E. H. Twizell has been Lecturer in Numerical Analysis at Brunel University since 1979. A former Senior Lecturer in Mathematics at Sunderland Polytechnic, he graduated in 1963 with a B.Sc. in Mathematics, an M.A. in Mathematics, adding a Ph.D. in Mathematics (Numerical Analysis) in 1977.

He was elected to Membership of the Canadian Mathematical Society in 1969, and has published over 40 papers in the numerical solution of ordinary and partial differential equations, some with appications in the biomedical sciences. He has also previously written a book *Computational Methods for Partial Differential Equations* (Ellis Horwood, 1984).

NUMERICAL METHODS, WITH APPLICATIONS IN THE BIOMEDICAL SCIENCES

MATHEMATICS AND ITS APPLICATIONS
Series Editor: G. M. BELL, Professor of Mathematics,
King's College London (KQC), University of London

NUMERICAL ANALYSIS, STATISTICS AND OPERATIONAL RESEARCH
Editor: B. W. CONOLLY, Professor of Operational Research,
Queen Mary College, University of London

Mathematics and its applications are now awe-inspiring in their scope, variety and depth. Not only is there rapid growth in pure mathematics and its applications to the traditional fields of the physical sciences, engineering and statistics, but new fields of application are emerging in biology, ecology and social organization. The user of mathematics must assimilate subtle new techniques and also learn to handle the great power of the computer efficiently and economically.

The need for clear, concise and authoritative texts is thus greater than ever and our series will endeavour to supply this need. It aims to be comprehensive and yet flexible. Works surveying recent research will introduce new areas and up-to-date mathematical methods. Undergraduate texts on established topics will stimulate student interest by including applications relevant at the present day. The series will also include selected volumes of lecture notes which will enable certain important topics to be presented earlier than would otherwise be possible.

In all these ways it is hoped to render a valuable service to those who learn, teach, develop and use mathematics.

Mathematics and its Applications
Series Editor: G. M. BELL, Professor of Mathematics, King's College London (KQC), University of London

Artmann, B. — **The Concept of Number**
Arczewski, K. & Pietrucha, J. — **Mathematical Modelling in Discrete Mechanical Systems**
Arczewski, K. and Pietrucha, J. — **Mathematical Modelling in Continuous Mechanical Systems**
Bainov, D.D. & Konstantinov, M. — **The Averaging Method and its Applications**
Balcerzyk, S. & Joszefiak, T. — **Commutative Rings**
Balcerzyk, S. & Joszefiak, T. — **Noetherian and Krull Rings**
Baldock, G.R. & Bridgeman, T. — **Mathematical Theory of Wave Motion**
Ball, M.A. — **Mathematics in the Social and Life Sciences: Theories, Models and Methods**
de Barra, G. — **Measure Theory and Integration**
Bell, G.M. and Lavis, D.A. — **Co-operative Phenomena in Lattice Models, Vols. I & II**
Berkshire, F.H. — **Mountain and Lee Waves**
Berry, J.S., Burghes, D.N., Huntley, I.D., James, D.J.G. & Moscardini, A.O. — **Mathematical Modelling Courses**
Berry, J.S., Burghes, D.N., Huntley, I.D., James, D.J.G. & Moscardini, A.O. — **Mathematical Methodology, Models and Micros**
Berry, J.S., Burghes, D.N., Huntley, I.D., James, D.J.G. & Moscardini, A.O. — **Teaching and Applying Mathematical Modelling**
Brown, R. — **Topology**
Burghes, D.N. & Borrie, M. — **Modelling with Differential Equations**
Burghes, D.N. & Downs, A.M. — **Modern Introduction to Classical Mechanics and Control**
Burghes, D.N. & Graham, A. — **Introduction to Control Theory, including Optimal Control**
Burghes, D.N., Huntley, I. & McDonald, J. — **Applying Mathematics**
Burghes, D.N. & Wood, A.D. — **Mathematical Models in the Social, Management and Life Sciences**
Butkovskiy, A.G. — **Green's Functions and Transfer Functions Handbook**
Butkovskiy, A.G. — **Structural Theory of Distributed Systems**
Cao, Z.-Q., Kim, K.H. & Roush, F.W. — **Incline Algebra and Applications**
Cerny, I. — **Complex Domain Analysis**
Chorlton, F. — **Textbook of Dynamics, 2nd Edition**
Chorlton, F. — **Vector and Tensor Methods**
Cohen, D.E. — **Computability and Logic**
Crapper, G.D. — **Introduction to Water Waves**
Cross, M. & Moscardini, A.O. — **Learning the Art of Mathematical Modelling**
Cullen, M.R. — **Linear Models in Biology**
Dunning-Davies, J. — **Mathematical Methods for Mathematicians, Physical Scientists and Engineers**
Eason, G., Coles, C.W. & Gettinby, G. — **Mathematics and Statistics for the Bio-sciences**
El Jai, A. & Pritchard, A.J. — **Sensors and Controls in the Analysis of Distributed Systems**
Exton, H. — **Handbook of Hypergeometric Integrals**
Exton, H. — **Multiple Hypergeometric Functions and Applications**
Exton, H. — ***q*-Hypergeometric Functions and Applications**

Series continued at back of book

NUMERICAL METHODS, WITH APPLICATIONS IN THE BIOMEDICAL SCIENCES

E. H. TWIZELL, B.Sc., M.A., Ph.D.
Reader in Computational Mathematics
Department of Mathematics and Statistics
Brunel University, Uxbridge

ELLIS HORWOOD LIMITED
Publishers · Chichester

Halsted Press: a division of
JOHN WILEY & SONS
New York · Chichester · Brisbane · Toronto

First published in 1988 by
ELLIS HORWOOD LIMITED
Market Cross House, Cooper Street,
Chichester, West Sussex, PO19 1EB, England
The publisher's colophon is reproduced from James Gillison's drawing of the ancient Market Cross, Chichester.

Distributors:
Australia and New Zealand:
JACARANDA WILEY LIMITED
GPO Box 859, Brisbane, Queensland 4001, Australia
Canada:
JOHN WILEY & SONS CANADA LIMITED
22 Worcester Road, Rexdale, Ontario, Canada
Europe and Africa:
JOHN WILEY & SONS LIMITED
Baffins Lane, Chichester, West Sussex, England
North and South America and the rest of the world:
Halsted Press: a division of
JOHN WILEY & SONS
605 Third Avenue, New York, NY 10158, USA
South-East Asia
JOHN WILEY & SONS (SEA) PTE LIMITED
37 Jalan Pemimpin # 05–04
Block B, Union Industrial Building, Singapore 2057
Indian Subcontinent
WILEY EASTERN LIMITED
4835/24 Ansari Road
Daryaganj, New Delhi 110002, India

British Library Cataloguing in Publication Data
Twizell, E. H.
Numerical methods, with applications in the biomedical sciences. —
(Ellis Horwood series in mathematics and its applications).
1. Medicine — Mathematical models
I. Title
610′.724 R853.M3

Library of Congress Card No. 87–29307

ISBN 0–7458–0027–0 (Ellis Horwood Limited)
ISBN 0–470–21002–8 (Halsted Press)

Phototypeset in Times by Ellis Horwood Limited
Printed and bound in Great Britain by Butler & Tanner Ltd, Frome and London

Table of Contents

To my mother
and in memory of
my father

Preface

This book has been written for undergraduate students of mathematics who are embarking on a first course in numerical methods, and senior undergraduate and postgraduate students and research workers in the biomedical sciences, biomathematics and bioengineering, who need to acquire knowledge of numerical methods for solving the algebraic and differential equations that arise from the mathematical modelling of biomedical systems.

It is assumed that the reader has undertaken first courses in calculus and algebra and can write computer programs in a high-level scientific language. The topics covered include numerical treatments of vector and matrix algebra, linear algebraic systems, eigenvalue problems, interpolation and approximation, differentiation and integration, initial and boundary value problems, and steady-state and time-dependent partial differential equations. There are two appendices, one giving an outline of some introductory mathematics which will benefit an understanding of the book, the other giving details of relevant software available from the NAG (Numerical Algorithms Group) library.

The text is liberally illustrated with case studies and worked examples from the biomedical sciences. These include mathematical models of predator–prey interactions, continuously fed biochemical reaction tanks, therapy of autoimmune diseases, growth of oncological diseases, arterial blood pressure, nerve impulse transmissions, percutaneous drug absorption, genetics, the otolith membrane, and heat flow in the human body.

I would like to thank Mrs B. Yates, Ms M. E. Demmar, Mrs M. Storey, Miss T. Brent and Mrs M. E. Saunders of Brunel University for typing the manuscript and Mrs V. R. Kessler also of Brunel University for help in the general preparation of the typescript. I am very pleased to acknowledge the contributions made by a number of my students, particularly Dr P. Smith and Dr A. Q. M. Khaliq, and the constant encouragement, whilst I was writing the book, of my wife Anne and our son Linus, who also helped with the computer programming.

January 1988 E. H. Twizell

1

Linear algebra

1.1 VECTORS AND VECTOR NORMS

A real *vector* $\mathbf{x}$ in N dimensions can be regarded as a quantity with N real *components* or *elements* $x_1, x_2, \ldots, x_N$. It is usual to write the elements of $\mathbf{x}$ column-wise so that

$$\mathbf{x} = \begin{bmatrix} x_1 \\ x_2 \\ \vdots \\ x_N \end{bmatrix} \tag{1.1}$$

The corresponding row vector $\mathbf{x}^{\mathrm{T}} = [x_1, x_2, \ldots, x_N]$ is called the *transpose of* $\mathbf{x}$. The vector $\mathbf{x} \in R^N$ is said to be of *order* or *dimension* N.

Definition 1.1 A set of vectors $\mathbf{v}_1, \mathbf{v}_2, \ldots, \mathbf{v}_m$ is said to be *linearly independent* if the only solution to the equation $c_1\mathbf{v}_1 + c_2\mathbf{v}_2 + \ldots + c_m\mathbf{v}_m = \mathbf{0}$ is $c_1 = c_2 = \ldots = c_m = 0$, where $\mathbf{0} = [0, 0, \ldots, 0]^{\mathrm{T}}$ is the *zero-vector*, and to be *linearly dependent* otherwise.

Definition 1.2 The *scalar* or *inner product* of two vectors $\mathbf{x}$ and $\mathbf{y}$ of order N is written as $(\mathbf{x}, \mathbf{y})$ and is defined by

$$(\mathbf{x}, \mathbf{y}) = \mathbf{x}^{\mathrm{T}}\mathbf{y} = \mathbf{y}^{\mathrm{T}}\mathbf{x} = x_1y_1 + x_2y_2 + \ldots + x_Ny_N \ . \tag{1.2}$$

The inner product of two vectors is a scalar quantity and it is clear that the distributive low holds, viz.:

$$(a\mathbf{x} + b\mathbf{y}, \mathbf{z}) = a(\mathbf{x}, \mathbf{z}) + b(\mathbf{y}, \mathbf{z}) \tag{1.3}$$

where a, b are scalars and $\mathbf{x}, \mathbf{y}, \mathbf{z}$ are vectors of the same dimension. Writing $\mathbf{z} = a\mathbf{x} + b\mathbf{y}$ in (1.3) and using (1.2) it is easy to show that

$$(\mathbf{x}, \mathbf{y})^2 \leqslant (\mathbf{x}, \mathbf{x})(\mathbf{y}, \mathbf{y}) \tag{1.4}$$

which if known as the *Cauchy–Schwarz inequality*.

Definition 1.3 A *vector norm* on R^N, the set of all N-dimensional column vectors with real components, is a function $\|\,.\,\|$, from R^N into R with the following properties:

(i) $\|\mathbf{x}\| \geqslant 0$ for all $\mathbf{x} \in R^N$,
(ii) $\|\mathbf{x}\| = 0$ if and only if $\mathbf{x} = \mathbf{0}$,
(iii) $\|c\mathbf{x}\| = |c|\,.\,\|\mathbf{x}\|$ for all $c \in R$ and $\mathbf{x} \in R^N$,
(iv) $\|\mathbf{x}+\mathbf{y}\| \leqslant \|\mathbf{x}\| + \|\mathbf{y}\|$ for all $\mathbf{x}, \mathbf{y} \in R^N$ (*the triangle inequality*).

There are three common vector norms; they are defined as follows:

Definition 1.4 The L_1, L_2 and L_∞ vector norms are defined by

$$\|\mathbf{x}\|_1 = \sum_{i=1}^{N} |x_i| \,, \tag{1.5}$$

$$\|\mathbf{x}\|_2 = \left\{\sum_{i=1}^{N} x_i^2\right\}^{\frac{1}{2}} , \tag{1.6}$$

$$\|\mathbf{x}\|_\infty = \max_{1 \leqslant i \leqslant N} |x_i| \,. \tag{1.7}$$

The L_1 and L_∞ vector norms are sometimes referred to as the *absolute* and *maximum* norms, respectively, while the L_2 vector norm is called the *Euclidean* norm of the vector $\mathbf{x}$ since it represents the familiar definition of distance from the origin in R_N.

Example 1.1 Consider the vector $\mathbf{x} = [-1, 3, -2]^{\mathrm{T}}$ in R^3. Its L_1, L_2 and L_∞ vector norms are

$$\|\mathbf{x}\|_1 = |-1| + |3| + |+2| = 1+3+2 = 6 \,,$$
$$\|\mathbf{x}\|_2 = \{(-1)^2 + 3^2 + (-2)^2)\}^{\frac{1}{2}} = \{1+9+4\}^{\frac{1}{2}} = \surd 14 \,,$$
$$\|\mathbf{x}\|_\infty = \max\{|-1|, |3|, |-2|\} = 3 \,.$$

This example illustrates that, in general, different vector norms are not equal.

Since the norm of a vector gives a measure of the distance between the vector and the origin, the *distance between two vectors* can be defined as the norm of the difference of the vectors, formalized as follows.

Definition 1.5 If $\mathbf{x} = [x_1, x_2, \ldots, x_N]^T$ and $\mathbf{y} = [y_1, y_2, \ldots, y_N]^T$ are two vectors in R^N, the distances between them in the L_1, L_2 and L_∞ norms are given by

$$L_1 \colon \|\mathbf{x}-\mathbf{y}\|_1 = \sum_{i=1}^{N} |x_i - y_i| \ , \tag{1.8}$$

$$L_2 \colon \|\mathbf{x}-\mathbf{y}\|_2 = \left\{\sum_{i=1}^{N} |x_i - y_i|^2\right\}^{\frac{1}{2}} \ , \tag{1.9}$$

$$L_\infty \colon \|\mathbf{x}-\mathbf{y}\|_\infty = \max_{1 \leq i \leq N} |x_i - y_i| \ . \tag{1.10}$$

These definitions are applicable in estimating the error in a computed solution obtained following the mathematical modelling of a given biomedical system, the error being the distance between the theoretical solution and a computed solution. In such applications, criteria which control the growth of the error or make it tend to zero as the computation proceeds will be sought: a number of these applications will occur in Chapters 5–9.

Definition 1.6 A sequence of vectors $\mathbf{x}^{(0)}, \mathbf{x}^{(1)}, \mathbf{x}^{(2)}, \ldots$, is said to *converge* to $\mathbf{x}$ with respect to the norm $\|\,.\,\|$, if, given any $\varepsilon > 0$, there exists an integer $M(\varepsilon)$ such that $\|\mathbf{x}^{(k)} - \mathbf{x}\|_. < \varepsilon$ for all $k \geqslant M(\varepsilon)$.

Theorem 1.1 All vector norms on R^N are equivalent with respect to convergence.

This theorem say that, if $\|\,.\,\|$ and $\|\,.\,\|'$ are any two norms on R^N and the sequence $\mathbf{x}^{(0)}, \mathbf{x}^{(1)}, \mathbf{x}^{(2)}, \ldots$ has the limit $\mathbf{x}$ with respect to $\|\,.\,\|$, then it also has the limit $\mathbf{x}$ with respect to $\|\,.\,\|'$. The proof of the theorem may be found in Ortega (1972, page 8).

1.2 MATRICES AND MATRIX NORMS

A *matrix* is defined as an ordered rectangular array of numbers. Suppose a matrix A has M rows and N columns. If the *element* in the ith row and jth column is denoted by a_{ij} the matrix may be written in the compact form $A = [a_{ij}]$: the whole matrix has the form

$$A = \begin{bmatrix} a_{11} & a_{12} & \dots & a_{1N} \\ a_{21} & a_{22} & \dots & a_{2N} \\ \dots & \dots & \dots & \dots \\ a_{M1} & a_{M2} & \dots & a_{MN} \end{bmatrix} . \tag{1.11}$$

This matrix is said to be of *order* $M \times N$; when $M = 1$, the matrix is a row vector, when $N = 1$ the matrix is a column vector and when $M = N$ the matrix is a *square* matrix of order N (when $M = N = 1$ the matrix is a scalar, identified with the quantity representing the single element). Some elementary properties of, operations on, and terminology relating to matrices are now revised.

Matrices of the same order may be added together or subtracted from each other. Thus, if $A = [a_{ij}]$ and $B = [b_{ij}]$ are two matrices of the same order, their sum/difference is a matrix $C = [c_{ij}]$ of the same order as A and B defined by $C = A \pm B$ with $c_{ij} = a_{ij} \pm b_{ij}$.

Definition 1.7 Two matrices A and B are said to be *equal* if they are of the same order $M \times N$ and if corresponding elements a_{ij} and b_{ij} ($i = 1,2,\dots,M$; $j = 1,2,\dots,N$) are equal.

If $A = [a_{ij}]$ is a matrix of order $M \times L$ and $B = [b_{ij}]$ is a matrix of order $L \times N$, their *product* $C = AB$ is a matrix of order $M \times N$ with c_{ij} found by taking the inner product of row i of matrix A with column j of matrix B, that is

$$c_{ij} = \sum_{r=1}^{L} a_{ir} b_{rj} \qquad (i = 1,2,\dots,M; j = 1,2,\dots,N) .$$

Clearly the product CB is not defined. In general, therefore, $BC \neq CB$ and indeed the product FG of some matrix F of order $M \times P$ with another matrix G of order $Q \times N$ is not defined unless $P = Q$: when P does equal Q the product matrix FG is of order $M \times N$.

Returning momentarily to vectors, the *outer product* of a vector $\mathbf{x} = [x_1, x_2, \dots, x_M]^T$ with a vector $\mathbf{y} = [y_1, y_2, \dots, y_N]^T$ is the matrix

$$\mathbf{x}\mathbf{y}^T = \begin{bmatrix} x_1y_1 & x_1y_2 & \dots & x_1y_N \\ x_2y_1 & x_2y_2 & \dots & x_2y_N \\ \dots & \dots & & \dots \\ x_My_1 & x_My_2 & \dots & x_My_N \end{bmatrix} .$$

Outer products of two vectors are rarely encountered. Note, $\mathbf{x}$ is of order $M \times 1$ and y^T is of order $1 \times N$; hence $\mathbf{x}\mathbf{y}^T$ *is* defined and is of order $M \times N$.

The *transpose* of a matrix A, denoted by A^{T}, is determined by interchanging rows and columns; hence, the transpose of A in (1.11) is the $N \times M$ matrix

$$A^{\mathrm{T}} = \begin{bmatrix} a_{11} & a_{21} & \dots & a_{M1} \\ a_{12} & a_{22} & \dots & a_{M2} \\ \dots & \dots & \dots & \dots \\ a_{1N} & a_{2N} & \dots & a_{MN} \end{bmatrix}.$$

Conjugation of a matrix $A = [a_{rs}]$, where $a_{rs} = u_{rs} + \mathrm{i}v_{rs}$ ($\mathrm{i} = +\sqrt{-1}$), gives the *conjugate matrix* $A^* = [a^*_{rs}]$ in which $a^*_{rs} = u_{rs} - \mathrm{i}v_{rs}$. If a matrix A is transposed and conjugated at the same time, the symbol A^{H} is used where $A^{\mathrm{H}} = (A^{\mathrm{T}})^* = (A^*)^{\mathrm{T}}$. The matrix A is said to be *Hermitian* if $A^{\mathrm{H}} = A$.

A square matrix $A = [a_{ij}]$ is said to be *symmetric* if $a_{ij} = a_{ji}$. The elements $a_{11}, a_{22}, \dots, a_{NN}$ of a square matrix $A = [a_{ij}]$ of order N form the *leading diagonal* of A and the quantity $a_{11} + a_{22} + \dots + a_{NN}$ is called the *trace* of A. The matrix A is said to be *diagonally dominant if* $|a_{ii}| \geqslant |a_{i1}| + |a_{i2}| + \dots + |a_{i,i-1}| + |a_{i,i+1}| + \dots + |a_{iN}|$ for all $i = 1, 2, \dots, N$. A real symmetric matrix A us said to be *positive definite* if $\mathbf{x}^{\mathrm{H}}A\mathbf{x} > 0$ for all vectors $\mathbf{x} \neq \mathbf{0}$ (of the same order as A) and to be *positive semi-definite* if $\mathbf{x}^{\mathrm{H}}A\mathbf{x} \geqslant 0$ (note: $\mathbf{x}^{\mathrm{H}}A\mathbf{x}$ is a scalar). A square matrix $A = [a_{ij}]$ of order N in which $a_{ij} = 0$ if $i > j$ is said to be *upper triangular*; if $a_{ij} = 0$ for $i < j$ the matrix is said to be *lower triangular*. A square matrix $A = [a_{ij}]$ of order N in which $a_{ij} = 0$ for $i \neq j$ is called a *diagonal matrix*; if, in addition, $a_{ii} = 1$ for $i = 1, 2, \dots, N$, A is called the *identity* or *unit matrix* of order N (usually I or I_N is used to denote this matrix). A matrix $A = [a_{ij}]$ of order $M \times N$ in which every element $a_{ij} = 0$ is called the *zero* or *null matrix* of order $M \times N$; if most, but not all, of the elements are zero, the matrix is said to be *sparse*.

The *inverse matrix* of a square matrix A, denoted by A^{-1}, is such that $AA^{-1} = A^{-1}A = I$. If the determinant of the matrix A is equal to zero, the inverse A^{-1} does not exist and A is said to be *singular*; if the inverse does exist, A is said to be *nonsingular*. The reader is referred to any of the numerous elementary texts on matrix algebra for revision examples on how to compute A^{-1} by elementary methods. The matrix A is *unitary* if $A^{\mathrm{H}} = A^{-1}$.

Given three matrices of orders such that addition, multiplication and inversion are defined, the following is a list of fundamental computing laws for matrices:

1. $(A + B) + C = A + (B + C)$ (the associative law for addition),
2. $A + B = B + A$ (the commutative law for addition),
3. $(AB)C = A(BC)$ (the associative law for multiplication),
4. $A(B + C) = AB + AC$ and $(A + B)C = AC + BC$ (the distributive laws for multiplication),
5. $(A + B)^{\mathrm{T}} = A^{\mathrm{T}} + B^{\mathrm{T}}$,
6. $(AB)^{\mathrm{T}} = B^{\mathrm{T}}A^{\mathrm{T}}$,
7. $(AB)^{-1} = B^{-1}A^{-1}$,

8. $(A^{\mathrm{T}})^{-1} = (A^{-1})^{\mathrm{T}}$.

A matrix is said to have *rank* r if it has r linearly independent rows or columns. If a matrix is of rank r, an arbitrarily chosen row (or column) can be written as a linear combination of at most r of the remaining rows (or columns); otherwise there are $r+1$ linearly independent rows (or columns) and the rank would be at least $r+1$.

Definition 1.8 A *matrix norm* on the set of all $N \times N$ matrices is a function $\|\,.\,\|$, defined on this set, satisfying for all $N \times N$ matrices A and B and scalars c the following properties:

(1) $\|A\| \geqslant 0$,
(2) $\|A\| = 0$ if and only if A is the zero matrix,
(3) $\|cA\| = |c|.\|A\|$,
(4) $\|A+B\| \leqslant \|A\| + \|B\|$ (the triangle inequality),
(5) $\|AB\| \leqslant \|A\|.\|B\|$ (the Schwarz inequality).

There are four common natural norms; they are defined as follows:

Definition 1.9 The L_1, L_2, L_∞ and *Euclidean* norms, respectively, are defined for a square matrix A of order N by

$$\|A\|_1 = \max_j \sum_{i=1}^{N} |a_{ij}| \, , \tag{1.12}$$

$$\|A\|_2 = \lambda_1^{\frac{1}{2}}, \text{ where } \lambda_1 \text{ is the largest eigenvalue of } A^{\mathrm{H}}A, \tag{1.13}$$

$$\|A\|_\infty = \max_i \sum_{j=1}^{N} |a_{ij}| \, , \tag{1.14}$$

$$\|A\|_{\mathrm{E}} = [\operatorname{trace}(A^{\mathrm{H}}A)]^{\frac{1}{2}} = [\operatorname{trace}(AA^{\mathrm{H}})]^{\frac{1}{2}} = \left\{ \sum_{i=1}^{N} \sum_{j=1}^{N} |a_{ij}|^2 \right\}^{\frac{1}{2}} \tag{1.15}$$

The L_2 norm is often called the *spectral* norm and occasionally the *Hilbert* norm. The *distance* between the matrices A and B is defined as $\|A - B\|$.

Since matrices and vectors frequently occur together, it is useful to be able to relate corresponding norms. A given vector norm is said to *induce* a matrix norm through the definition

$$\|A\| = \sup_{\mathbf{x}\neq\mathbf{0}} \|A\mathbf{x}\|/\|\mathbf{x}\| \tag{1.16}$$

or equivalently as

$$\|A\| = \sup \|A\mathbf{x}\| \text{ taken over } \|\mathbf{x}\| = 1 \;. \tag{1.17}$$

A matrix norm defined in this way has the property

$$\|A\mathbf{x}\| \leqslant \|A\|.\|\mathbf{x}\| \tag{1.18}$$

and a vector norm and a matrix norm satisfying (1.18) are said to be *compatible*.

Definition 1.10 If A is a real square matrix of order N, the polynomial defined by

$$p(\lambda) = \det(A - \lambda I) \tag{1.19}$$

is called the *characteristic polynomial* of A.

Clearly $p(\lambda)$ is a polynomial of degree N in λ and therefore has at most N distinct zeros. The coefficients are real (since A is real) and the coefficient of λ^N is $(-1)^N$.

Definition 1.11 If $p(\lambda)$ defined by (1.19) is the characteristic polynomial of the matrix A, the zeros $\lambda_1, \lambda_2, \ldots, \lambda_N$ of $p(\lambda)$ are known as the *eigenvalues* or *characteristic values* of A.

In the case when A is real, complex eigenvalues occur in complex conjugate pairs.

Definition 1.12 If λ is an eigenvalue of the matrix A and $\mathbf{x} \neq \mathbf{0}$ has the property that $A\mathbf{x} = \lambda\mathbf{x}$, then $\mathbf{x}$ is the *eigenvector* associated with λ.

Three very useful theorems, relating the the eigenvalues of A, are as follows:

Theorem 1.2 (*The Cayley–Hamilton Theorem*) Every square matrix satisfies its own characteristic equation.

Theorem 1.3 (*Geršgorin's First Theorem*) The eigenvalue of the square matrix A which has largest modulus cannot exceed the sum of the moduli of the elements along any row or down any column of A.

Theorem 1.4 (*Geršgorin's Circle Theorem or Brauer's Theorem*) Every eigenvalue of the matrix A lies inside at least one of the discs in the complex plane with centre a_{ii} and radius

$$r_i = \sum_{j \neq i} |a_{ij}| , \qquad i = 1,2,\ldots,N . \tag{1.20}$$

Definition 1.13 The *spectral radius* of a square matrix A, denoted by $\rho(A)$, is given by

$$\rho(A) = \max_i |\lambda_i|$$

where λ_i ($i = 1,2,\ldots,N$) are the eigenvalues of A.

The spectral radius is the radius of the smallest disc in the complex plane, with centre at the origin, which contains all of the eigenvalues of A. It follows from Theorems 1.3 and 1.4 that

$$\rho(A) \leqslant \min \left| \max_i \sum_{j=1}^{N} |a_{ij}| \;,\; \max_j \sum_{i=1}^{N} |a_{ij}| \right| .$$

The spectral radius is closely related to the norm of a matrix as is shown in the following theorem.

Theorem 1.5 If A is a real matrix of order N, then

(1) $[\rho(A^{\mathrm{T}}A)]^{\frac{1}{2}} = \|A\|_2$,
(2) $\rho(A) \leqslant \|A\|$ for any natural norm $\|.\|$.

In studying iterative matrix techniques, it is important to know when powers of a matrix become small; that is, when all of the elements approach zero.

Definition 1.14 A square matrix A of order N is said to be *convergent* if

$$\lim_{k \to \infty} (a_{ij})^k = 0 \text{ for each } i,j = 1,2,\ldots,N .$$

Definition 1.15 If $\|A\|$ denotes any matrix norm then the *convergence of a sequence* *of matrices* $A^{(1)}, A^{(2)}, A^{(3)}, \ldots$ in the sense of this norm is defined as follows:

$$\lim_{r\to\infty} A^{(r)} = A \text{ if and only if } \lim_{r\to\infty} \|A - A^{(r)}\| = 0 \ .$$

It is clear that, if $\lim_{r\to\infty} A^{(r)} = A$, then $\lim_{r\to\infty} \|A^{(r)}\| = \|A\|$.

Lemma 1.1 If, in any norm, $\|A\| < 1$, then $\lim_{r\to\infty} A^r = 0$.

Proof $\|A^r\| = \|AA^{r-1}\| \leqslant \|A\|\,\|A^{r-1}\|$, from the Schwarz inequality,

$$\leqslant \|A\|^2\|A^{r-1}\| \leqslant \ldots$$
$$\leqslant \|A\|^r \to 0 \text{ as } r\to\infty \text{ provided } \|A\| < 1.$$

Lemma 1.2 If λ is given eigenvalue of A then $|\lambda| \leqslant \|A\|$.

Proof Let $\mathbf{x} \neq \mathbf{0}$ be the nonzero vector such that $A\mathbf{x} = \lambda\mathbf{x}$. Then

$$|\lambda|\,.\,\|\mathbf{x}\| = \|\lambda\mathbf{x}\| = \|A\mathbf{x}\| \leqslant \|\mathbf{A}\|\,.\,\|\mathbf{x}\|$$

from which the hypothesis of the lemma follows.

Lemma 1.3 For any matrix A,

$$\|A\|_2^2 \leqslant \|\overline{A}^{\mathrm{T}}A\|_1 \leqslant \|\overline{A}^{\mathrm{T}}\|_1\|A\|_1 = \|A\|_\infty\,\|A\|_1.$$

The following theorem, the proof of which occurs in many specialist texts but requires theories not discussed in this section, has important applications in the numerical solution of partial differential equations:

Theorem 1.6 For a given matrix A, the necessary and sufficient condition for the series $I, A, A^2, \ldots$ to converge to the zero matrix is that every eigenvalue of A should be less than unity in modulus.

Theorem 1.7 A necessary and sufficient condition for the series $I + A + A^2 + A^3 + \ldots$ to converge is that $\lim_{r\to\infty} A^r = 0$.

Proof The necessity of the hypothesis is obvious. To establish sufficiency, Theorem

1.6 is used; this shows that if $A^r \to 0$ as $r \to \infty$, then all eigenvalues of A are less than unity in modulus. This shows that the determinant of the matrix $I - A$ is not zero and thus the matrix $(I-A)^{-1}$ exists.

Consider next the formula

$$(I-A)(I+A+A^2+\ldots+A^r) = I-A^{r+1} .$$

Since $(I-A)^{-1}$ exists, it follows that

$$\begin{aligned} I+A+A^2+\ldots+A^r &= (I-A)^{-1}(I-A^{r+1}) \\ &= (I-A)^{-1}-(I-A)^{-1}A^{r+1} \\ &\to (I-A)^{-1} \text{ since } A^{r+1} \to 0 \text{ as } r \to \infty . \end{aligned}$$

The series $I+A+A^2+\ldots$ thus converges to $(I-A)^{-1}$ if $A^r \to 0$ as $r \to \infty$.

Lemma 1.4 For any matrix norm with $\|I\| = 1$, $\|(I+A)^{-1}\| \leqslant (I - \|A\|)^{-1}$.

Proof The proof follows from Theorem 1.7 by replacing A with $-A$.

Lemma 1.5 A sufficient condition for the convergence of the series $I+A+A^2+A^3+\ldots$ is that $\|A\| < 1$ for any norm of A.

Proof If $\|A\| < 1$, then from Lemma 1.2 it follows that every eigenvalue of A is less than unity in modulus, and Theorem 1.6 then establishes the lemma.

Lemma 1.6 A sufficient condition for the convergence of the series $I+A+A^2+\ldots$ is that there exists a matrix similarity transform $A = HBH^{-1}$ for which $\|B\| < 1$ in some matrix norm.

Proof If $A = HBH^{-1}$ then

$$I+A+A^2+\ldots+A^r = H(I+B+\ldots+B^r)H^{-1}$$

which converges if $\|B\| < 1$ for some norm.

Lemma 1.7 Let A be some matrix; then, if $\|A\| < 1$ for some norm,

$$\|(I-A)^{-1}-(I+A+\ldots+A^r)\| \leqslant \|A\|^{r+1}/(1-\|A\|) .$$

Proof It is easy to see from Theorem 1.7 that

$$(I-A)^{-1}-(I+A+\ldots+A^r) = A^{r+1}+A^{r+2}+\ldots .$$

From property (4) of a matrix norm, listed in definition 1.8 it follows that

$$\begin{aligned}\|(I-A)^{-1}-(I+A+\ldots+A^r)\| &\leqslant \|A^{r+1}\|+\|A^{r+2}\|+\ldots\\ &\leqslant \|A\|^{r+1}+\|A\|^{r+2}+\ldots\\ &= \|A\|^{r+1}(1+\|A\|+\|A\|^2+\ldots)\\ &= \|A\|^{r+1}/(1-\|A\|)\end{aligned}$$

if $\|A\| < 1$.

Theorem 1.8 The following statements are equivalent:

(1) A is a convergent matrix,

(2) $\lim_{n\to\infty} \|A^n\| = 0$ for some norm $\|.\|$,

(3) $\rho(A) < 1$.

1.3 LINEAR ALGEBRAIC SYSTEMS: INTRODUCTION

The problem to be solved is that of solving the matrix–vector equation

$$A\mathbf{x} = \mathbf{b} \tag{1.21}$$

for the *solution vector* $\mathbf{x} = [x_1,x_2,\ldots,x_N]^{\mathrm{T}}$. The N-square coefficient matrix A in (1.21) has the general form

$$A = \begin{bmatrix} a_{11} & a_{12} & \ldots & a_{1N} \\ a_{21} & a_{22} & \ldots & a_{2N} \\ \ldots & \ldots & \ldots & \ldots \\ a_{N1} & a_{N2} & \ldots & a_{NN} \end{bmatrix} \tag{1.22}$$

while the vector $\mathbf{b} = [b_1,b_2,\ldots,b_N]^{\mathrm{T}}$ is the *constant vector*. The vector $\mathbf{x}$ is given by

$$\mathbf{x} - A^{-1}\mathbf{b} \tag{1.23}$$

from (1.21) and is determined uniquely provided $\det A$ (the determinant of A) is nonzero, for then A is nonsingular and A^{-1} exists. If $\det A = 0$ and $\mathbf{b} \neq \mathbf{0}$, there are in general no solutions: if $\det A \neq 0$ and $\mathbf{b} = \mathbf{0}$ there is only the trivial solution $\mathbf{x} = \mathbf{0}$.

The inverse matrix A^{-1} can be determined by any elementary method such as that which determines the adjoint matrix of A and then divides by $\det A$, or by Cramer's rule which states that $x_m = \det A_m/\det A$ where A_m is the matrix obtained by replacing column m of A by $\mathbf{b}$. However, for large systems of the form (1.21) which arise frequently in solving the partial differential equations associated with many biomedical systems, such elementary methods are impractical and a numerical method should be used instead.

The system of equations given in (1.21) consists of N linear algebraic equations in the N unknowns $x_1, x_2, \ldots, x_N$; this system is properly posed or properly determined. In systems which have M equations in N unknowns, so that A is an $M \times N$ matrix, $\mathbf{x}$ is an N-vector and $\mathbf{b}$ is an M-vector, the system is said to be overdetermined if it has more equations than unknowns (that is, $M > N$) and to be underdetermined if it has fewer equations than unknowns (that is, $M < N$). Overdetermined systems have in general no solution: in Chapter 2, however, the reader will see how to find an approximate 'best' solution by minimizing a suitably defined error. Underdetermined systems have in general infinitely many solutions. In this chapter only properly determined systems will be discussed.

The methods to be developed include direct methods and iterative methods. Systems in which the matrix A has a special form, as in linear partial differential equations, will also be discussed.

1.4 DIRECT METHODS

1.4.1 The method of Gaussian elimination

The method of Gaussian elimination or pivotal condensation has earned a well-deserved reputation and is particularly suited to solving systems of the form (1.21) when the matrix A given by (1.22) is *full*, that is, most of its elements are nonzero.

The method begins with the system (1.21) in the form

$$\begin{aligned}
a_{11}^{(0)}x_1 + a_{12}^{(0)}x_2 + a_{13}^{(0)}x_3 + \ldots + a_{1N}^{(0)}x_N &= b_1^{(0)} \;, \\
a_{21}^{(0)}x_1 + a_{22}^{(0)}x_2 + a_{23}^{(0)}x_3 + \ldots + a_{2N}^{(0)}x_N &= b_2^{(0)} \;, \\
a_{31}^{(0)}x_1 + a_{32}^{(0)}x_2 + a_{33}^{(0)}x_3 + \ldots + a_{3N}^{(0)}x_N &= b_3^{(0)} \;, \\
\ldots \quad \ldots \quad \ldots \quad \ldots \quad &\ldots \;, \\
a_{N1}^{(0)}x_1 + a_{N2}^{(0)}x_2 + a_{N3}^{(0)}x_3 + \ldots + a_{NN}^{(0)}x_N &= b_N^{(0)} \;,
\end{aligned} \tag{1.24}$$

where the superscript zero has been appended to the coefficients to indicate that (1.24) is the original form of the given system. If the coefficient $a_{11}^{(0)}$ is not the coefficient of x_1 with largest magnitude, row 1 of (1.24) should be interchanged with the row which does contain the coefficient of x_1 with largest magnitude. Suppose this row is row m, then the coefficient $a_{m1}^{(0)}$ is called the *pivot* or *pivotal element* of the given system. After interchanging rows 1 and m, the coefficients of the old row m are renamed $a_{1j}^{(0)}$ $(j = 1, 2, \ldots, N)$, $b_1^{(0)}$ while the elements of the old row 1 are renamed $a_{mj}^{(0)}$ $(j = 1, 2, \ldots, N)$, $b_m^{(0)}$. The rearranged system has the form (1.24).

For each of rows 2, 3, $\ldots$, N of the rearranged system, multipliers $m_i^{(0)} = a_{i1}^{(0)}/a_{11}^{(0)}$ are now defined $(i = 2, 3, \ldots, N)$ and the first equation multiplied by $m_i^{(0)}$ is subtracted from equation i $(i = 2, 3, \ldots, N)$. By construction, the coefficients of x_1 in each of equations $2, 3, \ldots, N$ are now equal to zero and the system (1.24) takes the form

$$
\begin{array}{llllll}
a_{11}^{(0)}x_1 + a_{12}^{(0)}x_2 + & a_{13}^{(0)}x_3 & + \ldots + a_{1N}^{(0)}x_N = b_1^{(0)} \ , & & & \\
- \quad - & - & - \quad - & - & & \\
\quad a_{22}^{(1)}x_2 & + a_{23}^{(1)}x_3 + \ldots + a_{12N}^{(1)}x_N = b_2^{(1)} \ , & & & & \\
\quad a_{32}^{(1)}x_2 & + a_{33}^{(1)}x_3 + \ldots + a_{3N}^{(1)}x_N = b_3^{(1)} \ , & & & & \\
\quad \ldots & \ldots \qquad \ldots \quad \ldots & & & & \\
\quad a_{N2}^{(1)}x_2 & + a_{N3}^{(1)}x_3 + \ldots + a_{NN}^{(1)}x_N = b_N^{(1)} \ , & & & &
\end{array}
\tag{1.25}
$$

which is known as the *first derived system*. The coefficients of the subsystem (below the broken line) are given by

$$
a_{ij}^{(1)} = a_{ij}^{(0)} - m_i^{(0)}a_{1j}^{(0)} \ , \qquad b_i^{(1)} = b_i^{(0)} - m_i^{(0)}b_1^{(0)} \tag{1.26}
$$

for $i,j = 2, 3, \ldots, N$.

The pivot of the subsystem is now located and the row in which it lies is interchanged with the first row of the *subsystem*. Thereafter, the coefficients of the (old) first row of the subsystem and the pivotal row are renamed as described above. The first row of the given system (1.24), which appears above the broken line in (1.25), plays no further part in the elimination process.

For each of rows 3, 4, ..., N in the rearranged system (1.25), multipliers $m_i^{(1)} = a_{i2}^{(1)}/a_{22}^{(1)}$ are now defined ($i = 3, 4, \ldots, N$) and the second equation of (1.25) (the first equation of the subsystem) multiplied by $m_i^{(1)}$ is subtracted from equation i ($i = 3, 4, \ldots, N$). By construction, the coefficients of x_2 in each of equations 3, 4, ..., N of (1.25) are now equal to zero and the original system (1.24) has the form

$$
\begin{array}{llll}
a_{11}^{(0)}x_1 & + a_{12}^{(0)}x_2 + & a_{13}^{(0)}x_3 + & \ldots + \ a_{1N}^{(0)}x_N = b_1^{(0)} \ , \\
 & \quad a_{22}^{(1)}x_2 & + a_{23}^{(1)}x_3 + & \ldots + \ a_{2N}^{(1)}x_N = b_2^{(1)} \ , \\
- & \quad - & \quad - & - \quad - \quad - \\
 & & \quad a_{33}^{(2)}x_3 & + \ldots + a_{3N}^{(2)}x_N = b_3^{(2)} \ , \\
 & & \quad \ldots & \qquad \ldots \quad \ldots \\
 & & + \quad a_{N3}^{(2)}x_3 & + \ldots + a_{NN}^{(2)}x_N = b_N^{(2)} \ ,
\end{array}
\tag{1.27}
$$

which is known as the *second derived system*. The coefficients of the subsystem (below the broken line in (1.27)) are given by

$$
a_{ij}^{(2)} = a_{ij}^{(1)} - m_i^{(1)}a_{2j}^{(1)} \ , \qquad b_i^{(2)} = b_i^{(1)} - m_i^{(1)}b_2^{(1)} \tag{1.28}
$$

for $i,j = 3, 4, \ldots, N$.

Continuing through this elimination process for a total of $N - 1$ times, the given system eventually takes the *upper triangular form*

$$
\begin{aligned}
a_{11}^{(0)}x_1 + a_{12}^{(0)}x_2 + a_{13}^{(0)}x_3 + \ldots + a_{1,N-1}^{(0)}x_{N-1} + a_{1N}^{(0)}x_N &= b_1^{(0)} \quad , \\
a_{22}^{(1)}x_2 + a_{23}^{(1)}x_3 + \ldots + a_{2,N-1}^{(1)}x_{N-1} + a_{2N}^{(1)}x_N &= b_2^{(1)} \quad , \\
a_{33}^{(2)}x_3 + \ldots + a_{3,N-1}^{(2)}x_{N-1} + a_{3N}^{(2)}x_N &= b_3^{(2)} \quad , \\
\ldots \qquad \ldots \qquad \ldots \qquad & \ldots \\
a_{N-1,N-1}^{(N-2)}x_{N-1} + a_{N-1,N}^{(N-2)}x_N &= b_{N-1}^{(N-2)} \quad , \\
a_{NN}^{(N-1)}x_N &= b_N^{(N-1)}
\end{aligned}
\tag{1.29}
$$

in which all the diagonal elements are nonzero and

$$
a_{ij}^{(k)} = a_{ij}^{(k-1)} - m_i^{(k-1)}a_{kj}^{(k-1)} \quad , \qquad b_i^{(k)} = b_i^{(k-1)} - m_i^{(k-1)}b_k^{(k-1)} \tag{1.30}
$$

for $k = 1, 2, \ldots, N-1$ and $i, j = k+1, k+2, \ldots, N$. This transformation into upper triangular form is always possible provided A is nonsingular.

The elements $x_N, x_{N-1}, \ldots, x_2, x_1$ are now obtained from (1.29) by the *method of back substitution*. The last equation in (1.29) gives

$$
x_N = b_N^{(N-1)}/a_{NN}^{(N-1)}
$$

which is used in the penultimate equation to give

$$
x_{N-1} = [b_{N-1}^{(N-2)} - a_{N-1,N}^{(N-2)}x_N]/a_{N-1,N-1}^{(N-2)}
$$

and so on. In general

$$
x_i = \left[b_i^{(i-1)} - \sum_{j=i+1}^{N} a_{ij}^{(i-1)}x_j \right] \Big/ a_{ii}^{(i-1)} \tag{1.31}
$$

for $i = N, N-1, \ldots, \ldots, 2, 1$.

The number of multiplications and divisions for the method is known to be $\frac{1}{3}N^3 + N^2 - \frac{1}{3}N$ (Burden and Faires, 1985) and the number of additions and subtractions to be $\frac{1}{3}N^3 + \frac{1}{2}N^2 - \frac{5}{6}N$.

Example 1.2 Use Gaussian Elimination to solve the system

$$
\begin{aligned}
x_1 + 2x_2 + x_3 &= -3 \; , \\
5x_1 - x_2 - x_3 &= 8 \; , \\
2x_1 - x_2 + 2x_3 &= -1 \; .
\end{aligned}
\tag{1.32}
$$

The first pivot is clearly $+5$, the coefficient of x_1 in the second equation, so that the first and second equations must be interchanged and the given system written in the form

$$\begin{aligned} 5x_1 - \ x_2 - \ x_3 &= \ \ 8 \ , \\ x_1 + 2x_2 + \ x_3 &= -3 \ , \\ 2x_1 - \ x_2 + 2x_3 &= -1 \ . \end{aligned}$$

The multipliers $m_i^{(0)}$, $i = 2,3$, are clearly $m_2^{(0)} = 0.2$ and $m_3^{(0)} = 0.4$ and substracting $m_2^{(0)}$ times the first equation of (1.33) (the pivotal equation) from the second, and $m_3^{(0)}$ times the first equation from the third, leads to the first derived system

$$\begin{aligned} 5x_1 - x_2 - x_3 &= 8 \ , \\ 2.2x_2 + 1.2x_3 &= -4.6 \ , \\ -0.6x_2 + 2.4x_3 &= -4.2 \ . \end{aligned} \tag{1.34}$$

No rearrangement is necessary to get the equation into the 'top' position of the subsystem formed by rows 2 and 3 of the first derived system (1.34). The multiplier $m_3^{(1)}$ is given by $m_3^{(1)} = -0.6/2.2 = -0.2727$ to four decimal places; subtracting $m_3^{(1)}$ times the second equation of (1.34) (the pivotal equation) from the third leads to the second, and final, derived system.

$$\begin{aligned} 5x_1 - x_2 - \qquad x_3 &= 8 \ , \\ 2.2x_2 + \quad 1.2x_3 &= -4.6 \ , \\ 2.7272x_3 &= -5.4544 \ , \end{aligned} \tag{1.35}$$

The method of back substitution gives

$$\begin{aligned} x_3 &= -5.4544/2.7272 = -2.000 \ , \\ x_2 &= [-4.6 - 1.2(-2.000)]/2.2 = -1.000 \ , \\ x_1 &= [8 + (-1.000) + (-2.000)]/5 = 1.000 \ . \end{aligned}$$

It is noted that the values of x_3, x_2, x_1 are given to three decimal places, the intermediate arithmetic having been carried out to four decimal places. The theoretical solution of (1.32) is $\mathbf{x} = (x_1, x_2, x_3)^{\mathrm{T}} = (1, -1, -2)^{\mathrm{T}}$.

1.4.2 *LU* decomposition methods

The matrix of coefficients obtained in the final derived system of the method of Gaussian Elimination is in upper triangular form. In solving a system of the form (1.21) by an *LU decomposition* method, the matrix A is decomposed in such a way that it is written as $A = LU$, where L is a lower triangular matrix and U is an upper triangular matrix.

The *Doolittle* method decomposes A in such a way that the diagonal elements of L are all unity and $A = LU$ takes form

$$\begin{bmatrix} a_{11} & a_{12} & \ldots & a_{1N} \\ a_{21} & a_{22} & \ldots & a_{2N} \\ \ldots & \ldots & & \ldots \\ a_{N1} & a_{N2} & \ldots & a_{NN} \end{bmatrix} = \begin{bmatrix} 1 & & & \\ l_{21} & 1 & 0 & \\ \ldots & \ldots & & \\ l_{N1} & l_{N2} & \ldots & 1 \end{bmatrix} \begin{bmatrix} u_{11} & u_{12} & \ldots & u_{1N} \\ & u_{22} & \ldots & u_{2N} \\ & & & \ldots \\ 0 & & & u_{NN} \end{bmatrix} . \quad (1.36)$$

It is easily seen from (1.36) that thc element a_{ij} of the matrix A is the inner product of row i of the matrix L with column j of matrix U, that is $a_{ij} = r_i(L)c_j(U)$.

The elements of L and U are calculated as follows:

1. $a_{11} = r_1(L)c_1(U) = U_{11}, a_{12} = r_1(L)c_2(U) = u_{12}, \ldots, a_{1N} = r_1(L)c_N(U) = u_{1N}$ from which it is seen that row 1 of U is the same as row 1 of A;
2. $a_{21} = r_2(L)c_1(U) = l_{21}u_{11}$ from which l_{21} is computed,
   $a_{22} = r_2(L)c_2(U) = l_{21}u_{12} + u_{22}$ from which u_{22} is computed,
   $a_{23} = r_2(L)c_3(U) = l_{21}u_{13} + u_{23}$ from which u_{23} is computed,
   $\ldots \quad \ldots \quad \ldots \quad \ldots$
   $a_{2N} = r_2(L)c_N(U) = l_{21}u_{1N} + u_{2N}$ from which u_{2N} is computed;
3. $a_{31} = r_3(L)c_1(U) = l_{31}u_{11}$ from which l_{31} is computed,
   $a_{32} = r_3(L)c_2(U) = l_{31}u_{12} + l_{32}u_{22}$ from which l_{32} is computed,
   $a_{33} = r_3(L)c_3(U) = l_{31}u_{13} + l_{32}u_{23} + u_{33}$ from which u_{33} is computed,
   $\ldots \quad \ldots \quad \ldots \quad \ldots$
   $a_{3N} = r_3(L)c_N(U) = l_{31}u_{1N} + l_{32}u_{2N} + u_{3N}$ from which u_{3N} is computed.

The decomposition is continued with steps 4, ..., N until L and U are fully determined. If they are used in the order given, each of these equations contains only one unknown. The decomposition may be summarized by the formulas

$$u_{ij} = a_{ij} - \sum_{k=1}^{i-1} l_{ij}u_{kj} \; ; \qquad i=1,\ldots,N \; , \qquad j=i,i+1,\ldots,N, \qquad (1.37)$$

$$l_{ij} = \left[a_{ji} - \sum_{k=1}^{i-1} l_{jk}u_{ki} \right] / u_{ii} \; ; \quad i=1,\ldots,N \; , \quad j=i+1,i+2,\ldots,N \qquad (1.38)$$

Having determined the decomposition matrices L and U, the system can be written as the pair of equations

$$L\mathbf{y} = \mathbf{b} \; , \qquad (1.39)$$

$$U\mathbf{x} = \mathbf{y} \; . \qquad (1.40)$$

The intermediate vector $\mathbf{y}$ is found by the method of forward substitution from (1.39) to give

$$y_i = b_i - \sum_{j=1}^{i-1} l_{ij} y_j \ , \qquad i = 1, \ldots, N \ . \tag{1.41}$$

The vector **y** is then used in (1.40) from which **x** is found by back substitution, giving

$$x_i = \left[b_i - \sum_{j=i+1}^{N} u_{ij} y_j \right] \Big/ u_{ii} \ ; \qquad i = N, N-1, \ldots, 1 \ . \tag{1.42}$$

Including the processes of forward and backward substitution, Doolittle's method also uses $\frac{1}{3}N^3 + N^2 - \frac{1}{3}N$ multiplications/divisions and $\frac{1}{3}N^3 + \frac{1}{2}N^2 - \frac{5}{6}N$ additions/subtractions.

Example 1.3 Solve the system of Example 1.2 using Doolittle's method.
Here,

$$\begin{bmatrix} 1 & 2 & 1 \\ 5 & -1 & -1 \\ 2 & -1 & 2 \end{bmatrix} = \begin{bmatrix} 1 & 0 & 0 \\ l_{21} & 1 & 0 \\ l_{31} & l_{32} & 1 \end{bmatrix} \cdot \begin{bmatrix} u_{11} & u_{12} & u_{13} \\ 0 & u_{22} & u_{23} \\ 0 & 0 & u_{33} \end{bmatrix} .$$

Then $u_{11} = 1, u_{12} = 2, u_{13} = 1$ since the first row of U is the same as the first row of A. Taking next the inner product of row 2 of L with each column of U in turn gives

$$\begin{aligned} 5 &= l_{21} u_{11}, && \text{and hence } l_{21} = 5; \\ -1 &= 10 + u_{22}, && \text{and hence } u_{22} = -11; \\ -1 &= 5 + u_{23}, && \text{and hence } u_{23} = -6. \end{aligned}$$

To complete the decomposition, take the inner product of row 3 of L with each column of U in turn; this gives

$$\begin{aligned} 2 &= l_{31} u_{11}, \text{ and hence } l_{31} = 2; \\ 1 &= 4 + l_{32}(-11), \text{ and hence } l_{32} = 0.4545; \\ 2 &= 2 + (0.4545)6 + u_{33}, \text{ and hence } u_{33} = 2.7270, \end{aligned}$$

working to four decimal places.

Equation (1.39) takes the form

$$\begin{bmatrix} 1 & 0 & 0 \\ 5 & 1 & 0 \\ 2 & 0.4545 & 1 \end{bmatrix} \begin{bmatrix} y_1 \\ y_2 \\ y_3 \end{bmatrix} = \begin{bmatrix} -3 \\ 8 \\ -1 \end{bmatrix}$$

and solving by forward substitution gives $y_1 = -3$, $y_2 = 23$, $y_3 = -5.4535$.

Equation (1.40) becomes

$$\begin{bmatrix} 1 & 2 & 1 \\ 0 & -11 & -6 \\ 0 & 0 & 2.7270 \end{bmatrix} \begin{bmatrix} x_1 \\ x_2 \\ x_3 \end{bmatrix} = \begin{bmatrix} -3 \\ 23 \\ -5.4535 \end{bmatrix}$$

and solving by backward substitution gives $x_3 = -2.000$, $x_2 = -1.000$, $x_1 = 1.000$, corrected to three decimal places, agreeing with the results obtained by Gaussian Elimination.

The Doolittle decomposition method fails if one of the diagonal elements of U, which are used as divisors in the back substitution (1.42), vanishes. This corresponds to the occurrence of a zero pivot in the method of Gaussian Elimination and may be avoided by introducing a modification analogous to the interchange of rows in Gaussian Elimination.

The *Crout* decomposition of A into the form LU imposes the value unity on the diagonal elements of U instead of L. Apart from this, the methods proceed in much the same way and there is little to distinguish between them in accuracy.

The *Choleski* decomposition of a symmetric matrix A decomposes the matrix to $A = LL^{\mathrm{T}}$, where L is lower triangular, so that

$$\begin{bmatrix} a_{11} & a_{12} & \dots & a_{1N} \\ a_{21} & a_{22} & \dots & a_{2N} \\ \dots & \dots & & \dots \\ a_{N1} & a_{N2} & \dots & a_{NN} \end{bmatrix} = \begin{bmatrix} l_{11} & & & \\ l_{21} & l_{22} & & 0 \\ \dots & \dots & & \\ l_{N1} & l_{N2} & \dots & l_{NN} \end{bmatrix} \begin{bmatrix} l_{11} & l_{21} & \dots & l_{N1} \\ & l_{22} & \dots & l_{N2} \\ & & & \dots \\ 0 & & & l_{NN} \end{bmatrix}.$$

The determination of each l_{ij} ($i = 1, \dots, N$; $j = 1, \dots, i$) is carried out as for the Doolittle method. The method uses the symmetry of A and is summarized by the formulas

$$l_{ii} = \left[a_{jj} - \sum_{k=1}^{j-1} l_{jk}^2 \right]^{\frac{1}{2}} ; \qquad i = 1, 2, \dots, N \quad \text{and} \quad j = i \tag{1.43}$$

$$l_{ij} = \left[a_{ji} - \sum_{k=1}^{j-1} l_{ik} l_{jk} \right] ; \qquad j = 1, \dots, N \quad \text{and} \quad i = j+1, j+2, \dots, N \,. \tag{1.44}$$

Including the forward and backward substitutions, Choleski's method uses $\frac{1}{6}N^3 +$

$\frac{3}{2}N^2 + \frac{1}{3}N$ multiplications/divisions, $\frac{1}{6}N^3 + N^2 - \frac{7}{6}N$ additions/subtractions and N square roots. This represents a significant saving in CPU time over Gaussian Elimination or Doolittle's method.

The obvious advantage of the Choleski method over the Doolittle method is that once L has been determined, $U = L^T$ follows immediately. For a large matrix A, the saving in computational effort is worthwhile. The biggest disadvantage of the method is that the method fails if any diagonal element of L is zero, and any interchange of rows to avoid this destroys symmetry. This difficulty does not arise, of course, for symmetric matrices which are positive definite, for then all diagonal elements of L are real and positive. An additional disadvantage of the method is that, in (1.43), it may be necessary to find the square root of a negative number, so that pure imaginary numbers will arise during the computation. No complex numbers will occur, however, and the decomposition is valid.

An important special matrix which arises frequently in the numerical solution of partial differential equations is the tridiagonal matrix defined in (1.45) below. Often, tridiagonal matrices associated with partial differential equations are positive definite so that Gaussian Elimination will not fail to give the unique solution of the related linear system.

The general form of a tridiagonal matrix of order N is

$$A = \begin{bmatrix} a_1 & c_1 & & 0 \\ b_2 & a_2 & c_2 & \\ \cdots & \cdots & \cdots & \\ & b_{N-1} & a_{N-1} & c_{N-1} \\ 0 & & b_N & a_N \end{bmatrix}, \tag{1.45}$$

though in the numerical solution of partial differential equations, the a_i are often equal to each other as are the b_i and the c_i. The LU decomposition of such a matrix may be carried out by decomposing A into

$$A = LU = \begin{bmatrix} 1 & & & 0 \\ l_2 & 1 & & \\ & l_3 & 1 & \\ 0 & \cdots\cdots & \cdots\cdots & \\ & & l_N & 1 \end{bmatrix} \begin{bmatrix} u_1 & c_1 & & 0 \\ & u_2 & c_2 & \\ & \cdots\cdots & \cdots\cdots & \\ 0 & & u_{N-1} & c_{N-1} \\ & & & u_N \end{bmatrix}$$

and using the following algorithm which is based on the Doolittle method:

(1) let $u_1 = a_1$;
(2) let $l_i = b_i/u_{i-1}$ and then $u_i = a_i - l_i c_{i-1}$ for $i = 2, \ldots, N$.

The solution vector $\mathbf{x}$ is then computed using forward and backward substitution, as in equations (1.41), (1.42). The algorithm must be used with care because it does not pivot and some of the u_i may be zero.

This algorithm requires only $5N-4$ multiplications/divisions and $3N-3$ additions/subtractions and obviously has a considerable advantage in CPU time over Gaussian Elimination, Doolittle's method and Choleski's method, which do not take into account the tridiagonality of the matrix. This advantage is particularly marked for large N.

1.5 ITERATIVE METHODS

When the linear system is small enough to be economically stored in the memory of a computer, it is generally most efficient to use a direct method to obtain the solution. Large sparse systems, on the other hand, which are frequently associated with the numerical solution of partial differential equations, can generally be solved more efficiently using an *iterative method*, provided that full advantage is taken of the sparseness of the coefficient matrix.

An iterative technique to solve a linear system of the form (1.21) starts with an initial approximation $\mathbf{x}^{(0)}$ to the solution $\mathbf{x}$ and generates a sequence of vectors $\mathbf{x}^{(0)}$, $\mathbf{x}^{(1)}$, $\mathbf{x}^{(2)}$, ... which converges to $\mathbf{x}$. Such iterative methods proceed to convert the given system $A\mathbf{x} = \mathbf{b}$ into an equivalent system of the form $\mathbf{x} = B\mathbf{x}+\mathbf{c}$ where B is a square matrix of order N and $\mathbf{c}$ is a vector with N components.

The choice of $\mathbf{x}^{(0)}$ is largely arbitrary (the zero-vector is the usual choice) and the sequence of iterates is generated from

$$\mathbf{x}^{(n+1)} = B\mathbf{x}^{(n)}+\mathbf{c} \; ; \qquad n = 0,1,2,\ldots \; , \tag{1.46}$$

Writing $\mathbf{x} = A^{-1}\mathbf{b}$ in $\mathbf{x} = B\mathbf{x}+\mathbf{c}$, it is easy to show that $\mathbf{c} = (I-B)A^{-1}\mathbf{b}$ and defining the error at the nth iterate to be

$$\mathbf{e}^{(n)} = \mathbf{x}-\mathbf{x}^{(n)} \tag{1.47}$$

it follows that

$$\begin{aligned}
\mathbf{e}^{(n+1)} &= \mathbf{x}-\mathbf{x}^{(n+1)} = \mathbf{x}-B\mathbf{x}^{(n)}-\mathbf{c} \\
&= \mathbf{x}-B(\mathbf{x}-\mathbf{e}^{(n)})-\mathbf{c} \\
&= B\mathbf{e}^{(n)}+(I-B)\mathbf{x}-\mathbf{c} \\
&= B\mathbf{e}^{(n)}+(I-B)A^{-1}\mathbf{b}-\mathbf{c} \\
&= B\mathbf{e}^{(n)} \; ,
\end{aligned}$$

that is at each iteration the error is premultiplied by the matrix B. Hence

$$\mathbf{e}^{(n+1)} = B^{n+1}\mathbf{e}^{(0)} \tag{1.48}$$

and the method converges if $\lim_{n \to \infty} \| \mathbf{e}^{(n)} \| = 0$. Further analysis of the convergence of the sequence of iterates requires the following lemma, which is given without proof:

Lemma 1.8 A necessary and sufficient condition for the convergence of (1.46) to the solution $\mathbf{x}$ from any initial vector $\mathbf{x}^{(0)}$ is that the series $I + B + B^2 + \ldots$ is convergent.

Lemma 1.5 then shows that (1.46) converges if $\| B \| < 1$ for some matrix norm. This follows also from (1.48) since

$$\| \mathbf{e}^{(n+1)} \| \leqslant \| B^{n+1} \| . \| \mathbf{e}^{(0)} \| = \| B \|^{n+1} \| \mathbf{e}^{(0)} \|$$

so that $\| B \| < 1$ is a sufficient condition for the convergence of (1.46) to $\mathbf{x}$. In particular, the sufficient condition for convergence is $\rho(B) < 1$.

The first iterative method of the type (1.46) to be discussed in this chapter is the *Jacobi* method in which the matrix A is split into a diagonal matrix, a lower triangular matrix and an upper triangular matrix as follows:

$$A = \begin{bmatrix} a_{11} & a_2 & \ldots & a_{1N} \\ a_{21} & a_{22} & \ldots & a_{2N} \\ \ldots & \ldots & & \ldots \\ a_{N1} & a_{N2} & \ldots & a_{NN} \end{bmatrix}$$

$$= \begin{bmatrix} a_{11} & & & 0 \\ & a_{22} & & \\ & & \ldots\ldots & \\ 0 & & & a_{NN} \end{bmatrix} + \begin{bmatrix} 0 & & & 0 \\ a_{21} & 0 & & \\ \ldots & & & \\ a_{N1} & \ldots & a_{N,N-1} & 0 \end{bmatrix} + \begin{bmatrix} 0 & a_{12} & \ldots & a_{1N} \\ & 0 & & \ldots \\ & & & a_{N-1,N} \\ 0 & & & 0 \end{bmatrix} \tag{1.49}$$

$$= D + L + U.$$

The system $A\mathbf{x} = \mathbf{b}$ then becomes $(D + L + U)\mathbf{x} = \mathbf{b}$, which may be arranged to $D\mathbf{x} = -(L + U)\mathbf{x} + \mathbf{b}$ and then to $\mathbf{x} = -D^{-1}(L + U)\mathbf{x} + D^{-1}\mathbf{b}$. This suggests the iterative scheme

$$\mathbf{x}^{(n+1)} = -D^{-1}(L + U)\mathbf{x}^{(n)} + D^{-1}\mathbf{b} \ ; \qquad n = 0, 1, 2, \ldots , \tag{1.50}$$

which is of the form (1.46) with $B = -D^{-1}(L + U)$ and $\mathbf{c} = D^{-1}\mathbf{b}$.

Clearly D must be nonsingular and the given system must be rearranged, if necessary, to ensure that the diagonal elements of A are nonzero. In practice, the system is rearranged to ensure that A is diagonally dominant, for then the elements of D^{-1}, and consequently of B, are as small as possible so that convergence, if it occurs at all, is as rapid as possible. A sufficient condition for convergence of the Jacobi method is

$$\| -D^{-1}(L+U) \| < 1 \ . \tag{1.51}$$

Example 1.4 Solve the following system using the Jacobi method

$$\begin{aligned} 10x_1 + x_2 - 2x_3 + 2x_4 &= 3, \\ 2x_1 - 8x_2 + x_3 - 3x_4 &= 15, \\ 4x_1 - x_2 + 7x_3 + 2x_4 &= 19, \\ 4x_1 + 3x_2 + x_3 - 9x_4 &= -22, \end{aligned} \tag{1.52}$$

The system has to be written in diagonally dominant form. The first, second, third and fourth equations of (1.52) are rearranged to express the unknowns x_1, x_2, x_3, x_4, respectively, in terms of the other variables. The superscripts relating to the iterations are appended as in (1.50) to give

$$\begin{aligned} x_1^{(n+1)} &= -\tfrac{1}{10}[\quad x_2^{(n)} - 2x_3^{(n)} + 2x_4^{(n)}] - \tfrac{3}{10} \ , \\ x_2^{(n+1)} &= \tfrac{1}{8}\ [2x_1^{(n)} \quad + x_3^{(n)} - 3x_4^{(n)}] - \tfrac{15}{8} \ , \\ x_3^{(n+1)} &= -\tfrac{1}{7}\ [4x_1^{(n)} - x_2^{(n)} \quad + 2x_4^{(n)}] - \tfrac{19}{7} \ , \\ x_4^{(n+1)} &= \tfrac{1}{9}\ [4x_1^{(n)} + 3x_2^{(n)} + x_3^{(n)} \quad] + \tfrac{22}{9} \ . \end{aligned} \tag{1.53}$$

The reader will note that, in (1.53), gaps have been left in all four equations, from where the terms on the left-hand side have come.

Using the initial approximation $\mathbf{x}^{(0)} = (0,0,0,0)^{\mathrm{T}}$, succeeding iterations are shown in Table 1.1. The iterates are given to four decimal places, and after 15 iterations from $\mathbf{x}^{(0)}$ the method has attained convergence of *all* of x_1, x_2, x_3, x_4 to three decimal places; the true solution is $x_1 = 0$, $x_2 = 1$, $x_3 = 2$, $x_4 = 3$.

The criterion used for stopping the iterations, namely $|x_m^{(n+1)} - x_m^{(n)}| < 10^{-3}$ for all $m = 1,2,3,4$, is an example of

$$\| \mathbf{x}^{(n+1)} - \mathbf{x}^{(n)} \| < \varepsilon \qquad \text{for some } n \ , \tag{1.54}$$

where $\varepsilon > 0$ is some *tolerance* and the usual norm used is the L_∞ norm for vectors given by

$$\| \mathbf{x} \| = \max_r |x_r| \ ; \qquad r = 1,2,\ldots,N \ .$$

Another criterion for stopping the iterations is

Table 1.1

Iterate n	$x_1^{(n)}$	$x_2^{(n)}$	$x_3^{(n)}$	$x_4^{(n)}$
0	0.0000	0.0000	0.0000	0.0000
1	0.3000	1.8750	2.7143	2.4444
2	0.1665	1.3726	2.1123	3.5044
3	−0.1157	0.8665	1.8140	3.2107
4	−0.0660	0.8688	1.9868	2.8834
5	0.0338	1.0256	2.0523	2.9255
6	0.0228	1.0429	2.0056	3.0294
7	−0.0090	0.9954	1.9847	3.0251
8	−0.0076	0.9864	1.9973	2.9928
9	0.0023	1.0005	2.0045	2.9918
10	0.0025	1.0042	2.0011	3.0017
11	−0.0005	1.0001	1.9987	3.0026
12	−0.0008	0.9987	1.9996	2.9997
13	0.0001	0.9999	2.0004	2.9992
14	0.0003	1.0004	2.0002	3.0000
15	−0.0000	1.0001	1.9999	3.0003

$$\frac{\|\mathbf{x}^{(n+1)} - \mathbf{x}^{(n)}\|}{\|\mathbf{x}^{(n)}\|} < \varepsilon \tag{1.55}$$

for some n and some tolerance $\varepsilon > 0$. The absolute error criterion given by (1.54) is suitable for use when the variables x_i ($i = 1, \ldots, N$) are all of the same order or magnitude, while (1.55), which calculates relative errors, is more suitable for use when the variables x_i ($i = 1, \ldots, N$) have widely differing orders of magnitude.

The second iterative method of the type (1.46) to be discussed is the method of *Gauss–Seidel* for which the matrix A is again split into the form (1.49). The system $A\mathbf{x} = \mathbf{b}$ is then rearranged in the form $(D+L)\mathbf{x} = -U\mathbf{x} + \mathbf{b}$, and subsequently to $\mathbf{x} = -(D+L)^{-1}U\mathbf{x} + (D+L)^{-1}\mathbf{b}$, suggesting the iterative scheme

$$\mathbf{x}^{(n+1)} = -(D+L)^{-1}U\mathbf{x}^{(n)} + (D+L)^{-1}\mathbf{b} \tag{1.56}$$

which is of the form (1.46) with $B = -(D+L)^{-1}U$ and $\mathbf{c} = (D+L)^{-1}\mathbf{b}$. A sufficient condition for the convergence of the Gauss–Seidel method from some initial vector $\mathbf{x}^{(0)}$ to the solution $\mathbf{x}$ of (1.21) is clearly

$$\| -(D+L)^{-1}U\| < 1$$

The criteria (1.54), (1.55) for stopping the iterations may still be used.

Returning to Example 1.4, the rearrangement (1.53) must be augmented to

$$
\begin{aligned}
x_1^{(n+1)} &= -\tfrac{1}{10}[\qquad\quad x_2^{(n)} - 2x_3^{(n)} + 2x_4^{(n)}] - \tfrac{3}{10}\,,\\
x_2^{(n+1)} &= \quad -\tfrac{1}{8}[2x_1^{(n+1)} \qquad + \; x_3^{(n)} - 3x_4^{(n)}] - \tfrac{15}{8}\,,\\
x_3^{(n+1)} &= -\tfrac{1}{7}[4x_1^{(n+1)} - \; x_2^{(n+1)} \qquad + 2x_4^{(n)}] - \tfrac{19}{7}\,,\\
x_4^{(n+1)} &= \quad \tfrac{1}{9}[4x_1^{(n+1)} + 3x_2^{(n+1)} + \; x_3^{(n+1)} \qquad] + \tfrac{22}{9}\,.
\end{aligned}
\tag{1.57}
$$

$n = 0, 1, 2, \ldots$, to be consistent with (1.46). It is easy to see that, in the Gauss–Seidel method, an iterate is used as soon as it is calculated, whereas in the Jacobi method a new set of iterates $x_m^{(n+1)}$ ($m = 1, \ldots, N$; $n = 0, 1, 2, \ldots$) was not used until the iteration procedure had completed its sweep through the system

Using $\mathbf{x}^{(0)} = (0, 0, 0, 0)^{\mathrm{T}}$, succeeding iterates for the Gauss–Seidel method are shown in Table 1.2.

Table 1.2

Iterate n	$x_1^{(n)}$	$x_2^{(n)}$	$x_3^{(n)}$	$x_4^{(n)}$
0	0.0000	0.0000	0.0000	0.0000
1	0.3000	1.9500	2.8214	3.5413
2	− 0.0390	0.8900	1.8519	2.9295
3	− 0.0045	1.0068	2.0237	3.0029
4	0.0035	1.0028	1.9976	3.0022
5	− 0.0012	0.9986	1:9999	2.9990
6	0.0003	1.0004	2.0002	3.0003
7	− 0.0001	0.9999	1.9999	2.9999

In this example, the Gauss–Seidel method has converged in less than half as many iterations as the Jacobi method; this is fairly typical of the behaviour of the two methods. A general comparison of them produces the following observations:

(1) the Gauss–Seidel method invariably converges when the Jacobi method does;
(2) the Gauss–Seidel method may converge when the Jacobi method diverges;
(3) generally, when both methods converge, the Gauss–Seidel method converges more quickly than the Jacobi method.

These observations are no more than general guidelines; in fact, there are linear systems for which the Jacobi method converges while the Gauss–Seidel method does not (see Varga (1962, p. 74)).

It is not necessary to complete a sweep of the Gauss–Seidel method through the linear system before concluding that convergence has taken place. For instance, examination of Table 1.2 leads to the conclusion that convergence to two decimal places has been reached for all variables as soon as $x_3^{(5)}$ has been calculated; convergence of x_4 had been achieved after four iterations. For the Jacobi method, on the other hand, complete sweeps of the system are required, convergence to two decimal places having been reached after 11 complete iterations. In practice, of course, it is quicker to carry out complete sweeps with the Gauss–Seidel method than to check for convergence after individual iterates.

It was noted earlier that the equations of the given linear system should be arranged in such a way that the coefficient matrix A is diagonally dominant. Suppose, now, that the first and second equations and the third and fourth equations of Example 1.4 are interchanged. The Gauss–Seidel formulation (1.57) thus becomes

$$\begin{aligned}
x_1^{(n+1)} &= \tfrac{1}{2}[8x_2^{(n)} - x_3^{(n)} + 3x_4^{(n)}] + \tfrac{15}{2} \ , \\
x_2^{(n+1)} &= -10x_1^{(n+1)} + 2x_3^{(n)} - 2x_4^{(n)} + 3 \ , \\
x_3^{(n+1)} &= -4x_1^{(n+1)} - 3x_2^{(n+1)} + 9x_4^{(n)} - 22 \ , \\
x_4^{(n+1)} &= \tfrac{1}{2}[-4x_1^{(n+1)} + x_2^{(n+1)} - 7x_3^{(n+1)}] + \tfrac{19}{2} \ .
\end{aligned}$$

Starting with the initial approximation $\mathbf{x}^{(0)} = (0, 0, 0, 0)^{\mathrm{T}}$ as before, succeeding iterates are shown in Table 1.3 where it is obvious that divergence has taken place.

Table 1.3

Iterate n	$x_1^{(n)}$	$x_2^{(n)}$	$x_3^{(n)}$	$x_4^{(n)}$
0	0.0000E + 00	0.0000E + 00	0.0000E + 00	0.0000E + 00
1	0.7500E + 01	− 0.7200E + 02	0.1640E + 03	− 0.6155E + 03
2	− 0.1286E + 04	0.1442E + 05	− 0.4368E + 05	0.1627E + 06
3	0.3235E + 06	− 0.3648E + 07	0.1111E + 08	− 0.4137E + 08
4	− 0.8220E + 08	0.9270E + 09	− 0.2824E + 10	0.1051E + 11

The reader will be able to demonstrate for himself or herself the failure of the Jacobi method following a similar rearrangement of (1.52). These failures emphasize the need to have a coefficient matrix which is diagonally dominant.

It is easy to show that the Gauss–Seidel method may be written in the form

$$x_i^{(n+1)} = x_i^{(n)} + \left[b_i - \sum_{j=1}^{i-1} a_{ij} x_j^{(n+1)} - \sum_{j=1}^{N} a_{ij} x_j^{(n)} \right] \Big/ a_{ii} \tag{1.58}$$

for every $i = 1, \ldots, N$. The second term on the right-hand side may be regarded as the correction which is added to $x_i^{(n)}$ by one Gauss–Seidel iteration. For the linear systems which arise following approximations to many partial differential equations, the corrections all have the same sign and convergence of the Gauss–Seidel method can be accelerated by multiplying the correction term by the *acceleration parameter* ω which generally lies in the interval $1 < \omega < 2$.

Methods involving such acceleration factors, which are sometimes referred to as *relaxation factors*, are known as *relaxation methods*. For choices of ω in the range $0 < \omega < 1$, the methods are called *under-relaxation methods* and can be used to obtain convergence of some systems for which the Gauss–Seidel method does not converge. For choices of $\omega > 1$, the methods are called *over-relaxation methods* and can be used to accelerate convergence for systems that are convergent for the Gauss–Seidel method. Such methods are usually described as *SOR methods* or *successive over-relaxation methods*.

In general, the SOR iteration is defined by

$$\begin{aligned} x_i^{(n+1)} &= x_i^{(n)} + \omega\left[b_i - \sum_{j=1}^{i-1} a_{ij}x_j^{(n+1)} - \sum_{j=1}^{N} a_{ij}x_j^{(n)}\right]\Big/ a_{ii} \\ &= \omega\left[b_i - \sum_{j=1}^{i-1} a_{ij}x_j^{(n+1)} - \sum_{j=i+1}^{N} a_{ij}x_j^{(n)}\right]\Big/ a_{ii} - (\omega - 1)x_i^{(n)} \end{aligned} \tag{1.59}$$

for each $i = 1, \ldots, N$, from which it is seen that the SOR method is directly related to the Gauss–Seidel method (1.56) and is exactly the Gauss–Seidel method for $\omega = 1$.

In vector form, (1.59) becomes

$$\mathbf{x}^{(n+1)} = \omega D^{-1}\mathbf{b} - \omega D^{-1}L\mathbf{x}^{(n+1)} - \omega D^{-1}U\mathbf{x}^{(n)} - (\omega - 1)\mathbf{x}^{(n)}$$

so that

$$(I + \omega D^{-1}L)\mathbf{x}^{(n+1)} = [(1-\omega)I - \omega D^{-1}U]\mathbf{x}^{(n)} + \omega D^{-1}\mathbf{b} \; .$$

It follows that the SOR method may be written in the form (1.46) as

$$\mathbf{x}^{(n+1)} = (I + \omega D^{-1}L)^{-1}[(1-\omega)I - \omega D^{-1}U]\mathbf{x}^{(n)} + \omega(I + \omega D^{-1}L)^{-1}D^{-1}\mathbf{b} \; .$$

This may be simplified to

$$\mathbf{x}^{(n+1)} = (D + \omega L)^{-1}[(1-\omega)D - \omega U]\mathbf{x}^{(n)} + \omega(D + \omega L)^{-1}\mathbf{b} \tag{1.60}$$

which is of the form (1.46) with $B = (D + \omega L)^{-1}[(1-\omega)D - \omega U]$ and $\mathbf{c} = \omega(D + \omega L)^{-1}\mathbf{b}$. A sufficient condition for the convergence of the SOR method is

$$\| (D + \omega L)^{-1}[(1-\omega)D - \omega U] \| < 1 \; . \tag{1.61}$$

The following lemmas relating to the relaxation matrices B_J, B_G, B_ω of the Jacobi, Gauss–Seidel, SOR methods give conditions for the convergence of these methods:

Lemma 1.9 If $a_{ii} \neq 0$ for each $i = 1, \ldots, N$ then $\rho(B_\omega) \geqslant |\omega - 1|$, where $\rho(B_\omega)$ is the spectral radius of B_ω.

Corollary The SOR method converges only if $0 < \omega < 2$.

Proof A sufficient condition for convergence is $\| B_\omega \| < 1$. Considering the spectral radius, it follows that $\rho(B_\omega) < 1$, so that $|\omega - 1| < 1$, and the hypothesis of the corollary follows.

Lemma 1.10 If A is a positive definite matrix and $0 < \omega < 2$, then the SOR method converges to the solution $\mathbf{x}$ of the system $A\mathbf{x} = \mathbf{b}$ from any initial vector $\mathbf{x}^{(0)}$.

Lemma 1.11 If the matrix A is positive definite and tridiagonal, then $\rho(B_G) = [\rho(B_J)]^2 < 1$ and the optimal choice of ω for the SOR method is

$$\omega = 2[1 + \{1 - \rho(B_J)^2\}^{\frac{1}{2}}]^{-1}$$

and with this choice of ω, $\rho(B_\omega) = \omega - 1$.

Returning once again to the numerical example given by (1.52), the solution was obtained using the SOR method with $\omega = 0.9$. Using $\mathbf{x}^{(0)} = (0,0,0,0)^T$, succeeding iterates are shown in Table 1.4 from which it is seen that convergence of all variables

Table 1.4

Iterate n	$x_1^{(n)}$	$x_2^{(n)}$	$x_3^{(n)}$	$x_4^{(n)}$
0	0.0000	0.0000	0.0000	0.0000
1	0.2700	1.7482	2.5288	3.0854
2	0.0395	1.1144	2.0253	3.0612
3	−0.0128	0.9908	1.9922	2.9974
4	−0.0014	0.9987	2.0004	2.9989
5	0.0003	1.0004	2.0003	3.0001
6	0.0000	1.0000	2.0000	3.0000

to three decimal places has occurred after six iterations of the method. With just one decimal place, $\omega = 0.9$ gives the quickest convergence for this example taking one iteration fewer than the Gauss–Seidel method for which $\omega = 1.0$.

1.6 CONJUGATE GRADIENTS

When the matrix of coefficients in (1.21) is symmetric, the method of *conjugate gradients* may be used to determine $\mathbf{x}$. This method, which uses the best features of both direct and iterative methods, needs a *residual vector* $\mathbf{r} = \mathbf{b} - A\mathbf{x}$ and a scalar quantity $s = \mathbf{r}^{\mathrm{T}}\mathbf{r}$ which is the sum of the squares of the components of $\mathbf{r}$. The aim of the method is clearly to minimize s.

The iterates $\mathbf{x}^{(k)}$ $(k = 0, 1, 2, \ldots)$ are determined from

$$\mathbf{x}^{(k+1)} = \mathbf{x}^{(k)} + \alpha_k \mathbf{w}^{(k)} \tag{1.62}$$

where α_k is a scalar and $\mathbf{w}^{(k)}$ is a vector to be determined. The scalar α_k is chosen to make $s_{k+1} = \mathbf{r}^{(k)^{\mathrm{T}}}\mathbf{r}^{(k)}$ a minimum, in which case

$$\alpha_k = \mathbf{w}^{(k)^{\mathrm{T}}}\mathbf{r}^{(k)}/\mathbf{w}^{(k)^{\mathrm{T}}}A\mathbf{w}^{(k)} \tag{1.63}$$

(see, for example, Greig (1980)). In determing $\mathbf{w}^{(k)}$, the combination of $\mathbf{r}^{(k)}$ and $\mathbf{w}^{(k-1)}$ given by

$$\mathbf{w}^{(k)} = \mathbf{r}^{(k)} + \beta_{k-1}\mathbf{w}^{(k-1)} \tag{1.64}$$

is chosen. The successive constants β_k can be chosen so that the process terminates in exactly N steps. To this end, $\mathbf{w}^{(k)}$ is made *conjugate* (that is orthogonal with respect to the matrix A) to $\mathbf{w}^{(k-1)}$, that is $\mathbf{w}^{(k)^{\mathrm{T}}}A\mathbf{w}^{(k-1)} = 0$. The corresponding value of β is thus given by

$$\beta_{k-1} = -\mathbf{r}^{(k)^{\mathrm{T}}}A\mathbf{w}^{(k-1)}/\mathbf{w}^{(k-1)^{\mathrm{T}}}A\mathbf{w}^{(k-1)} . \tag{1.65}$$

The full computational algorithm is given by

$$\beta_{k-1} = -\mathbf{r}^{(k)^{\mathrm{T}}}A\mathbf{w}^{(k-1)}/\mathbf{w}^{(k-1)^{\mathrm{T}}}A\mathbf{w}^{(k-1)} ,$$

$$\mathbf{w}^{(k)} = \mathbf{r}^{(k)} + \beta_{k-1}\mathbf{w}^{(k-1)} ,$$

$$\alpha_k = \mathbf{r}^{(k)^{\mathrm{T}}}\mathbf{w}^{(k)}/\mathbf{w}^{(k)^{\mathrm{T}}}A\mathbf{w}^{(k)} , \tag{1.66}$$

$$\mathbf{x}^{(k+1)} = \mathbf{x}^{(k)} + \alpha_k\mathbf{w}^{(k)} ,$$

$$\mathbf{r}^{(k+1)} = \mathbf{b} - A\mathbf{x}^{(k+1)} = \mathbf{r}^{(k)} - \alpha_k A\mathbf{w}^{(k)} , \tag{1.67}$$

with $\mathbf{x}^{(0)}$ arbitrary and $\mathbf{w}^{(-1)} = \mathbf{0}$ so that $\mathbf{w}^{(0)} = \mathbf{r}^{(0)}$.

It is well beyond the intended scope of the present text to undergo a study of the advanced numerical methods for solving equations of the form (1.21). Such methods include band and envelope methods, the minimum degree ordering algorithm for use when A is sparse, quotient tree methods for finite difference and finite element problems, one-way dissection methods for finite element problems, and nested dissection methods. For a detailed study of these methods the reader is referred to an advanced text, such as that by George and Liu (1981) which contains a number of FORTRAN subroutines.

1.7 THE NUMERICAL EIGENVALUE PROBLEM

1.7.1 Introduction

The solution of many problems in the mathematical modelling of biomedical systems requires the estimation of the eigenvalues of a matrix associated with a system of linear algebraic equations. It has been noted already that a square matrix A of order N has exactly N eigenvalues, not necessarily distinct, which may be found by solving the *characteristic equation* $p(\lambda) = \det(A - \lambda I) = 0$ for λ. When N is large, finding the characteristic polynomial $p(\lambda)$ is not itself an easy exercise, while finding the zeros of a polynomial of large degree also poses problems as will be seen in Chapter 3.

The numerical eigenvalue problem attempts to overcome these difficulties by providing the user with approximation techniques for finding the eigenvalues. In subsection 1.7.2 the *Krylov* method of finding the characteristic equation is described: this direct method is useful for small N, as it is relatively easy to find the zeros of a low-degree polynomial. The *power* method is described in subsection 1.7.3: this iterative method locates the eigenvalue of largest modulus which is assumed to be distinct. In subsection 1.7.4, *transformation* methods are discussed briefly.

An understanding of these techniques requires the stating of the following definitions and results.

Definition 1.16 Two matrices A and B are said to be *similar* if a nonsingular matrix Q exists with $A = Q^{-1}BQ$.

Theorem 1.9 Suppose that A and B are similar matrices and that λ is an eigenvalue of A with associated eigenvalue $\mathbf{x}$. Then λ is also an eigenvalue of B and if $A = Q^{-1}AQ$, then $Q\mathbf{x}$ is an eigenvector associated with the eigenvalue λ of the matrix B.

Definition 1.17 A set of vectors $\mathbf{y}_1, \mathbf{y}_2, \ldots, \mathbf{y}_N$ is said to be *orthogonal* if $\mathbf{y}_r^T\mathbf{y}_s = 0$ for all $r \neq s$. If, in addition $\mathbf{y}_r^T\mathbf{y}_r = 1$ *for all* $r = 1,2,\ldots, N$ then the set is said to be *orthonormal*.

The reader should note that in testing for orthonormality, it is necessary to check for orthogonality first.

Theorem 1.10 If A is a square matrix of order N and $\lambda_1, \lambda_2, \ldots, \lambda_m$ $(m \leq N)$ are distinct eigenvalues of A with associated eigenvectors $\mathbf{x}_1, \mathbf{x}_2, \ldots, \mathbf{x}_k$, then the set $\{\mathbf{x}_1, \mathbf{x}_2, \ldots, \mathbf{x}_k\}$ is linearly independent.

Theorem 1.11 An orthogonal set of vectors which does not contain the zero-vector is linearly independent.

Definition 1.18 A matrix P is said to be an *orthogonal matrix* if $\mathrm{P}^{-1} = P^{\mathrm{T}}$.

Theorem 1.12 If A is a Hermitian matrix, then there exists a unitary matrix U such that $U^{-1}AU$ is a diagonal matrix.

Theorem 1.13 If A is a square symmetric matrix, then there exists an orthogonal matrix P such that $D = P^{-1}AP$ where D is a diagonal matrix, the diagonal elements of which are the eigenvalues of A.

Theorem 1.14 If A is a square symmetric matrix of order N, then these exist N eigenvectors of A that form an orthonormal set.

Theorem 1.15 The eigenvectors of a symmetric matrix are all real.

1.7.2 The Krylov method

Suppose that the characteristic polynomial of a matrix A has the form

$$p(\lambda) = b_0 + b_1\lambda + b_2\lambda^2 + \ldots + b_{N-1}\lambda^{N-1} + (-1)^N\lambda^N , \tag{1.68}$$

where the b_i ($i = 0, 1, \ldots, N-1$) are constants. The characteristic equation is therefore

$$b_0 + b_1\lambda + \ldots + b_{N-1}\lambda^{N-1} + (-1)^N\lambda^N = 0 . \tag{1.69}$$

Using the Cayley–Hamilton Theorem (Theorem 1.2) it follows that

$$b_0 I + b_1 A + b_2 A^2 + \ldots + b_{N-1}A^{N-1} + (-1)^N A^N = 0 \tag{1.70}$$

where 0 is the zero-matrix of order N. Post-multiplication of (1.70) by an arbitrary vector $\mathbf{y} \neq \mathbf{0}$ gives

$$b_0\mathbf{y} + bA\mathbf{y} + b_2A^2\mathbf{y} + \ldots + b_{N-1}A^{N-1}\mathbf{y} + (-1)^N A^N\mathbf{y} = \mathbf{0} \tag{1.71}$$

which is, in fact, a system of N linear equations in the N unknowns $b_0, b_1, \ldots, b_{N-1}$. This system may be solved using one of the methods described in sections 1.4 or 1.5. Having found the b_i, the characteristic polynomial $p(\lambda)$ may be written down immediately.

The expensive feature of the Krylov method is the need to calculate the powers of A. Expense notwithstanding, some of the higher powers of A may be very large in size and some may be very small, implying that (1.71) may be an ill-conditioned system.

Example 1.5 A matrix which occurs frequently in the differential equations arising from the mathematical modelling of numerous biomedical systems (see Chapter 8) is the tridiagonal matrix A_N of order N defined by $a_{ii} = -2\,(i = 1,2,\ldots,N)$, $a_{r,r+1} = 1\,(r = 1,2,\ldots,N-1)$, $a_{s,s-1} = 1\,(s = 2,3,\ldots,N)$. Use the Krylov method to find the characteristic polynomial of A_3.

$$\text{Clearly, } A_3 = \begin{bmatrix} -2 & 1 & 0 \\ 1 & -2 & 1 \\ 0 & 1 & 2 \end{bmatrix} \text{ so that } A_3^2 = \begin{bmatrix} 5 & -4 & 1 \\ -4 & 6 & -4 \\ 1 & 4 & 5 \end{bmatrix}$$

$$\text{and } A_3^2 = \begin{bmatrix} -14 & 14 & -6 \\ 14 & -18 & 14 \\ -6 & -18 & 14 \end{bmatrix}.$$

The reader will note that the elements of successive powers of A_3 are indeed increasing. Choosing $\mathbf{y} = [1,0,0]^{\mathrm{T}}$ and noting that N is odd, equation (1.71) becomes

$$b_0 \begin{bmatrix} 1 \\ 0 \\ 0 \end{bmatrix} + b_1 \begin{bmatrix} -2 & 1 & 0 \\ 1 & -2 & 1 \\ 0 & 1 & -2 \end{bmatrix} \begin{bmatrix} 1 \\ 0 \\ 0 \end{bmatrix} + b_2 \begin{bmatrix} 5 & -4 & 1 \\ -4 & 6 & -4 \\ 1 & -4 & 5 \end{bmatrix} \begin{bmatrix} 1 \\ 0 \\ 0 \end{bmatrix}$$

$$- \begin{bmatrix} -14 & 14 & -6 \\ 14 & -18 & 14 \\ -6 & 14 & -14 \end{bmatrix} \begin{bmatrix} 1 \\ 0 \\ 0 \end{bmatrix} = \begin{bmatrix} 0 \\ 0 \\ 0 \end{bmatrix},$$

which gives

$$\begin{aligned} b_0 - 2b_1 + 5b_2 + 14 &= 0\,, \\ b_1 + 4b_2 - 14 &= 0\,, \\ b_2 + 6 &= 0\,. \end{aligned}$$

so that $b_2 = -6$, $b_1 = -10$, $b_0 = -4$, giving the characteristic equation

$$p(\lambda) = -4 - 10\lambda - 6\lambda^2 - \lambda^3\,.$$

The reader can verify that this is indeed the characteristic polynomial by simplifying the determinant $\det(A_3 - \lambda I)$. Clearly, Krylov's method would not, in practice, be chosen in preference to this elementary method for matrices of low order.

1.7.3 The power method

In using numerical methods to solve the partial differential equations arising from the mathematical modelling of many biomedical systems, knowledge of the eigenvalue of largest modulus of a matrix is often required. This eigenvalue can be found using the *power* method, which starts with an initial guess to the eigenvalue and follows an iterative procedure to obtain the eigenvalue to the required accuracy.

In order to apply the power method, it must be assumed that the matrix A of order N has N eigenvalues $\lambda_1, \lambda_2, \ldots, \lambda_N$ with an associated set of eigenvectors $\mathbf{x}_1, \mathbf{x}_2, \ldots, \mathbf{x}_N$ which is linearly independent. It is also assumed that A has an eigenvalue which is larger in modulus than all the others. The eigenvalues may then, without loss of generality, be assumed to follow the ordering given by

$$|\lambda_1| > |\lambda_2| \geqslant |\lambda_3| \geqslant \ldots \geqslant |\lambda_{N-1}| \geqslant |\lambda_N| \geqslant 0 \,. \tag{1.72}$$

The power method allows A to operate repeatedly on the vector $\mathbf{x}$, which is a linear combination of the eigenvectors $\mathbf{x}_1, \mathbf{x}_2, \ldots, \mathbf{x}_N$, that is to say

$$\mathbf{x} = c_1\mathbf{x}_1 + c_2\mathbf{x}_2 + \ldots + c_N\mathbf{x}_N \,. \tag{1.73}$$

Then,

$$\begin{aligned} A\mathbf{x} &= c_1A\mathbf{x}_1 + c_2A\mathbf{x}_2 + \ldots + c_NA\mathbf{x}_N = c_1\lambda_1\mathbf{x}_1 + c_2\lambda_2\mathbf{x}_2 + \ldots + c_N\lambda_N\mathbf{x}_N \\ &= \lambda_1\left(c_1\mathbf{x}_1 + c_2\frac{\lambda_2}{\lambda_1}\mathbf{x}_2 + \ldots + c_N\frac{\lambda_N}{\lambda_1}\mathbf{x}_N\right) \end{aligned}$$

and thus

$$A^k\mathbf{x} = \lambda_1^k\left\{c_1\mathbf{x}_1 + c_2\left(\frac{\lambda_2}{\lambda_1}\right)^k\mathbf{x}_2 + \ldots + c_N\left(\frac{\lambda_N}{\lambda_1}\right)^k\mathbf{x}_N\right\} .$$

For large values of k, the vector

$$c_1\mathbf{x}_1 + \left(\frac{\lambda_2}{\lambda_1}\right)^k\mathbf{x}_2 + \ldots + \left(\frac{\lambda_N}{\lambda_1}\right)^k\mathbf{x}_N$$

tends to $c_1\mathbf{x}_1$ (because of (1.72)), that is the eigenvector corresponding to λ_1. This eigenvalue is actually obtained as

$$\lambda_1 = \lim_{k\to\infty} (A^{k-1}\mathbf{x})_r/(A^k\mathbf{x})_r \,, \qquad r = 1,2,\ldots,n$$

where the subscript r signifies the rth component in the corresponding vector. It is clear that the rate of convergence depends on the quotient λ_2/λ_1; convergence is faster the smaller $|\lambda_2/\lambda_1|$ is.

For computational purposes, the power method may be applied in the way indicated in the following algorithm for $k = 0, 1, \ldots$:

(1) given a vector $\mathbf{y}_k$, form two other vectors $\mathbf{z}_{k+1}$ and $\mathbf{y}_{k+1}$ from

$$\mathbf{z}_{k+1} = A\mathbf{y}_k \;,$$

$$\alpha_{k+1} = \max_r |(z_{k+1})_r| = \text{largest (modulus) component of } \mathbf{z}_{k+1}$$

(note α_{k+1} is not the modulus; it is the algebraic value of the element),

$$\mathbf{y}_{k+1} = \mathbf{z}_{k+1}/\alpha_{k+1} \text{ (this step normalizes } \mathbf{z}_{k+1});$$

(2) α_{k+1} is the current approximation to λ_1 and $\mathbf{y}_{k+1}$ the current approximation to $\mathbf{x}_1$;

(3) next k.

The initial approximation to x_1 is chosen in a convenient way such as $\mathbf{y}_0 = [1, 1, \ldots, 1]^T$ or $\mathbf{y}_0 = [0, 0, \ldots, 1, 0, \ldots, 0]^T$ (here, the lone 1 is in position i or j where a_{ij} is the element of A of largest modulus).

Example 1.6 Starting with the initial vector $\mathbf{y}_0 = [1, 0, 0]^T$, use the power method to find, to two decimal places, the eigenvalue of the matrix A in Example 1.5 which has largest modulus.

Taking $\mathbf{y}_0 = [1, 0, 0]^T$, the computation proceeds as in Table 1.5. The largest modulus eigenvalue is therefore $\lambda_1 = -3.41$ to two decimal places and the corresponding eigenvector is $[-0.71, 1.00, -0.70]^T$. The reader can check by putting $h = 1$ and $N = 3$ in equation (8.12) of Chapter 8 that $\lambda_3 = -3.4142\ldots$.

The reader will note that, although the required eigenvalue had converged to two decimals of accuracy after seven iterations, it was not until the eleventh iteration that all three components of the associated eigenvector had converged to the same accuracy.

1.7.4 Transformation methods

All transformation methods are based on the fact that the eigenvalues of any matrix are invariant under any similarity transformation (Theorem 1.9). The approach used in transformation methods is to construct a series of simple transformations which will transform the given matrix into a special form, such as diagonal, tridiagonal or triangular form (upper or lower). Often the transformation process is slow.

In the case of real symmetric matrices, Theorem 1.13 shows that the original matrix A can be diagonalized by orthogonal transformations; the eigenvalues will then appear on the main diagonal. The *Jacobi* method for finding the eigenvalues of such a matrix is based on this principle. The *Householder* method brings the real symmetric matrix A to tridiagonal form. The eigenvalues are then computed using a

Table 1.5

r	Components of $\mathbf{z}_r$ 1	2	3	α_{r+1}	Components of $\mathbf{y}_r$ 1	2	3
0	—	—	—		1.000	0.000	0.000
1	−2.000	1.000	0.000	2.000	−1.000	0.500	0.000
2	2.500	−2.000	0.500	2.500	1.000	−0.800	0.200
3	−2.800	2.800	−1.200	−2.800	1.000	−1.000	0.429
4	−3.000	3.429	−1.858	3.429	−0.875	1.000	−0.542
5	2.750	−3.417	2.084	−3.417	−0.805	1.000	−0.610
6	2.610	−3.415	2.220	−3.415	−0.764	1.000	−0.650
7	2.528	−3.414	2.300	−3.414	−0.740	1.000	−0.674
8	2.480	−3.414	2.348	−3.414	−0.726	1.000	−0.688
9	2.452	−3.414	2.376	−3.414	−0.718	1.000	−0.696
10	2.436	−3.414	2.392	−3.414	−0.713	1.000	−0.701
11	2.426	−3.414	2.402	−3.414	−0.711	1.000	−0.704

simple recursive formula. Hermitian matrices, which can be diagonalized by unitary transformations (Theorem 1.12), can be treated by variations of these methods.

In trying to determine the eigenvalues of matrices without symmetry properties, it is better first of all to transform them into Hessenberg form.

Definition 1.19 The square matrix $A = [a_{ij}]$ of order N is said to be *upper Hessenberg* form if $a_{ij} = 0$ for $i > j+1$ and to be of *lower Hessenberg* form if $a_{ij} = 0$ for $j > i+1$.

Having been transformed into Hessenberg form, the original matrix is transformed in the *QR* method to triangular form by a sequence of unitary transformations. The *QR* method is efficient and robust and has good convergence properties. It can be used on arbitrary matrices but is most effective when used to determine the eigenvalues of Hessenberg or symmetric band matrices which occur in the numerical solution of second-order parabolic partial differential equations in two space variables (see Chapter 8).

Usually, only the eigenvalue of largest modulus will be needed in analysing the numerical solution of the time-dependent partial differential equations of medical and biological systems; the power method discussed in subsection 1.7.3 can be relied on to obtain this eigenvalue. The transformation methods outlined above will not be discussed further: the interested reader should refer to any of the specialist texts, aimed at advanced levels of readership, such as Burden and Faires (1985), Fröberg (1985), Gerald and Wheatley (1984) and Wilkinson (1965).

PROBLEMS

1. Given a square matrix A of order N whose eigenvalues are $\lambda_1, \lambda_2, \ldots, \lambda_N$ with corresponding eigenvectors $\mathbf{v}_1, \mathbf{v}_2, \ldots, \mathbf{v}_N$. Prove that the eigenvalues of the matrix A^2 are $\lambda_1^2, \lambda_2^2, \ldots, \lambda_N^2$.

2. Use Gaussian elimination to solve the system of equations

$$\begin{aligned} 2x_1 + 6x_2 + 2x_3 &= -2 \,, \\ x_1 + 6x_2 + 4x_3 &= 5 \,, \\ -2x_1 - 5x_2 + 2x_3 &= 1 \,. \end{aligned}$$

3. Use Gaussian elimination to solve the system of equations

$$\begin{aligned} 2x_1 + x_2 \qquad\qquad &= 4 \,, \\ x_1 + 2x_2 + x_3 \qquad &= 5 \,, \\ 0.5x_2 + 2x_3 + x_4 &= 2 \,, \\ x_3 + x_4 &= 1 \,. \end{aligned}$$

4. Use Doolittle's method to solve the system of equations

$$\begin{aligned} 4x_1 + x_2 + x_3 &= 8 \,, \\ -2x_1 + 1.5x_2 + 0.5x_3 &= 2 \,, \\ 2x_1 + 1.5x_2 - x_3 &= -1 \,. \end{aligned}$$

5. Use the Doolittle method to solve the tridiagonal system

$$\begin{aligned} -2x_1 + x_2 \qquad &= 2 \,, \\ x_1 - 2x_2 + 2x_3 &= -4 \,, \\ x_2 - 2x_3 &= -2 \,, \end{aligned}$$

retaining exact arithmetic throughout the computation.

6. Write the following system of equations in matrix form and use Doolittle's LU Decomposition Method to calculate the solution for this set

$$\begin{aligned} 2x_1 + x_2 &= 1 , \\ x_1 + 2x_2 + x_3 &= 0 , \\ x_2 + 2x_3 + x_4 &= 0 , \\ x_3 + 2x_4 + x_5 &= 0 , \\ x_4 + 2x_5 &= 0 . \end{aligned}$$

7. Use the Doolittle decomposition method to solve the system

$$\begin{aligned} 2x_1 + x_2 + x_3 &= 2 , \\ x_1 - 2x_2 + x_3 &= -2 , \\ 3x_1 - 3x_2 - x_3 &= 1 . \end{aligned}$$

8. Write an algorithm, or program, to find the solution $x_1, \ldots, x_n$ of the upper triangular system of equations

$$\begin{array}{ccccccc} a_{1,1}x_1 & + \ldots & & \ldots + & a_{1,n}x_n & = & b_1 \\ & \ldots & & & \ldots & & \ldots \\ & & a_{i,i}x_i & + \ldots + & a_{i,n}x_n & = & b_i \ , \\ & & & \ldots & \ldots & & \ldots \\ & & & & a_{n,n}x_n & = & b_n \ . \end{array}$$

9. A system of equations has the following LU decomposed form

$$\begin{bmatrix} 1 & & & \\ -2 & 1 & & \\ 0 & 0 & 0 & \\ 1 & -1 & 1 & 1 \end{bmatrix} \begin{bmatrix} 3 & 0 & 3 & 1 \\ & 2 & 1 & 1 \\ & & 2 & 2 \\ & & & 1 \end{bmatrix} \begin{bmatrix} x_1 \\ x_2 \\ x_3 \\ x_4 \end{bmatrix} = \begin{bmatrix} 1 \\ 0 \\ 0 \\ 0 \end{bmatrix} .$$

Find the solution x_1, x_2, x_3, x_4.

10. An $n \times n$ tridiagonal system of equations has the decomposed form $LU\,\mathbf{x} = \mathbf{b}$, where

$$L = \begin{bmatrix} 1 & & & & & \\ l_2 & 1 & & & & \\ & & \cdots & & & \\ & & l_i & 1 & & \\ & & & \cdots & & \\ & & & & l_n & 1 \end{bmatrix}, \qquad U = \begin{bmatrix} u_1 & d_1 & & & & \\ & u_2 & d_2 & & & \\ & & \cdots & & & \\ & & & u_i & d_i & \\ & & & \cdots & & \\ & & & & & u_n \end{bmatrix}.$$

Write forward and backward substitution algorithms to solve the system as

(a) $L\mathbf{y} = \mathbf{b}$ (b) $U\mathbf{x} = \mathbf{y}$.

11. If

$$M = \begin{bmatrix} m_{1,1} & m_{1,2} & m_{1,3} \\ m_{2,1} & m_{2,2} & m_{2,3} \\ m_{3,1} & m_{3,2} & m_{3,3} \end{bmatrix} \quad \text{and } \mathbf{x} = \begin{bmatrix} x_1 \\ x_2 \\ x_3 \end{bmatrix}$$

show that

$$\|M\mathbf{x}\| \leqslant \|M\| . \|x\|$$

where

$$\|M\| = \max_{1 \leqslant i \leqslant 3} \{|m_{i,1}| + |m_{i,2}| + |m_{i,3}|\}$$

$$\|\mathbf{x}\| = \max_{1 \leqslant i \leqslant 3} |x_i| .$$

Hence show that the iterative process

$$\mathbf{x}^{(k+1)} = M\mathbf{x}^{(k)} + \mathbf{c} ; \qquad k = 0, 1, 2, \ldots ,$$

converges to a vector $\mathbf{x}$ such that

$$\mathbf{x} = M\mathbf{x} + \mathbf{c}$$

if $\|M\| < 1$.

Define the Jacobi iterative method for the system of equations

$$\begin{aligned} 4x_1 + x_2 + x_3 &= 1 \ , \\ x_1 + 3x_2 + x_3 &= 2 \ , \\ x_1 + x_2 + 5x_3 &= 1 \ . \end{aligned}$$

12. Define the Jacobi and the Gauss–Seidel iterative methods for the system of equations

$$\begin{aligned} x_1 + 3x_2 + x_3 &= 1 \ , \\ x_1 \qquad + 3x_3 &= 1 \ , \\ 4x_1 + 2x_2 + x_3 &= 1 \ . \end{aligned}$$

Perform *two* steps of *each* iteration using $[0, 0, 0]^{\mathrm{T}}$ as the initial approximation.- Show that each method can be represented by vector iterations of the form

$$\mathbf{x}^{(k+1)} = M\mathbf{x}^{(k)} + \mathbf{c} \ , \qquad k = 0, 1, 2, \ldots \ ,$$

where M is a 3×3 matrix and $\mathbf{c}$ is a 3×1 vector.

13. Define the Jacobi and Gauss–Seidel iterative methods for the system

$$\begin{aligned} 3x_1 + x_2 + x_3 &= 1 \ , \\ x_1 + 4x_2 + 2x_3 &= 1 \ , \\ 2x_1 + x_2 + 4x_3 &= 2 \ , \end{aligned}$$

and prove that the Jacobi method converges.

14. Starting with $\mathbf{x}^{(0)} = (0, 0, 0)^{\mathrm{T}}$, use the Gauss–Seidel method to solve the system

$$\begin{aligned} 10x_1 + x_2 + x_3 - 1 &= 0 \ , \\ x_1 + x_2 + 10x_3 - 1 &= 0 \ , \\ x_1 + 10x_2 + x_3 - 1 &= 0 \ , \end{aligned}$$

giving the answers to two decimal plases.

15. Use the Power method to find the spectral radius of the matrix

$$B = \begin{bmatrix} -9 & -1 \\ -10 & 0 \end{bmatrix} .$$

Start with the vector $(1,0)^T$ and work to three decimal places, giving the answer to two decimal places. Find a starting vector which gives immediate convergence.

16. Use the Power Method to find the spectral radius of the matrix

$$B = \begin{bmatrix} -8 & -2 \\ -10 & 0 \end{bmatrix};$$

start with the vector $(1,0)^T$ and work to three decimal places, giving the answer to two decimal places.

Write down a starting vector which will give immediate convergence.

2

Interpolation and approximation

2.1 POLYNOMIAL APPROXIMATION

Suppose that a function $y(x)$ represents a dependent variable y expressed in terms of an independent variable x. Suppose, further, that the function $y(x)$ is unknown, but that its numerical value, y_i, is known at $N+1$ discrete values, $x_i(i=0,1,\ldots,N)$, of x. The y_i might well represent observed values in some laboratory experiment measured at various times represented by x_i.

Polynomial interpolation is concerned with passing a polynomial through the $N+1$ data pairs (x_i,y_i), $i = 0,1,\ldots,N$, and then using the polynomial to estimate the numerical value, $y(X)$, of y at some point X which lies in the interval bounded by the smallest and the greatest, algebraically, of the $x_i(i=0,1,\ldots,N)$.

Suppose that the polynomial is $p(x)$; then $p(X)$ is an approximation to $y(X)$. In many situations $p(X)$ will be an adequate approximation to $y(X)$, but it will be useful to be able to estimate the error, $y(X)-p(X)$, in $p(X)$ and then to decide if another technique such as trigonometric or exponential interpolation or a function approximation technique will be more useful.

Algebraic polynomials are one of the most useful, well known, and well understood classes of functions of some independent variable x (in the terminology of abstract algebra, polynomials map the real line into itself). They take the form

$$p_N(x) = a_0 + a_1x + \ldots + a_Nx^N \tag{2.1}$$

where N is a non-negative integer and the $a_i(i=0,1,\ldots,N)$ are real constants. Algebraic polynomials uniformly approximate continuous functions; that is, given a function $y(x)$, defined and continuous on a closed interval of x, there exists a polynomial which is as close, geometrically, to the function $y(x)$ as is desired.

This phenomenon is formalized in the following theorem, the proof of which may be found in the text by Bartle (1976), pages 165–172.

Theorem 2.1 (*The Weierstrass Approximation Theorem*) If $y(x)$ is a defined and continuous function on the finite interval $[a,b]$ of x, and $\varepsilon > 0$ is given, then there exists a polynomial $p(x)$, defined on $[a,b]$, such that $|y(x)-p(x)| < \varepsilon$ for all x in $[a,b]$.

2.2 THE TAYLOR POLYNOMIALS

The Taylor polynomial of degree m approximates the function $y(x)$ near the point $x = 0$. It is derived in such a way that it agrees with $y(x)$ and its first m derivatives at the point $x = 0$, thus requiring the curve $p(x)$ to be close to the curve $y(x)$ near $x = 0$ in the (x,y) plane.

Example 2.1 (a) find the fourth-degree Taylor polynomial about $x = 0$ for $y(x) = \cos x$; (b) use the polynomial in part (a) to approximate $\cos(0.1)$ and estimate the error involved; (c) use the polynomial in part (a) to estimate $\int_0^{0.1} \cos x \, dx$ and estimate the error involved.

(a) Set $p_4(x) = a_0 + a_1x + a_2x^2 + a_3x^3 + a_4x^4$; then, since $y(0) = \cos 0 = 1$, $a_0 = 1$.

Differentiating $p(x)$ successively gives

$$\begin{aligned} p_4'(x) &= a_1 + 2a_2x + 3a_3x^2 + 4a_4x^3 \\ p_4''(x) &= 2a_2 + 6a_3x + 12a_4x^2, \\ p_4'''(x) &= 6a_3 + 24a_4x, \\ p_4^{(iv)}(x) &= 24a_4. \end{aligned}$$

Then, since the derivatives of $y(x) = \cos x$ at $x = 0$ are successively $0, -1, 0, 1$, it follows that $a_1 = 0$, $a_2 = \frac{1}{2}$, $a_3 = 0$, $a_4 = \frac{1}{24}$ so that the fourth-degree Taylor polynomial for $\cos x$ near $x = 0$ is

$$p_4(x) = 1 - \frac{x^2}{2} + \frac{x^4}{24} \, . \tag{2.2}$$

(b) substituting $x = 0.1$ in (2.2) gives $p_4(0.1) = 0.99500$. The error associated with this result may be estimated from section A.2 of Appendix A and is given by

$$R_4(x) = \frac{d^5}{dx^5}[\cos(0.1)](0.1)^5/5! = -0.83 \times 10^{-8} \, .$$

(c) $$\int_0^{0.1} \cos x \, dx \simeq \int_0^{0.1} p_4(x)dx = \int_0^{0.1} \left(1 - \frac{x^2}{2} + \frac{x^4}{24}\right) dx$$

$$= \left[x - \frac{x^3}{6} + \frac{x^5}{120}\right]_0^{0.1} = 0.09983 \text{ (to five decimal places).}$$

The error is estimated from $\int_0^{0.1} R_4(x)dx$ and is thus $\sin(0.1).(0.5)^5/5!$ which has the value 0.83×10^{-8}.

2.3 LAGRANGIAN INTERPOLATION

The Taylor polynomials have the property that all the information used in the approximation originates from one point and using these polynomials to approximate functions far removed from this point will incur large errors. The reader will have noticed that the Taylor approximating polynomials of section 2.2 are in fact the Taylor expansions of Appendix A with $x_0 = 0$.

Clearly, for more general purposes, it is better to use polynomials which use information at various points. One such polynomial is the *Lagrange interpolating polynomial*, which is considered now.

Suppose, to begin with, that two distinct points (x_0,y_o) and (x_1,y_1) with $x_0 < x_1$ are given and that it is required to find the interpolating polynomial $p_1(x)$ of first degree which passes through these points. The polynomial $p_1(x)$ is thus an approximation to $y(x)$ in the interval $x_0 \leqslant x \leqslant x_1$; clearly $p_1(x_0) = y(x_0) = y_0$ and $p_1(x_1) = y(x_1) = y_1$.

Suppose, next, that the polynomial has the form

$$p_1(x) = a_0 + a_1 x \ , \tag{2.3}$$

where the parameters a_0 and a_1 are to be determined. The *method of undetermined coefficients*, so named for a reason which will become clear to the reader, will be used to find a_0 and a_1. The polynomial $p_1(x)$ passes through the two given points so that

$$y_0 = a_0 + a_1 x_0 \ ,$$
$$y_1 = a_0 + a_1 x_1 \, .$$

It is then a very easy exercise to show that

$$a_0 = \frac{x_1 y_0 - x_0 y_1}{x_1 - x_0}, \quad a_1 = \frac{y_0 - y_1}{x_0 - x_1}$$

and substituting these values of a_0 and a_1 into (2.3) gives

$$p_1(x) = \frac{x-x_1}{x_0-x_1}y_0 + \frac{x-x_0}{x_1-x_0}y_1 \tag{2.4}$$

which is known as the *Lagrange interpolating polynomial of degree* 1.

The principle may easily be extended to find the second-degree polynomial

$$p_2(x) = a_0 + a_1x + a_2x^2$$

which passes through the three known points (x_i,y_i), $i = 0,1,2$, giving

$$y_i = a_0 + a_1x_i + a_2x_i^2 \quad (i=0,1,2). \tag{2.5}$$

There are three equations in (2.5) so that, assuming $x_0 < x_i < x_2$, the three parameters a_0,a_1,a_2 (the 'undetermined coefficients') may be determined uniquely. The *Lagrange interpolating polynomial of degree* 2 may subsequently be shown to take the form

$$p_2(x) = \frac{(x-x_1)(x-x_2)}{(x_0-x_1)(x_0-x_2)}y_0 + \frac{(x-x_0)(x-x_2)}{(x_1-x_0)(x_1-x_2)}y_1 + \frac{(x-x_0)(x-x_1)}{(x_2-x_0)(x_2-x_1)}y_2 \,. \tag{2.6}$$

The Lagrange interpolating polynomial of degree N passes through the $N+1$ data points (x_i,y_i) for $i = 0,1,\ldots,N$ with $x_0 < x_1 < \ldots < x_N$. Its form is a generalization of (2.4) and (2.6) and is given by

$$p_N(x) = \sum_{i=0}^{N}\left[\prod_{\substack{j=0\\ j\neq i}}^{N}\frac{(x-x_j)}{(x_i-x_j)}y_i\right] . \tag{2.7}$$

One of the most useful features of (2.7) is that the given data points need not be equally spaced with regard to the x_i $(i=0,1,\ldots,N)$.

The next step is to calculate a bound on the error involved in approximating a function by (2.7). The error bound which will now be found is restricted to those functions having derivatives with known bounds. This is the case when trigonometric exponential or logarithmic functions are to be interpolated using tabulated values.

Consider the function

$$g(x) = y(x) - p_N(x) - \frac{y(X)-p_N(x)}{(X-x_0)(X-x_1)\ldots(X-x_N)} \,.$$

$$(x-x_0)\,(x-x_1)\ldots(x-x_N)\ , \tag{2.8}$$

where X is some point in the open interval (x_0,x_N) at which the function $y(x)$ is to be approximated by the polynomial value $p_N(X)$ calculated from (2.7). Assume that $y(x)$ possesses $N+1$ continuous derivatives in (x_0,x_N).

It is clear that $g(x_i) = 0(i=0,1,\ldots,N)$ and that $g(X) = 0$; that is, $g(x)$ vanishes $N+2$ times in the closed interval $[x_0,x_N]$. It follows, therefore, that $g'(x)$ vanishes $N+1$ times, $g''(x)$ vanishes N times, etc., in $[x_0,x_N]$. In particular, $g^{(N+1)}(x)$ must vanish once in $[x_0,x_N]$; that is, there exists a number r in the interval $[x_0,x_N]$ such that $g^{(N+1)}(r) = 0$.

Differentiating $g(x)$ $N+1$ times gives

$$g^{(N+1)}(x) = y^{(N+1)}(x) - \frac{y(X)-p_N(X)}{(X-x_0)\,(X-x_1)\ldots(X-x_N)}\cdot(N+1)\ ! \tag{2.9}$$

since $p_N^{(n+1)}(x) = 0$, $p_N(x)$ being a polynomial of degree N. Putting $x = r$ in (2.9) gives

$$0 = y^{(N+1)}(r) - \frac{y(X)-p_N(X)}{(X-x_0)\,(X-x_1)\ldots(X-x_N)}\cdot(N+1)\ !$$

which leads to

$$y(X)-p_N(X) = (X-x_0)\,(X-x_1)\ldots(X-x_N)\,.\frac{y^{(N+1)}(r)}{(N+1)\ !}\ . \tag{2.10}$$

This result is formalized by the following theorem:

Theorem 2.2 If $x_0 < x_1 < \ldots < x_N$, if $y(x)$ is a function with $N+1$ continuous derivatives in $[x_0,x_N]$, and if $p(x)$ is a polynomial of degree N or less such that $p(x_i) = y(x_i)$, $i = 0,1,\ldots,N$, then

$$y(X)-p(X) = (X-x_0)\,(X-x_1)\ldots(X-x_N)y^{(N+1)}(r)/(N+1)\ !$$

where X is an arbitrary value in (x_0,x_N) and $r = r(X)$ lies in $[x_0,x_N]$.

In order to use (2.10) it is necessary to know the value of $y^{N+1)}(r)$ or to be able to estimate its value. Often it is possible to place an upper bound on $|y^{(N+1)}(r)|$ and thus on the modulus of the eror $y(X)-p(X)$.

Example 2.2 Consider the function $y(x) = \sin x$ and the passing of the Lagrange interpolating polynomial $p_6(x)$ of degree six through the tabulated values of $\sin x_i$, where $x_i = i\pi/12$ and $i = 0,1,\ldots,6$. Estimate the error in $p_6(5\pi/24)$.

For the function $y(x) = \sin x$, $y^{(N+1)}(x) = y^{(\text{vii})}(x) = -\cos x$ so that $|y^{(\text{vii})}(r)| \leqslant 1$ for any r in the interval $[x_0,x_6] = [0,\frac{1}{2}\pi]$. Now,

$$y(X) - p(X) = (X - x_0)\,(X - x_1)\ldots(X - x_6).\,y^{(\text{vii})}(\text{r})/7!\ ,$$

and for $X = 5\pi/24$,

$$|y(5\pi/24) - p_6(5\pi/24| = \left|\left(\frac{5\pi}{24} - 0\right)\left(\frac{5\pi}{24} - \frac{\pi}{12}\right)\ldots\left(\frac{5\pi}{24} - \frac{\pi}{2}\right)\cdot y^{(\text{vii})}(r)/7\,!\right|$$

$$\leqslant \frac{1}{5040}\,.5.3.1.1.3.5.7.\,\frac{\pi^7}{24^7}\ .$$

This gives the error bound

$$|y(5\pi/24) - p_6(5\pi/24)| \leqslant 0.206 \times 10^{-6}\ .$$

2.4 OTHER INTERPOLATING FUNCTIONS

Numerous other interpolating formulas have been developed and used for many decades. Many were developed primarily for use with hand calculators and consequently their popularity has waned with the advent of high-speed digital computers. This has not proved to be so with the Lagrange interpolating polynomials, which are amenable to use with computers. Some such interpolating functions (in alphabetic order) are Bernstein polynomials, Bessel's formula, cubic spline interpolation, Everett's forward difference method, the Gauss backward and Gauss forward formulas, iterated interpolation, Neville's method, Newton's backward difference, backward divided difference, forward difference and forward divided difference formulas, and Stirling's formula.

Hermite interpolating polynomials (also known as osculatory polynomials) must also be mentioned. The development of these polynomials is similar to the development of the Lagrange interpolating polynomial except that a polynomial $p_{2N+1}(x)$ of degree $2N+1$ is fitted to the data set $\{(x_0,y(x_0)), (x_1,y(x_1)),\ldots,(x_N,y(x_N))\}$, such that $p_{2N+1}(x_i) = y(x_i)$ and $p'_{2N+1}(x_i) = y'(x_i)$ for $i = 0,1,\ldots,N$.

Inverse interpolation is, as might be expected from its name, concerned with finding x_k corresponding to a numerical value y_k and is formalized as follows. Suppose $y(x)$ and $y'(x)$ exist and are continuous, with $y'(x) \neq 0$, for x in some interval $[a, b]$. Suppose further that $y(x)$ has one zero q in $[a, b]$ and let $x_0, x_1,\ldots, x_N$ be $N+1$ distinct numbers in $[a, b]$ with $y(x_i) = y_i\ (i = ,1,\ldots,N)$. To approximate q, construct the interpolating polynomial of degree N on the numerical values $y_0,\ y_1,\ldots,\ y_N$ for $y^{-1}(x)$. Since $y_i = y(x_i)$ and $0 = y(q)$, it follows that $y^{-1}(y_i) = x_i$ for $i = 0,1,\ldots,N$, and $q = y^{-1}(0)$.

Details of the interpolation procedures named above may be found, in varying amounts of detail, in the more advanced texts of Burden and Faires (1985), Fröberg (1985), Gerald and Wheatley (1984), Morris (1983) and Ralston (1965).

2.5 DISCRETE L_p APPROXIMATION

The first four sections of this chapter were concerned with fitting polynomials to discrete data pairs relating to some known or unknown function, the polynomials then being used to find approximate values of the function.

The problem to be considered in the present section of the chapter is that of fitting functions to given data and finding the 'best' function in a certain class that is to be used to present the data.

The general discrete approximation problem, given a real-valued function $f(x)$ defined on a finite set of points $X = \{x_1,x_2,\ldots,x_N\}$ of an interval I of the real number line and a set of M real-valued continuous functions $\phi_j(x)$ $(j = 1,\ldots, M)$ defined on I, is concerned with forming the approximating function

$$F(A,x) = \sum_{j=1}^{M} a_j\phi_j(x) \tag{2.11}$$

for any real set $A = \{a_1,a_2,\ldots,a_M\}$. If $\rho\{f(x), F(A,x)\}$ is a *distance function* or *norm* which measures the distance of $F(A,x)$ from $f(x)$, then the discrete approximation problem is to determine the set of optimal parameters $A^* = \{a_1^*,a_2^*,\ldots,a_M^*\}$ such that

$$\rho\{f(x), F(A^*,x)\} \leqslant \rho\{f(x), F(A, x)\} \tag{2.12}$$

for all $A \subset E^M$. The set A^* is said to determine a *best approximation* to $f(x)$.

Given a function $f(x)$ to approximate, there are a number of steps to consider:

(1) the choice of approximating function and norm,
(2) proving the existence of a solution,
(3) a discussion of uniqueness,
(4) characterization of the solution,
(5) the numerical computation of the solution.

Lengthy discussion of all these points may be found in the literature on the approximation of functions, including the well-known texts by Cheney (1966) and Rice (1964, 1968).

The most widely used norm before the advent of high-speed digital computers was the L_2 norm. Since 1960, however, algorithms have appeared in the literature which give efficient computer codes for determining best approximations in the L_∞ and L_1 norms. These include works by Barrodale and Young (1966), Spyropoulos *et al.* (1973) and Shrager and Hill (1980). The L_∞ and L_1 norms call for a knowledge of mathematical programming whereas the L_2 norm calls for no more than elementary calculus and it is consequently this norm which will be emphasized in later sections of the chapter. In practice, the approximating function is frequently chosen from algebraic polynomials, when it must be remembered that, although the degree of the polynomial may have to be large for reasonable accuracy, the sequence of monomials $\{x^j\}$ becomes almost linearly dependent for large j, leading to numerical instability.

If F depends linearly on its parameters $a_1,a_2,\ldots,a_M$, it is said to be a *linear approximating function*; otherwise F is *nonlinear*. The inequality (2.12) is usually written in the alternative form

$$\| f(x) - F(A^*,x)\| \leqslant \| f(x) - F(A,x)\| \tag{2.13}$$

for some given norm $\| . \|$ and the L_p norm of the function $f(x)$ defined on the point set X_N is given by

$$L_p(f) = \left[\sum_{i=1}^{N} |f(x_i)|^p\right]^{1/p} , \quad p \geqslant 1 . \tag{2.14}$$

The pth root can be omitted in any minimization process and the sequence of best L_p approximations ($p \geqslant 1$) converges to a best approximation in the L_∞ norm defined by

$$L_\infty(f) = \max_i |f(x_i) - F(A,x_i)| . \tag{2.15}$$

The three commonly used norms, with associated prescribed weights $w(x_i)$, are

$$\begin{aligned} L_\infty &: \| w(x)\{f(x) - F(A,x)\} \|_\infty = \max_i \{w(x_i)|f(x_i) - F(A,x_i)|\} , \\ L_1 &: \| w(x)\{f(x) - F(A,x)\} \|_1 = \sum_{i=1}^{N} w(x_i)|f(x_i) - F(A,x_i)| , \\ L_2 &: \| w(x)\{f(x) - F(A,x)\} \|_2 = \sum_{i=1}^{N} w(x_i)\{f(x_i) - F(A,x)\}^2 ; \end{aligned} \tag{2.16}$$

often, in practice, $w(x) \equiv 1$.

2.6 LINEAR DISCRETE L_2 APPROXIMATION

2.6.1 The least-squares straight line

The least-squares or L_2 straight-line problem is concerned with fitting the straight line

$$F(A,x) = a_1 + a_2 x \tag{2.17}$$

to the data points (x_i, y_i), $i = 1,2,\ldots,N$, such that the sum of squares residual function

$$S(a_1,a_2) = \sum_{i=1}^{N} \{y_i - (a_1 + a_2 x_i)\}^2 \tag{2.18}$$

is minimized. That is, the optimal values a_1^*, a_2^* of the parameters a_1, a_2 are sought, these optimal values giving S its minimum value (Note: a_1 and a_2 occur linearly in (2.17)).

The necessary conditions for S to have a minimum value are

$$\partial S/\partial a_1 = 0 \quad \text{and} \quad \partial S/\partial a_2 = 0 \tag{2.19}$$

simultaneously. Differentiating $S(a_1,a_2)$ with respect to a_1 and a_2 gives

$$\frac{\partial S}{\partial a_1} = -2 \sum_{i=1}^{N} \{y_i - (a_1 + a_2 x_1)\} , \tag{2.20}$$

$$\frac{\partial S}{\partial a_2} = -2 \sum_{i=1}^{N} x_i\{y_i - (a_1 + a_2 x_i)\} \tag{2.21}$$

so that (2.19), (2.20), (2.21) give

$$N a_1 + a_2 \sum_{i=1}^{N} x_i = \sum_{i=1}^{N} y_i , \tag{2.22}$$

$$a_1 \sum_{i=1}^{N} x_i + a_2 \sum_{i=1}^{N} x_i^2 = \sum_{i=1}^{N} x_i y_i . \tag{2.23}$$

Equations (2.22) and (2.23) are known as the *normal equations* and their solution gives the optimal values a_1^*, a_2^* which minimize $S(a_1,a_2)$. Their substitution in (2.17) gives the least-squares line to the data set $(x_1y_1),(x_2,y_2),\ldots,(x_N,y_N)$.

2.6.2 The least-squares parabola

Here the problem is concerned with fitting the parabola

$$F(a,x) = a_1 + a_2x + a_3x^2 \tag{2.24}$$

to the data points (x_i,y_i), $i = 1,2,\ldots,N$, in such a way that the least-squares residual function

$$S(a_1,a_2,a_3) = \sum_{i=1}^{N} \{y_i - (a_1 + a_2x_i + a_3x_i)\}^2 \tag{2.25}$$

is minimized. (Note: a_1,a_2,a_3 occur linearly in (2.24).)

The necessary conditions for S to have a minimum value are that

$$\partial S/\partial a_1 = 0\ , \quad \partial S/\partial a_2 = 0 \quad \text{and} \quad \partial S/\partial a_3 = 0$$

simultaneously. Differentiating (2.25) with respect to a_1,a_2,a_3 gives

$$\frac{\partial S}{\partial a_1} = -2\sum_{i=1}^{N} \{y_i - (a_1 + a_2x_i + a_3x_i^2)\}\ ,$$

$$\frac{\partial S}{\partial a_2} = 2\sum_{i=1}^{N} x_i\{y_i - (a_1 + a_2x_i + a_3x_i^2)\}\ ,$$

$$\frac{\partial S}{\partial a_3} = -2\sum_{i=1}^{N} x_i^2\{y_i - (a_1 + a_2x_i + a_3x_i^2)\}$$

from which it is easy to show that the normal equations for this least-squares model are

$$Na_1 + a_2 \sum_{i=1}^{N} x_i + a_3 \sum_{i=1}^{N} x_i^2 = \sum_{i=1}^{N} y_i \ , \tag{2.26}$$

$$a_1 \sum_{i=1}^{N} x_i + a_2 \sum_{i=1}^{N} x_i^2 + a_3 \sum_{i=1}^{N} x_i^3 = \sum_{i=1}^{N} x_i y_i \ , \tag{2.27}$$

$$a_1 \sum_{i=1}^{N} x_i^2 + a_2 \sum_{i=1}^{N} x_i^3 + a_3 \sum_{i=1}^{N} x_i^4 = \sum_{i=1}^{N} x_i^2 y_i \ . \tag{2.28}$$

Solving (2.26), (2.27), (2.28) simultaneously gives a_1^*, a_2^*, a_3^*, substitution of which into (2.24) gives the (best) least-squares parabola to the data set $(x_1,y_1),(x_2,y_2),\ldots,(x_N,y_N)$.

Example 2.3 The percentage of sucrose released, $R\%$, at time t hours, observed in a cell kinetics experiment reported by Dean and Jessup (1982), is recorded in Table 2.1. Find the least-squares parabola of the form $R = a_1 + a_2 t + a_3 t^2$ which fits these data, giving a_1, a_2 and a_3 to two decimal places.

Table 2.1

t	0.00	2.25	4.33	6.00	18.00	42.00
R	3.11	10.33	10.00	13.93	20.98	33.11

The summations required for the normal equations (2.26), (2.27), (2.28), noting that x is replaced by t and y is replaced by R, are given in Table 2.2.

Table 2.2

	t_i	R_i	t_i^2	t_iR_i	t_i^3	$t_i^2R_i$	t_i^4
1	0.00	3.11	0.000	0.000	0.000	0.000	0.000
2	2.25	10.33	5.063	23.243	11.391	52.301	25.629
3	4.33	10.00	18.749	43.300	81.183	187.490	351.521
4	6.00	13.93	36.000	83.580	216.000	501.480	1296.000
5	18.00	20.98	324.000	377.640	5832.000	6797.520	104976.000
6	42.00	33.11	1764.000	1390.620	74088.000	58406.040	3111696.000
Sum	72.58	91.46	2147.812	1918.383	80228.574	65944.831	3218345.200

The normal equations are therefore

$$6a_1 + 72.58a_2 + 2147.812a_3 = 91.46 \ ,$$

$$72.58a_1 + 2147.812a_2 + 80228.574a_3 = 1918.383 \ ,$$

$$2147.812a_1 + 80228.574a_2 + 3218345.200a_3 = 65944.831 \ ,$$

which have solution $a_1 = 5.723$, $a_2 = 1.120$, $a_3 = -0.011$. The least-squares parabola is thus

$$R(t) = 5.72 + 1.12t - 0.01t^2 \ , \tag{2.29}$$

and this curve and the data points of the laboratory experiment are depicted in Fig. 2.1. As the coefficient of t^2 in (2.29) is very small, it may be concluded that the data

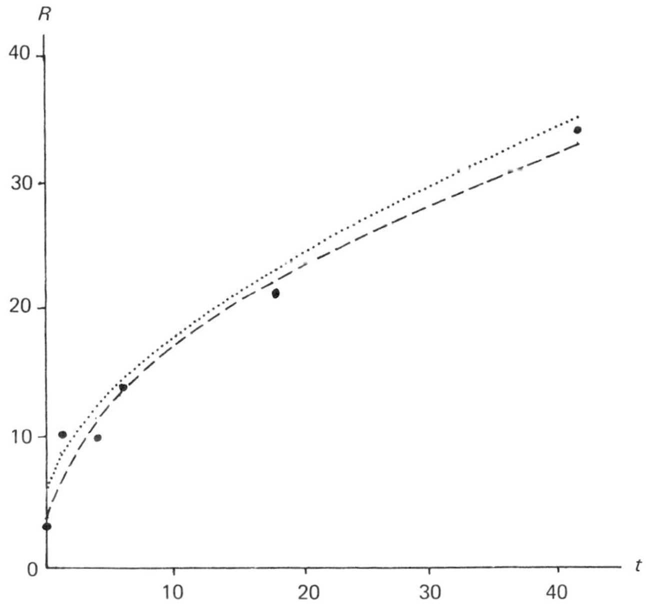

Fig. 2.1 — . . . Data points of Dean and Jessup (1982), … L_2 parabola and --- nonlinear L_2 model.

may be represented, with reasonable accuracy, by a straight line for small values of t.

2.6.3 Linearization of exponential curve fitting problems

Least-squares polynomials, in particular straight lines and parabolas, are arguably the most extensively used procedures for fitting curves to experimental data. Occasionally, however, it is more appropriate to assume that the data are related by an exponential function such as

$$Y = ae^{bX} \tag{2.30}$$

or

$$Y = aX^b \ . \tag{2.31}$$

Taking natural logarithms of each side of (2.30) gives

$$\ln Y = \ln a + bX \ , \tag{2.32}$$

which is of the linear form (2.17) with $a_1 = \ln a$, $a_2 = b$ and $x = X$. The procedure for fitting a least-squares straight line is described in subsection 2.6.1, the x_i and y_i of equation (2.18) being X_i and $\ln Y_i$ respectively. Having found a_1 and a_2, the parameters a and b of (2.30) are given by $a = e^{a_1}$ *and* $b = a_2$.

Taking base ten logarithms of each side of (2.31) gives

$$\log Y = \log a + b \log X \tag{2.33}$$

which is the form (2.17) with $a_1 = \log a$, $a_2 = b$ and $x = \log X$. The procedure described in subsection 2.6.1 for fitting the least-squares straight line to the data set (x_i, y_i), where $x_i = \log X_i$ and $y_i = \log Y_i$, may now be followed. Having calculated a_1 and a_2, the parameters A and B of (2.31) are given by $a = 10^{a_1}$ and $b = a_2$.

Example 2.4 In an experiment in which macrophages were incubated with increasing concentrations of acetyl-LDL in the presence of [^{14}C]oleate, Brown *et al.* (1979) reported a marked enhancement in the rate of [^{14}C]oleate incorporation into cholesteryl [^{14}C]oleate. Figures connecting [^{14}C]oleate→cholesteryl [^{14}C]oleate, measured in nmol.h^{-1}.mg protein^{-1} and denoted by Y, and Lipoprotein, measured in μg protein/ml and denoted by X, are given in Table 2.3. Fit a least-squares curve of

Table 2.3

X	0	4.17	8.33	29.17	50.00	147.25	245.83
Y	0	4.00	7.00	10.00	12.00	13.33	13.11

the form (2.31) to these data, giving a and b to two decimal places.

As $\log 0 = -\infty$, the data point $(X,Y) = (0,0)$ may not be used in (2.33) and must be ignored; it will, of course, satisfy (2.31) provided $b \neq 0$. Writing $y = \log Y$ and $x = \log X$, the summations for the normal equations are given in Table 2.4.

The normal equations are therefore

$$\begin{aligned} 6a_1 + \ 9.264a_2 &= 5.769 \ , \\ 9.264a_1 + 16.682a_2 &= 9.561 \end{aligned}$$

which have solution $a_1 = 0.537$ and $a_2 = 0.275$. Thus, $a = 10^{0.537} \simeq 3.44$ and $b = 0.27$ and the least-squares model of the form (2.31) is

$$Y = 3.44 \ X^{0.27} \ ;$$

note this equation is satisfied by the ignored data point $(X,Y) = (0,0)$. This curve and the data points of Table 2.3 are shown in Fig. 2.2.

2.7 LEAST-SQUARES APPROXIMATION OF FUNCTIONS

Suppose $y(x)$ is a continuous function in some interval $[a,b]$ of x and that a polynomial $p_N(x)$ of degree (at most) N of the form

$$p_N(x) = a_0 + a_1x + a_2x^2 + \ldots + a_{N-1}x^{N-1} + a_Nx^N \tag{2.34}$$

is to be found which will minimize the function

$$S(a_0,a_1,\ldots,a_N) = \int_a^b \left(y(x) - \sum_{i=0}^{N} a_ix^i \right)^2 \mathrm{d}x \ . \tag{2.35}$$

The least-squares problem is thus to find the parameters $a_0,a_1,\ldots,a_N$ which achieve this.

The necessary conditions for S to have a minimum value are

$$\partial S/\partial a_r = 0 \ ; \qquad r = 0,1,\ldots,N \tag{2.36}$$

simultaneously. Noting that

$$S = \int_a^b [y(x)]^2 \mathrm{d}x - 2\sum_{i=0}^{N} a_i \int_a^b x^i y(x)\mathrm{d}x + \int_a^b \left(\sum_{i=0}^{N} a_ix^i \right)^2 \mathrm{d}x \ ,$$

Table 2.4

i	x_i	y_i	x_i^2	$x_i y_i$
1	0.620	0.620	0.384	0.373
2	0.921	0.845	0.848	0.778
3	1.465	1.000	2.146	1.465
4	1.699	1.079	2.887	1.833
5	2.168	1.125	4.700	2.439
6	2.391	1.118	5.717	2.673
Sum	9.264	5.769	16.682	9.561

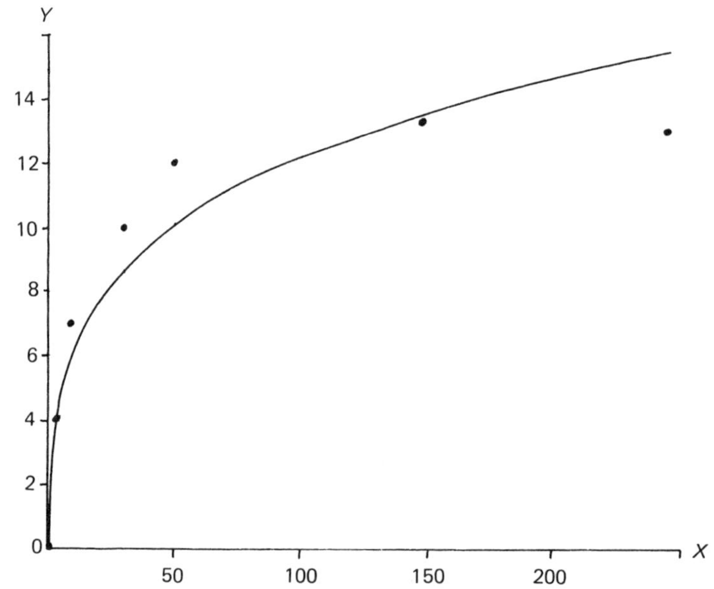

Fig. 2.2 — . . . Data points of Brown *et al.* (1979); ——— nonlinear L_2 model.

it follows that

$$\frac{\partial S}{\partial a_r} = -2\int_a^b x^r y(x)\mathrm{d}x + \sum_{i=0}^{N} a_i \int_a^b x^{r+i}\mathrm{d}x$$

for $r = 0,1,\ldots,N$, and the normal equations are thus

$$\sum_{i=0}^{N} a_i \int_a^b x^{r+i}\mathrm{d}x = \int_a^b x^r y(x)\mathrm{d}x \ .$$

This system of $N+1$ linear equations, which always has a unique solution provided by $a \neq b$ and $y(x)$ is continuous in the interval $[a,b]$, is solved for its unknowns $a_0, a_1, \ldots, a_N$.

2.8 NONLINEAR DISCRETE L_2 APPROXIMATION

2.8.1 The Gauss–Newton method

In Example 2.4 of sub-section 2.6.3, one of the data points was the origin and (2.31) was therefore an appropriate model. Often, however, the origin is not a data point and a more general nonlinear model must be fitted to the data points. Unlike (2.30) and (2.31) it is usually not possible to linearize such nonlinear models, so that a system of linear normal equations cannot be found with which to determine the parameters of the model.

There are many differences between linear and nonlinear curve fitting and, in keeping with the literature, a number of changes in notation will now be introduced.

Suppose that (ε_i,σ_i), $i = 1,2,\ldots,N$, are N data pairs which are suspected to satisfy a nonlinear model equation of the form

$$\sigma = s(\varepsilon) \tag{2.37}$$

where $s = (s_1(\varepsilon), s_2(\varepsilon), \ldots, s_M(\varepsilon))$ is some nonlinear function of ε and $N \geqslant M$ (the number of parameters in s). Such a nonlinear function could take the form

$$\sigma = a + b\varepsilon + c\varepsilon^2 \cos \varepsilon$$

for which $M=3$, $s_1 = a$, $s_2 = b\varepsilon$, $s_3 = c\varepsilon^2 \cos\varepsilon$ and $s = s_1 + s_2 + s_3$.

Let $f_i = \sigma_i - s(\varepsilon_i)$ be the error in σ_i $(i = 1,2,\ldots,N)$, then f is a function of s. Such errors could be experimental errors or errors of measurement. It is convenient to write $\mathbf{f} = \mathbf{f}(\mathbf{x})$ where $\mathbf{f}$ is a vector of order N and $\mathbf{x}$ is a vector of order M having, as its elements, the parameters of s. The least-squares criterion requires

$$S = \sum_{i=1}^{N} f_i^2 = \sum_{i=1}^{N} [\sigma_i - s(\varepsilon_i)]^2 \tag{2.38}$$

to be minimized for the best least-squares fit. The necessary conditions for this are

$$\partial S/\partial x_j = 0 \tag{2.39}$$

simultaneously for $j = 1,2,\ldots,M$. Suppose that the minimum value of S occurs when $\mathbf{x} = \mathbf{x}^*$.

Iterative algorithms for the minimization of S calculate a sequence $\mathbf{x}^{(n)} = [x_1^{(n)}, x_2^{(n)}, \ldots, x_M^{(n)}]^T$, $n = 0,1,2,\ldots$, with $\mathbf{x}^{(0)}$ some initial value chosen so that the sequence $\{\mathbf{x}^{(n)}\}$ will converge to $\mathbf{x}^*$. Consider the linear approximations to the error functions $f_i(\mathbf{x})$, $i = 1,2,\ldots,N$, afforded by Taylor's series and given by

$$L_i^{(n)}(\mathbf{x}) = f_i(\mathbf{x}^{(n)}) + \sum_{j=1}^{M} (x_j - x_j^{(n)}) p_{ij}^{(n)} \ , \qquad i = 1,\ldots,N \tag{2.40}$$

where $P^{(n)} = [p_{ij}^{(n)}]$ is the first derivative matrix of order $N \times M$ at the nth iterate; the element $p_{ij}^{(n)}$ is given by

$$p_{ij}^{(n)} = \partial f_i(\mathbf{x})/\partial \mathbf{x}_j \,|_{\mathbf{x} = \mathbf{x}^{(n)}} \tag{2.41}$$

with $i = 1,2,\ldots,N$, $j = 1,2,\ldots,M$ and $n = 0,1,2,\ldots$. Then

$$S = \sum_{i=1}^{N} [L_i^{(n)}(\mathbf{x})]^2 + O(\|\mathbf{x} - \mathbf{x}^{(n)}\|^2) \ . \tag{2.42}$$

Differentiating this form of S twice gives

$$\frac{\partial S}{\partial x_i \partial x_j} \simeq 2 \sum_{t=1}^{N} \frac{\partial L_t}{\partial x_j} \cdot \frac{\partial L_t}{\partial x_i} \simeq 2 \sum_{t=1}^{N} \frac{\partial f_t}{\partial x_j} \cdot \frac{\partial f_t}{\partial x_i} \tag{2.43}$$

so that $2P^{\mathrm{T}}P$ is an approximation to the Hessian matrix G. This approximation to G is positive definite or positive semi-definite.

Let $\mathbf{h} = \mathbf{x}^{(n)} - \mathbf{x}^* = [h_1^{(n)}, h_2^{(n)}, \ldots, h_M^{(n)}]^T$. Then

$$\begin{aligned} S^{(n)} &= S(\mathbf{x}^{(n)}) = S(\mathbf{x}^* + \mathbf{h}^{(n)}) \\ &= S(\mathbf{x}^*) + \sum_{j=1}^{M} h_j^{(n)} \left.\frac{\partial S}{\partial x_j}\right|_{\mathbf{x} = \mathbf{x}^*} + \text{higher-order terms} \end{aligned}$$

and so

$$\frac{\partial S}{\partial x_r} = \left.\frac{\partial S}{\partial x_r}\right|_{\mathbf{x}=\mathbf{x}^*} + \sum_{j=1}^{M} h_j^{(n)} \left.\frac{\partial^2 S}{\partial x_j \partial x_r}\right|_{\mathbf{x}=\mathbf{x}^*} + \text{higher-order terms}$$

at the nth iterate, where $r = 1,2,\ldots,M$. Using (2.39) it follows that

$$g_r^{(n)} = \sum_{j=1}^{M} h_j^{(n)} \left.\frac{\partial^2 S}{\partial x_j \partial x_r}\right|_{x=x^*} , \qquad r = 1,\ldots,M \tag{2.44}$$

where $g_r = \partial S/\partial x_r$ so that $\mathbf{g} = [g_1,g_2,\ldots,g_M]^T$ is the gradient vector of S. Thus $\mathbf{g}^{(n)} = G^{(n)}\mathbf{h}^{(n)}$ and, therefore, $\mathbf{h}^{(n)} = (G^{(n)})^{-1}\mathbf{g}^{(n)}$ and

$$\mathbf{x}^* = \mathbf{x}^{(n)} - \mathbf{h}^{(n)} = \mathbf{x}^{(n)} - (G^{(n)})^{-1}\mathbf{g}^{(n)}$$

which suggests the iterative scheme

$$\mathbf{x}^{(n+1)} = \mathbf{x}^{(n)} - (G^{(n)})^{-1}\mathbf{g}^{(n)} , \qquad n = 0,1,2,\ldots . \tag{2.45}$$

The drawback associated with (2.45) is that convergence is only guaranteed if $G^{(n)}$ is positive definite. So, G is approximated by $Q = 2P^{\mathrm{T}}P$. Noting also that

$$g_r = \frac{\partial S}{\partial x_r} = 2\sum_{i=1}^{N} f_i \frac{\partial f_i}{\partial x_r} , \qquad r = 1,2,\ldots,M , \tag{2.46}$$

it follows that

$$\mathbf{g} = 2P^{\mathrm{T}}\mathbf{f} . \tag{2.47}$$

Hence, in (2.45),

$$\mathbf{x}^{(n+1)} = \mathbf{x}^{(n)} - 2(Q^{(n)})^{-1}(P^{(n)})^{\mathrm{T}}\mathbf{f}(\mathbf{x}^{(n)}) . \tag{2.48}$$

Equation (2.48) gives the *Gauss–Newton method* for minimizing S in (2.38). The need to determine $(Q^{(n)})^{-1}$ is obviated by writing

$$\mathbf{z}^{(n)} = -2(Q^{(n)})^{-1}(P^{(n)})^{\mathrm{T}}\mathbf{f}(\mathbf{x}^{(n)}) \ , \tag{2.49}$$

solving the linear system

$$Q^{(n)}\mathbf{z}^{(n)} = -2(P^{(n)})^{T}\mathbf{f}(\mathbf{x}^{(n)}) \tag{2.50}$$

for $\mathbf{z}^{(n)}$ and then by finding the next iterate $\mathbf{x}^{(n+1)}$ from

$$\mathbf{x}^{(n+1)} = \mathbf{x}^{(n)} + \mathbf{z}^{(n)} \ . \tag{2.51}$$

The reader will note that, if the number of data pairs N is equal to the number of functions x_i $(i = 1,2,\ldots,M)$, the minimum value of S is zero and all matrices are square and of order N, $\mathbf{f}$ and $\mathbf{x}$ are N-vectors and the least-squares problem is equivalent to solving the nonlinear algebraic system

$$\mathbf{f}(\mathbf{x}) = \mathbf{0} \ . \tag{2.52}$$

The numerical solution of (2.52) is dealt with in further detail in Chapter 3 where, it suffices to say for now, the Gauss–Newton method (2.48) becomes the Newton–Raphson method for a nonlinear system.

2.8.2 The Levenberg–Marquardt method

It was noted in subsection 2.8.1 that a more general form of (2.31) may need to be fitted to a set of data points which does not include the origin. Retaining the notation (ε_i,σ_i), $i = 1,2,\ldots,N$, to denote these data points, suppose that a nonlinear model of one of the forms

$$\sigma = a(B \pm \varepsilon)^{b} \tag{2.53}$$

is to be fitted in which a, B and b are parameters.

The least-squares criterion requires that

$$S(a,B,b) = \sum_{i=1}^{N} f_i^2 = \sum_{i=1}^{N} \{\sigma_i - a(B \pm \varepsilon_i)^{b}\}^2 \tag{2.54}$$

be minimized, this minimum of S being reached by choosing optimal values of the parameters a, B and b. Adapting the notation of subsection 2.8.1 gives $\mathbf{x}^{(n)} = [x_1^{(n)}, x_2^{(n)}, x_3^{(n)}]^T \equiv [a^{(n)}, B^{(n)}, b^{(n)}]^T$, $n = 0,1,2,\ldots$, with $\mathbf{x}^{(0)}$ some initial vector chosen so that the sequence $\{\mathbf{x}^{(n)}\}$ will converge to a point $\mathbf{x}^* = [a^*, B^*, b^*]^T$ that minimizes S. The elements of the matrix $P^{(n)}$ are given, from (2.41), by

$$\begin{aligned} p_{i1}^{(n)} &= -(B^{(n)} \pm \varepsilon_i)^{b^{(n)}} , \\ p_{i2}^{(n)} &= -a^{(n)} b^{(n)} (B^{(n)} \pm \varepsilon_i)^{b^{(n)}-1} , \\ p_{i3}^{(n)} &= -a^{(n)} (B^{(n)} \pm \varepsilon_i)^{b^{(n)}} \ln(B^{(n)} \pm \varepsilon_i) \end{aligned} \tag{2.55}$$

for $i = 1,2,\ldots,N$ and $n = 0,1,2,\ldots$.

Often the Gauss–Newton method converges slowly for (2.53) from some initial vector $\mathbf{x}^{(0)}$ as happens when the *search direction* $-2(Q^{(n)})^{-1}(P^{(n)})^T\mathbf{f}(\mathbf{x}^{(n)})$ is almost perpendicular to the gradient vector

$$\begin{aligned} \mathbf{g}^{(n)} &= [\partial S/\partial x_1, \partial S/\partial x_2, \partial S/\partial x_3]^T_{\mathbf{x}=\mathbf{x}^*} \\ &= [\partial S/\partial a, \partial S/\partial B, \partial S/\partial b]^T_{a=a^{(n)}, B=B^{(n)}, b=b^{(n)}} . \end{aligned}$$

Marquardt (1963) has suggested an algorithm, similar to that proposed by Levenberg (1944), in which the search direction is biased towards the direction of steepest descent. The algorithm is closely related to the Gauss–Newton method and is given by

$$\mathbf{x}^{(n+1)} = \mathbf{x}^{(n)} - 2[Q^{(n)} + \lambda^{(n)} I]^{-1}(P^{(n)})^T\mathbf{f}(\mathbf{x}^{(n)}) \; ; \quad n = 0,1,2,\ldots . \tag{2.56}$$

In the form (2.56), the algorithm has become as the *Levenberg–Marquardt method.*

Clearly $\lambda^{(n)}$ must be selected so that

$$S^{(n+1)} < S^{(n)} \tag{2.57}$$

where $S^{(n)}$ denotes the value of S at the nth iteration. It is shown in Marquardt's paper that a sufficiently large $\lambda^{(n)}$ always exists such that (2.57) is satisfied (unless $\mathbf{x}^{(n)} = \mathbf{x}^*$) and the strategy of the algorithm is as follows.

(a) Let $\upsilon = 10$ (it is sufficient that $\upsilon > 1$ and the choice of υ is otherwise arbitrary); let $\lambda^{(0)} = 0.01$.

(b) Let $R(\lambda^{(n)})$, $R(\lambda^{(n)}/\upsilon)$ be the values of S when $\lambda^{(n)}$ and $\lambda^{(n)}/\upsilon$, respectively, are used in (2.56).

(c) Compute $S^{(n+1)}$, $R(\lambda^{(n)})$ and $R(\lambda^{(n)}/\upsilon)$. Then
 (i) if $R(\lambda^{(n)}/\upsilon) \leqslant S^{(n+1)}$, let $\lambda^{(n+1)} = \lambda^{(n)}/\upsilon$,

(ii) if $R(\lambda^{(n)}/\upsilon) > S^{(n+1)}$ and $R(\lambda^{(n)}) \leqslant S^{(n+1)}$, let $\lambda^{(n+1)} = \lambda^{(n)}$,

(iii) if $R(\lambda^{(n)}/\upsilon) > S^{(n+1)}$ and $R(\lambda^{(n)}) > S^{(n+1)}$, increase λ by successive multiplication by υ until, for some smallest positive constant, w, $R(\lambda^{(n)}\upsilon^{w}) \leqslant S^{(n+1)}$; let $\lambda^{(n+1)} = \lambda^{(n)}\upsilon^{w}$.

Marquardt gives the convergence criterion

$$|\alpha_j^{(n)}|/[\tau + |\mathbf{x}_j^{(n)}|] < \xi \ , \qquad j = 1,2,3$$

for some suitable $\xi > 0$ (say 0.000001) and some suitable τ (say 0.003) where

$$\mathbf{d}^{(n)} = -2[Q^{(n)} + \lambda^{(n)}I]^{-1}(P^{(n)})^{\mathrm{T}}\mathbf{f}(\mathbf{x}^{(n)}) \ . \tag{2.58}$$

It is easy to see that the convergence tests described in this strategy do increase computer time, and from good initial values $a^{(0)}$, $B^{(0)}$ and $b^{(0)}$ the Gauss–Newton method may well converge faster than the Levenberg–Marquardt method. From poor initial values, however, the Gauss–Newton method may diverge while the Levenberg–Marquardt method will converge. This highlights the superior reliability of the Levenberg–Marquardt algorithm. The method does require a very small amount more of core storage than the Gauss–Newton method but, in practice, this requirement does not detract from the superiority of the better convergence properties of the Levenberg–Marquardt method.

The Levenberg–Marquardt algorithm is available to IBM users through the IBM Share Library as *Share Program* #1428. The algorithm is also available from early versions of the NAG (Numerical Algorithm Group) subroutine library where it was implemented in Fortran as E04GAF and in Algol as E04GAA.

Example 2.5 Returning to the cell kinetics experiment of Dean and Jessup (1982) considered in Example 2.3, the least-squares model of the form $R = a(B + t)^b$ as suggested by (2.54), gives, for the data of Table 2.1,

$$R = 5.46(0.37 + t)^{0.48} \ . \tag{2.59}$$

The curve given by (2.59) is included in Fig. 2.1.

Example 2.6 In a steady-state finite element model of heat distribution in the human torso, Smith (1981) and Smith and Twizell (1980) fitted a curve of the form $u = a(B - r)^b$ to the data of Table 2.5, where u represents temperature and r $(0 \leqslant r \leqslant 1)$ represents the proportional distance from the centre of the torso, assumed to have a circular cross-section, to the skin surface. The curve, which has the form

$$u = 36.93(1 + 1 \times 10^{-16} - r)^{0.00356} \ , \tag{2.60}$$

Table 2.5

r	0	0.25	0.50	0.75	1.00
u	36.90	36.85	36.75	36.55	32.35

is shown, together with the data points of Table 2.5, in Fig. 2.3.

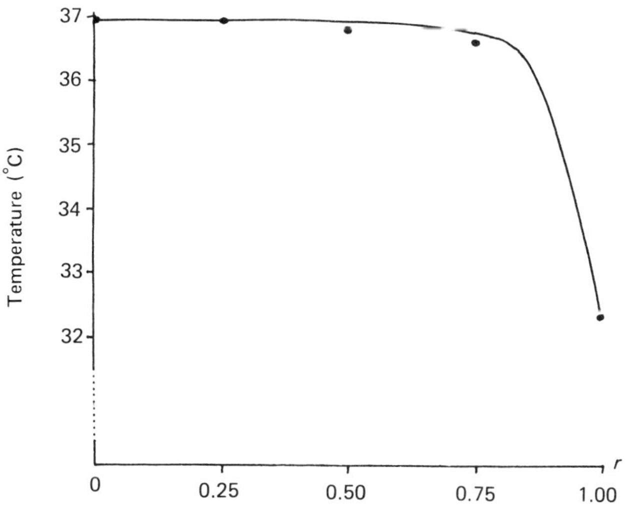

Fig. 2.3 — . . . Data points of Smith (1981) and Smith and Twizell (1980); ——— nonlinear L_2 model.

PROBLEMS

1. Let

$$p(x) \equiv a_0 + a_1 x + a_2 x^2 + a_3 x^3$$

be a cubic polynomial. Determine the coefficients of this polynomial so that

$$p(0) = f(0) \;, \qquad p(1) = f(1) \;,$$
$$p'(0) = f'(0), \qquad p'(1) = f'(1) \;,$$

where f is given function (*Cubic Hermite interpolation* on $0 \leqslant x \leqslant 1$).

2. Consider the function of the variable t defined by

$$F(t) = f(t) - p(t) - [f(x) - p(x)]\,\frac{t^2(t-1)^2}{x^2(x-1)^2}\,,$$

where $x \neq 0,1$ is considered fixed. Verify that $F(0) = F(x) = F(1) = F'(0) = F'(1) = 0$.

Hence prove that $F^{(iv)}(t)$ has at least one zero ξ, say, and thus derive an expression for the Hermite error $f(x) - p(x)$ in terms of $f^{(iv)}(\xi)$.

3. Let $p_0(t) \equiv a_0 + a_1 t + a_2 t^2 + a_3 t^3$ be a cubic polynomial on $0 \leqslant t \leqslant 1$. Determine the coefficients of this polynomial so that

$$p(0) = f_0\,, \qquad p(1) = f_1\,,$$
$$p'(0) = d_0\,, \qquad p'(1) = d_1\,,$$

where f_0, f_1, d_0, and d_1 are given values.

Write down the cubic polynomial $p_1(t)$ on $1 \leqslant t \leqslant 2$ which is such that

$$p(1) = f_1\,, \qquad p(2) = f_2\,,$$
$$p'(1) = d_1\,, \qquad p'(2) = d_2\,,$$

where f_1, f_2, d_1, and d_2 are given values.

If d_1 is not given, derive an equation for d_1 such that piecewise defined cubic function on $0 \leqslant t \leqslant 2$ is twice continuously differentable.

4. State the transformation for t which would map $p(t)$ to $x_i \leqslant x \leqslant x_{i+1}$.

The function $p(t)$ is piecewise defined on $0 \leqslant t \leqslant 2$ by the cubic polynomials

$$p(t) = \begin{cases} P_0(t) = 3t^3 - 5t^2 + t + 1 + (t^3 - t^2)K\,; & 0 \leqslant t \leqslant 1 \\ P_1(t) = (t^3 - 5t^2 + 8t - 4)K\,; & 1 \leqslant t \leqslant 2. \end{cases}$$

Determine and justify the continuity class of $p(t)$ on $0 \leqslant t \leqslant 2$ for an arbitrary K, and show that there exists a value of K so that $p(t)$ is twice continuously differentiable (*Cubic spline*).

5. Define the functions $l_0(x)$, $l_1(x)$, $l_2(x)$ in the Lagrange representation

$$p(x) \equiv l_0(x) f_0 + l_1(x) f_1 + l_2(x) f_2$$

where $p(x)$ is the interpolation polynomial of degree $\leqslant 2$ such that

$$p(x_i) = f_i\,; \qquad i = 0,1,2$$

and x_0, x_1, x_2 are distinct points.

Prove that $p(x)$ is the unique polynomial of degree $\leqslant 2$ with this interpolation property.

6. The quadratic function $p(x)$ interpolates $f(x)$ at the distinct points x_0, x_1, and x_2. By considering the function of the variable t

$$F(t) = f(t) - p(t) - [f(x) - p(x)] \frac{(t-x_0)(t-x_1)(t-x_2)}{(x-x_0)(x-x_1)(x-x_2)},$$

where x is considered fixed, prove that

$$f(x) - p(x) = \frac{(x-x_0)(x-x_1)(x-x_2)}{6} f'''(\xi)$$

where f is assumed to be three times continuously differentiable.

7. State the Lagrange representation of the quadratic polynomial which interpolates $f(x)$ at $x = -h, x = 0$, and $x = +h$, where h is a given interval length.

   Hence derive approximations to $f'(0)$, $f''(0)$ and $\int_{-h}^{h} f(x)\,\mathrm{d}x$.

8. Approximate $f(x) \equiv |x|$ by $p(x) \equiv a + bx^2$ such that

$$\int_{-1}^{1} [f(x) - p(x)]^2 \,\mathrm{d}x$$

is a minimum.

9. A function $f(x)$ is defined by

$$\begin{cases} x^2 \text{ for } -1 \leqslant x \leqslant 0\,, \\ -x^2 \text{ for } \quad 0 \leqslant x \leqslant 1\,. \end{cases}$$

Approximate $f(x)$ by $p(x) \equiv a + bx + cx^2$ such that

$$\int_{-1}^{1} [f(x) - p(x)]^2 \,\mathrm{d}x$$

is a minimum.

10. A function $f(x)$ is defined by

$$f(x) = \begin{cases} x^2 \text{ for } -1 \leqslant x \leqslant 0 \ , \\ x^3 \text{ for } \quad 0 \leqslant x \leqslant 1 \ . \end{cases}$$

Approximate $f(x)$ by $p(x) = a + bx$ such that

$$\int_{-1}^{1} [f(x) - p(x)]^2 \, dx$$

is a minimum.

11. Approximate $f(x) \equiv x^4$ by $p(x) \equiv a + bx + cx^2$ such that

$$\int_{-1}^{1} [f(x) - p(x)]^2 \, dx$$

is a minimum.

12. Approximate $f(x) \equiv x^5$ by $p(x) \equiv a + bx^3$ such that

$$\int_{0}^{1} [f(x) - p(x)]^2 \, dx$$

is a minimum.

3

Nonlinear equations and nonlinear systems

3.1 FIXED-POINT ITERATION

3.1.1 Introduction

In the first three sections of this chapter the problem to be addressed is that of finding the values of x which satisfy the algebraic equation

$$f(x) = 0 \tag{3.1}$$

for a given function $f = f(x)$. A solution of this problem is called a *zero* of f or a *root* of (3.1).

Fixed-point iteration methods are based upon changing the equation $f(x) = 0$ into the form

$$x = g(x) \ . \tag{3.2}$$

Clearly, there may be many possible rearrangements of (3.1) into the form (3.2) and these rearrangements are made so that a root of (3.2) is also a root of (3.1).

Definition 3.1 If $g(x)$ is defined on some interval $[a,b]$ of x and $g(\alpha) = \alpha$ for some $\alpha \in [a,b]$, then the function g is said to have a *fixed point* α in $[a,b]$.

Theorem 3.1 If $g(x)$ is continuous for $x \in [a,b]$ and $g(x) \in [a,b]$ for all $x \in [a,b]$, then g has a fixed point in $[a,b]$.

Theorem 3.2 Suppose, in addition to the hypotheses of Theorem 3.1, that $g'(x)$ exists on (a,b) and

$$|g'(x)| \leq k < 1 \text{ for all } x \in [a,b].$$

Then g has a unique fixed point α in $[a,b]$.

The proofs of these theorems rely on the Mean Value Theorem and the Intermediate Value Theorem (Theorems A.4 and A.5 of Appendix A).

In order to approximate the fixed point α of a function $g(x)$, an initial approximation x_0 to α is chosen. Then, the sequence of iterates $x_1,x_2,x_3,\ldots$ is generated using the *one-point iteration algorithm*

$$x_{n+1}=g(x_n);\quad n=0,1,2,\ldots\ . \tag{3.3}$$

Under the conditions to be derived in subsection 3.1.2, this sequence converges to the root $x=\alpha$ of (3.2) and, consequently, of (3.1). The algorithm given in (3.3) is also called the *fixed-point iterative technique* or *functional iteration*.

3.1.2 Convergence

Example 3.1 Consider the equation $f(x)\equiv x^2+3x-4=0$ which obviously has roots at $x=-4$ and $x=1$. Rearranging gives

(1) $x=-(4-3x)^{\frac{1}{2}}$ so that $g(x)=-(4-3x)^{\frac{1}{2}}$. Starting with $x_0=-5$ gives $x_1=-19^{\frac{1}{2}}=-4.359$, $x_2=-17.077^{\frac{1}{2}}=-4.132$, $x_3=-16.396^{\frac{1}{2}}=-4.049$, $x_4=-16.147^{\frac{1}{2}}=-4.018$, $x_5=-16.054^{\frac{1}{2}}=-4.007$, $x_6=-16.021^{\frac{1}{2}}=-4.003$, $x_7=-16.009^{\frac{1}{2}}=-4.001$. To two decimal places, $x_5=-4.01$ while $x_6=x_7=-4.00$ to two decimal places. The reader should note, therefore, that *seven* iterations were necessary to obtain convergence to two decimal places of the zero of $f(x)$ at $x=-4$;

(2) $x=4/(x+3)$ so that $g(x)=4/(x+3)$. Starting with $x_0=-5$ this time gives $x_1=-2$, $x_2=4$, $x_3=0.571$, $x_4=1.120$, $x_5=0.971$, $x_6=1.007$, $x_7=0.998$, $x_8=1.001$, $x_9=1.000$. Clearly the iterates have converged to the zero of $f(x)$ at $x=1$, but this time the iterates have oscillated about the zero whereas in (1) convergence was monotonic;

(3) $x=(4-x^2)/3$ so that $g(x)=(4-x^2)/3$. Starting again with $x_0=-5$ gives $x_1=-7$, $x_2=-15$, $x_3=-73.667$ and this rearrangement is obviously diverging.

It was seen in this simple example that if a method converges it may do so monotonically or oscillatorily. The conditions under which convergence or divergence occurs are now discussed. Let $x=\alpha$ be a solution to $f(x)=0$, so that $f(\alpha)=0$ and $\alpha=g(\alpha)$. Subtracting α from each side of (3.3) then multiplying and dividing by $x_n-\alpha$ gives

$$x_{n+1}-\alpha=g(x_n)-\alpha=\frac{g(x_n)-\alpha}{x_n-\alpha}(x_n-\alpha)\ .$$

If $g(x)$ and $g'(x)$ are continuous on the interval bounded by α and x_n, the Mean Value Theorem (Theorem A.4 of Appendix A) gives

$$x_{n+1}-\alpha=g'(r_n)(x_n-\alpha),$$

where r_n lies between x_n and α.

Defining the error of the nth iterate as $e_n = x_n - \alpha$ so that $x_{n+1} - \alpha = e_{n+1}$, it follows that

$$e_{n+1} = g'(r_n)e_n \quad . \tag{3.4}$$

Thus

$$|e_{n+1}| \leqslant |g'(r_n)|.|e_n|$$

and, provided $|g'(r_n)| \leqslant k < 1$, the error at each iteration will not grow in magnitude; that is, the iteration procedure will converge. In other words, if $|g'(x)| \leqslant k < 1$ for all values of x in an interval of radius h about α, and x_0 is chosen in this interval, x_1 will also be in this interval and the algorithm will converge since

$$|e_{n+1}| \leqslant k|e_n| \leqslant k^2|e_{n-1}| \leqslant \ldots \leqslant k^{n+1}|e_0| \quad .$$

This discussion is formalized in the following theorem.

Theorem 3.3 Let $g(x) \in C[a,b]$ and suppose that $g(x) \in [a,b]$ for all $x \in [a,b]$. Suppose also that $g'(x)$ exists and is continuous for $x \in (a,b)$ with $g'(x) \leqslant k < 1$ for all $x \in (a,b)$. If x_0 is any number in $[a,b]$ then the sequence $x_{n+1} = g(x_n)$, $n = 0,1,2,\ldots$, will converge to the unique fixed point $x = \alpha$ in $[a,b]$.

Corollary 3.3.1 A bound for the error $e_n = x_n - \alpha$ incurred by using x_n to approximate α is given by

$$|e_n| \leqslant k^n \max\{x_0 - a, b - x_0\} \text{ for all } n \geqslant 0 \quad .$$

Corollary 3.3.2 A bound for e_n is given by

$$|e_n| \leqslant k^n|x_1 - x_0|/(1-k) \text{ for all } n \geqslant 1 \quad .$$

Theorem 3.1 shows that the rate of convergence depends on $|g'(x)|$, convergence being rapid if k is small. If the derivative is positive, the errors will be of the same sign and convergence will be monotonic, while convergence will be oscillatory if $g'(x)$ is negative.

It must be noted that the condition $|g'(x)| < 1$ is sufficient only; it is not necessary as the following example shows.

Example 3.2 Consider the equation $f(x) \equiv x^2 - x - 2$; then clearly $f(x)$ has zeros at $x = 2$ and $x = -1$. Let $x = g(x) = x^2 - 2$. Using the one-point iteration function $x_{n+1} = x_n^2 - 2$, $n = 0,1,2,\ldots$, with $x_0 = 1$ gives $x_1 = -1$ so that convergence to the zero at $x = -1$ is immediate. However, $g'(1) = 2 \nless 1$ so that the condition $|g'(x)| < 1$

is violated yet convergence has occurred.

Definition 3.2 If the sequence of iterates $x_0, x_1, x_2, \ldots$ tends to the limit α in such a way that

$$\lim_{x_n \to \alpha} \frac{x_{n+1} - \alpha}{(x_n - \alpha)^p} = C$$

for some constant $C \neq 0$ and for $p \geq 1$, then C is called the *asymptotic error constant* of the method and p is the *order of convergence*.

The constants C and p are connected by the following theorem.

Theorem 3.4 The order of convergence of a one-point iteration function is p if and only if

$$g'(\alpha) = g''(\alpha) = \ldots = g^{(p-1)}(\alpha) = 0 \text{ and } g^{(p)}(\alpha) \neq 0 \ .$$

The asymptotic error constant has the value $C = g^{(p)}(\alpha)/p!$

If $p = 1$, the one-point iteration function (3.3) is said to be *linear* and if $p = 2$, it is said to be *quadratic*. Fixed-point iteration exhibits linear convergence if $g'(\alpha) \neq 0$; higher-order convergence can only occur when $g'(\alpha) = 0$ as indicated in the following theorem.

Theorem 3.5 Let α be a fixed point of $x = g(x)$. Suppose that $g'(\alpha) = 0$ and $g''(x)$ is continuous for all x in some open interval containing α. Then there exists a number $\delta > 0$ such that, for $x_0 \in [\alpha - \delta, \alpha + \delta]$, fixed-point iteration becomes a second-order convergent process. Also, if $|g''(x)| \leq 2M$, where M is some constant, for $x \in [\alpha - \delta, \alpha + \delta]$ and $M|e_0| < 1$, then fixed-point iteration will converge quadratically to the zero at $x = \alpha$.

Example 3.3 Consider again the equation $f(x) \equiv x^2 + 3x - 4 = 0$ of Example 3.1 and the rearrangement $x = g(x) = (x^2 + 4)/(2x + 3)$. Starting with $x_0 = -5$ as in Example 3.1, succeeding iterates are $x_1 = -4.143$, $x_2 = -4.004$, $x_3 = -4.000$.

Convergence to the zero of $f(x)$ at $x = -4$ has been very rapid, requiring only three iterations to obtain convergence to two decimal places compared to seven iterations for rearrangement (1) of Example 3.1.

Example 3.4 Use Theorem 3.4 to show that (i) $x = g(x) = -(4 - 3x)^{\frac{1}{2}}$ and (ii) $x = g(x) = (x^2 + 4)/(2x + 3)$ are, respectively, linear and quadratic one-point iteration functions for obtaining the zero at $x = -4$ of $f(x) = x^2 + 3x - 4$. (The solution to an exercise such as that of Example 3.4 should begin with the verification that the iteration functions are, indeed, rearrangements of $f(x) = 0$. Examples 3.1 and 3.3 render this part of the solution unnecessary on this occasion.) In part (i) it is easy to show that $g'(x) = 3/\{2(4 - 3x)^{\frac{1}{2}}\}$ giving $g'(-4) = 0.375 \neq 0$; hence $p = 1$ and convergence is linear. In part (ii) it may be shown that $g'(x) = 2(x^2 + 3x - 4)/(2x + 3)^2$ and that $g''(x) = 2(-4x^2 - 12x + 41)/(2x + 3)^3$ from which it follows that $g'(-4) = 0$ and

$g''(-4) = -0.4 \neq 0$, so that convergence is quadratic.

It is easy to see that quadratic convergence is to be preferred to linear convergence. One quadratically convergent iteration function is Newton's method, which is derived in the following section.

3.2 NEWTON'S METHOD (THE NEWTON–RAPHSON METHOD)

3.2.1 Derivation and convergence

Newton's method, also known as the *Newton–Raphson method*, is one of the most powerful, and arguably the most well-known, numerical methods for finding the roots of equations of the form (3.1).

Suppose that the function $f(x)$ is twice continuously differentiable on the interval $[a,b]$, that is $f(x) \in C^2[a,b]$ for all $x \in [a,b]$. Let $x_n \in [a,b]$ be an approximation to α, a zero of $f(x)$, such that $f'(x) \neq 0$ and $|\alpha - x_n|$ is small. The second-degree Taylor polynomial may be written as

$$f(x) = f(x_n) + (x - x_n)f'(x_n) + \tfrac{1}{2}(x - x_n)^2 f''(r(x)) \tag{3.5}$$

where $r(x)$ lies between x and x_n. Since α is a zero of $f(x)$, equation (3.5) with $x = \alpha$ becomes

$$0 = f(x_n) + (\alpha - x_n)f'(x_n) + \tfrac{1}{2}(\alpha - x_n)^2 f''(r(\alpha)) \tag{3.6}$$

and, since $\alpha - x_n$ is small so that $(\alpha - x_n)^2$ is even smaller and may be neglected, equation (3.6) gives

$$\alpha \simeq x_n - \frac{f(x_n)}{f'(x_n)} . \tag{3.7}$$

The right-hand side of (3.7) is a better approximation to α than x_n itself provided certain conditions are met and (3.7) suggests the iterative scheme

$$x_{n+1} = x_n - \frac{f(x_n)}{f'(x_n)}, \qquad n = 0,1,2,\ldots \tag{3.8}$$

which is the *Newton–Raphson method* for finding each root of (3.1).

It is clear that (3.8) is an example of (3.2) with $g(x) = x - f(x)/f'(x)$. It follows, therefore, that $g'(x) = f(x)f''(x)/[f'(x)]^2$ and so, if

$$|f(x)f''(x)/[f'(x)]^2| < 1 \tag{3.9}$$

on an interval about the zero $x = \alpha$ of $f(x)$, Newton's method will converge for any initial value x_0 in the interval. This condition is sufficient only and is formalized in the following theorem.

Theorem 3.6 Let $f(x)\in C^2[a,b]$ for all $x\in[a,b]$. If $\alpha\in[a,b]$ is such that $f(\alpha)=0$ and $f'(\alpha)\neq 0$, then there exists a number $\delta>0$ such that Newton's method generates a sequence $x_0,x_1,x_2,\ldots$ which converges to α for any initial approximation $x_0\in[\alpha-\delta,\alpha+\delta]$.

It is easy to see that the Newton–Raphson method exhibits quadratic convergence by writing

$$g(x_n)=g(\alpha)+(x_n-\alpha)g'(\alpha)+\tfrac{1}{2}(x_n-\alpha)^2g''(s_n)$$

where s_n lies in the interval from x_n to α. Then, since $g'(\alpha)=f(\alpha)f''(\alpha)/[f'(\alpha)]^2=0$ (because $f(\alpha)=0$),

$$g(x_n)=g(\alpha)+\tfrac{1}{2}(x_n-\alpha)^2g''(s_n)\ ,$$

that is,

$$x_{n+1}=\alpha+\tfrac{1}{2}(x_n-\alpha)^2g''(s_n)\ , \tag{3.10}$$

and so

$$\frac{x_{n+1}-\alpha}{(x_n-\alpha)^2}=\tfrac{1}{2}g''(s_n)$$

and Definition 3.2 establishes second-order convergence. Alternatively, equation (3.10) shows that

$$e_{n+1}=\tfrac{1}{2}g''(s_n).e_n^2\ ,$$

that is, the error in the iterate x_{n+1} is proportional to the square of the error in the iterate x_n, and so convergence is quadratic.

Example 3.5 In section 9.4 of Chapter 9, where a finite difference model of the otolith membrane is discussed, the solution of the polynomial equation

$$f(y)\equiv 54.2535y^3+2.1701y^2-0.5y-0.02$$

is required. It follows that

$$f'(y)=162.7605y^2+4.3402y-0.5$$

and the Newton–Raphson method

$$y_{n+1} = y_n - f(y_n)/f'(y_n) \ , \qquad n = 0,1,2,\ldots \ ,$$

from the initial approximation $y_0 = -0.064$, yields the iterates given in Table 3.1.

Table 3.1

Iterate number	Iterate
0	− 0.06400
1	− 0.00400
2	− 0.03891
3	− 0.03999
4	− 0.04000
5	− 0.04000

3.2.2 Multiple roots

In the above discussion of Newton's method it was assumed that $f'(\alpha) \neq 0$. Suppose, however, that the zero α of $f(x)$ has multiplicity m. Then $f(x)$ takes the form

$$f(x) = (x - \alpha)^m q(x) \ , \tag{3.11}$$

where α is not a zero of $q(x)$. It may be shown that, for $m \neq 1$, convergence of Newton's method (3.8) is only linear but that convergence of its modified form

$$x_{n+1} = x_n - mf(x_n)/f'(x_n) \ , \qquad n = 0,1,2,\ldots$$

is quadratic.

3.3 OTHER METHODS

These include the interval bisection method, the *regula falsi*, Horner's method, Bairstow's method for quadratic factors, the Graeffe and Muller interpolation methods for real roots and Laguerre's method which, for polynomial equations, gives cubic convergence to a simple root and linear convergence to a multiple root. Laguerre's method will also converge to nonreal roots from real starting values. The quotient-difference (QD) algorithm is an efficient method for determining all the roots of a polynomial equation. The reader is referred to the specialist texts by Henrici (1964), Ralston (1965), Morris (1983), Gerald and Wheatley (1984), Burden and Faires (1985), and Fröberg (1985) for further discussions of these methods.

3.4 FIXED-POINT ITERATION FOR NONLINEAR SYSTEMS

In this section, many of the definitions and theorems of section 3.1 will be extended to the system of nonlinear algebraic equations

$$f_i(x_1,x_2,\ldots,x_N)=0\ ,\qquad i=1,2,\ldots,N \tag{3.12}$$

which may be written in vector form as

$$\mathbf{f}(\mathbf{x})=\mathbf{0} \tag{3.13}$$

where $\mathbf{x}=[x_1,x_2,\ldots,x_N]^T$, $\mathbf{f}=[f_1,f_2\ldots,f_N]^{\mathrm{T}}$ and $\mathbf{0}$ is the zero-vector of order N. Each function f_i maps N-dimensional space R^N into the real line R and the vector $\mathbf{f}$ maps R^N into R^N.

Definition 3.3 Let f be a function defined on a set $S\subset R^N$ and mapping into R. The function f is said to have the limit q at the point $\mathbf{x}=\mathbf{x}_0$, written

$$\lim_{\mathbf{x}\to\mathbf{x}_0} f(\mathbf{x})=q$$

if, given a number $\varepsilon>0$, a number $\delta>0$ exists such that

$$|f(\mathbf{x})\to q|<\varepsilon$$

whenever $\mathbf{x}\in S$ and

$$0<\|\mathbf{x}-\mathbf{x}_0\|<\delta\ ,$$

where $\|.\|$ is any vector norm.

Definition 3.4 Let f be a function from a set $S\subset R^N$ into R. The function f is said to be *continuous* at $\mathbf{x}=\mathbf{x}_0\in S$ provided

$$\lim_{\mathbf{x}\to\mathbf{x}_0} f(\mathbf{x}) \text{ exists and } \lim_{\mathbf{x}\to\mathbf{x}_0} f(\mathbf{x})=f(\mathbf{x}_0)\ .$$

Definition 3.5 The function f is said to be *continuous on a set S* provided f is continuous at every point of S (written $f\in C(S)$).

Definition 3.6 Let $\mathbf{f}$ be a function from $S\subset R^N$ into R where $\mathbf{f}(\mathbf{x})=[\mathbf{f}_1(\mathbf{x}),f_2(\mathbf{x}),\ldots,f_N(\mathbf{x})]^{\mathrm{T}}$ with f_i a mapping from R^N into R $(i=1,2,\ldots,N)$. Then

$$\lim_{\mathbf{x}\to\mathbf{x}_0} \mathbf{f}(\mathbf{x})=\mathbf{q}=[q_1,q_2,\ldots,q_N]^{\mathrm{T}}$$

if and only if

$$\lim_{\mathbf{x}\to\mathbf{x}_0} f_i(\mathbf{x}) = q_i \quad (i = 1,2,\ldots,N) \ .$$

Definition 3.7 Suppose **f** is a function from $S \subset R^N$ into R^N with $\mathbf{f}(\mathbf{x})$ as given in Definition 3.6. The function **f** is said to be *continuous* at $\mathbf{x}_0 \in S$ provided

$$\lim_{\mathbf{x}\to\mathbf{x}_0} \mathbf{f}(\mathbf{x}) \text{ exists and } \lim_{\mathbf{x}\to\mathbf{x}_0} \mathbf{f}(\mathbf{x}) = \mathbf{f}(\mathbf{x}_0) \ .$$

Definition 3.8 The function **f** is said to be *continuous on the set S* if **f** is continuous at each point in S (written $\mathbf{f} \in C(S)$).

Theorem 3.7 Let f be a function from $S \subset R^N$ into R and suppose that $\mathbf{x}_0 \in S$. Suppose that $\delta > 0$ and $k > 0$ are constants such that

$$|\partial f(\mathbf{x})/\partial x_j| \leqslant k \ , \quad j = 1,2,\ldots,N$$

whenever $\|\mathbf{x} - \mathbf{x}_0\| < \delta$ and $\mathbf{x} \in S$. Then f is continuous at $\mathbf{x}_0$.

Theorem 3.7 connects the continuity of the function f of N variables, such as those in (3.12), with the partial derivatives of the function at a given point. In section 3.1, equation (3.1) was solved by transforming it into the form (3.2), equations (3.1) and (3.2) having the same fixed points. A similar procedure may be adopted for vector equations of the form (3.13), the following definition and theorem completing the preliminary theory.

Definition 3.9 A function **g** from $S \subset R^N$ into R^N is said to have a *fixed point* at $\boldsymbol{\alpha} \in S$ if $\boldsymbol{\alpha} = \mathbf{g}(\boldsymbol{\alpha})$.

Theorem 3.8 (*The Contraction Mapping Theorem*) Let $S = \{(x_1, x_2, \ldots, x_N)$: $a_i \leqslant x_i \leqslant b_i$ for each $i = 1,2,\ldots,N\}$, where the a_i, $b_i (i = 1,2,\ldots,N)$ are constants and suppose that **g** is a continuous function with continuous first partial derivatives from $S \subset R^N$ into R^N with $\mathbf{g}(\mathbf{x}) \in S$ whenever $x \in S$. Then **g** has a fixed point in S.

Theorem 3.9 If, in addition to the hypotheses of Theorem 3.8, a constant $k < 1$ exists and

$$|\partial g_i(\mathbf{x})/\partial x_j| \leqslant k/N \ ,$$

whenever $\mathbf{x} \in S$, for each $i,j = 1,2,\ldots,N$, then the fixed point is unique.

Each component $f_i(\mathbf{x})$ of the vector $\mathbf{f}(\mathbf{x})$ of (3.13) can be rearranged to express $x_i (i = 1,2,\ldots,N)$ as some function of the variables $x_i, x_2, \ldots, x_N$. Component-wise, therefore, (3.13) may be written as

$$x_i = g_i(x_1, x_2, \ldots, x_N) \ , \quad i = 1, 2, \ldots, N \ . \tag{3.14}$$

In vector form, (3.13) may thus be rearranged to give

$$\mathbf{x} = \mathbf{g}(\mathbf{x}) \ , \tag{3.15}$$

where $\mathbf{g} = [g_1(\mathbf{x}), g_2(\mathbf{x}), \ldots, g_N(\mathbf{x})]^{\mathrm{T}}$, which suggests the iteration scheme

$$\mathbf{x}^{(n+1)} = \mathbf{g}(\mathbf{x}^{(n)}) \ , \quad n = 0, 1, 2, \ldots \ . \tag{3.16}$$

This is known as a *Gauss–Jacobi* type of iteration scheme.

In practice, an initial vector $\mathbf{x}^{(0)} = [x_1^{(0)}, x_2^{(0)}, \ldots, x_N^{(0)}]^{\mathrm{T}}$ is selected and used to generate the sequence of vectors $\mathbf{x}^{(1)}, \mathbf{x}^{(2)}, \mathbf{x}^{(3)}, \ldots$ defined by (3.16). Under the hypotheses of the following theorem, which is the multi-dimensional case of Theorem 3.3, the sequence will converge to a unique fixed point $\boldsymbol{\alpha} \in S$. The proof of the theorem can be found on page 153 of the specialist text of Ortega (1972).

Theorem 3.10 Let S be the set of points of Theorem 3.8 and suppose that $\mathbf{g}(\mathbf{x})$ is a continuous function with continuous first partial derivatives from $S \subset R^N$ into R^N with the property that $\mathbf{g}(\mathbf{x}) \in S$ whenever $\mathbf{x} \in S$. If a constant k $(0 < k < 1)$ exists with

$$|\partial g_i(\mathbf{x})/\partial x_j| \leqslant k/N \text{ whenever } \mathbf{x} \in S$$

for each $i,j = 1, 2, \ldots, N$, then the sequence defined by (3.16) converges to the unique fixed point $\boldsymbol{\alpha} \in S$ from any initial point $\mathbf{x}^{(0)} \in S$.

Corollary 3.10.1 The error in the nth iterate $\mathbf{x}^{(n)}$ is given by

$$\|\mathbf{x}^{(n)} - \boldsymbol{\alpha}\|_\infty \leqslant k^n \|\mathbf{x}^{(1)} - \mathbf{x}^{(0)}\| / (1 - k) \ .$$

If instead of the Gauss–Jacobi scheme (3.16), the *Gauss–Seidel* type of scheme described by

$$x_i^{(n+1)} = g_i(x_1^{(n+1)}, x_2^{(n+1)}, \ldots, x_{i-1}^{(n+1)}, x_i^{(n)}, \ldots, x_N^{(n)}) \tag{3.17}$$

is used, convergence may be faster. The conditions for the convergence of (3.17) are more complicated than for (3.16) and details will be omitted here. It suffices to say that the user will be aware of divergence if it happens.

3.5 THE NEWTON–RAPHSON METHOD FOR A SYSTEM

Consider again equation (3.13) and suppose that $\mathbf{x} = \boldsymbol{\alpha}$ is a fixed point. Suppose, further, that the current approximation $\mathbf{x}^{(n)}$ *to* $\boldsymbol{\alpha} = [\alpha_1, \alpha_2, \ldots, \alpha_N]^T$ can be written as

$\mathbf{x}^{(n)} = \boldsymbol{\alpha} + \mathbf{h}^{(n)}$ where $\mathbf{h}^{(n)} = [h_1^{(n)}, h_2^{(n)}, \ldots, h_N^{(n)}]^{\mathrm{T}}$. The components $f_i(\mathbf{x}^{(n)})$ of $\mathbf{f}$ can be computed: let them be called $\phi_i^{(n)}$ $(i = 1,2,\ldots,N)$, then

$$f_i(\alpha_1 + h_1^{(n)}, \alpha_2 + h_2^{(n)}, \ldots, \alpha_N + h_N^{(n)}) = \phi_i^{(n)}$$

and so $\mathbf{f}(\boldsymbol{\alpha} + \mathbf{h}^{(n)}) = \boldsymbol{\phi}^{(n)} = \mathbf{f}(\mathbf{x}^{(n)})$.

It then follows from Appendix A that

$$f_i(\alpha_1, \alpha_2, \ldots, \alpha_N) + \sum_{j=1}^{N} h_i^{(n)} \partial f_i / \partial \alpha_j + \text{higher-order terms} \;= \phi_i^{(n)} \tag{3.18}$$

for $i = 1,2,\ldots,N$. Neglecting the higher-order terms and noting that $\mathbf{f}(\boldsymbol{\alpha}) = \mathbf{0}$, equation (3.18) gives

$$J\mathbf{h}^{(n)} = \boldsymbol{\phi}^{(n)} \text{ and } \mathbf{h}^{(n)} = J^{-1}\boldsymbol{\phi}^{(n)} = J^{-1}\mathbf{f}(\mathbf{x}^{(n)})$$

where it is assumed that the *Jacobian* J is nonsingular. Thus

$$\boldsymbol{\alpha} = \mathbf{x}^{(n)} - \mathbf{h}^{(n)} = \mathbf{x}^{(n)} - J^{-1}\mathbf{f}(\mathbf{x}^{(n)})$$

which suggests the iterative scheme

$$\mathbf{x}^{(n+1)} = \mathbf{x}^{(n)} - [J^{(n)}]^{-1}\mathbf{f}(\mathbf{x}^{(n)}) \tag{3.19}$$

where the matrix $J^{(n)}$, assumed to be nonsingular, is the Jacobian at $\mathbf{x} = \mathbf{x}^{(n)}$.

Equation (3.19) is the *Newton* or *Newton–Raphson method for a nonlinear system*. In the form (3.19) the method has a definite drawback, namely the apparent need to invert the Jacobian at each iteration. In practice, the method is usually implemented in a two-step fashion by defining a vector $\mathbf{z}^{(n)} = -[J^{(n)}]^{-1}\mathbf{f}(\mathbf{x}^{(n)})$, finding $\mathbf{z}^{(n)}$ by solving the *linear* system $J^{(n)}\mathbf{z}^{(n)} = -\mathbf{f}(\mathbf{x}^{(n)})$, and then calculating $\mathbf{x}^{(n+1)}$ from the equation $\mathbf{x}^{(n+1)} = \mathbf{x}^{(n)} + \mathbf{z}^{(n)}$. Newton's method for a system gives quadratic convergence to $\boldsymbol{\alpha}$ from a sufficiently accurate starting vector $\mathbf{x}^{(0)}$ provided the Jacobian $J^{(n)}$ is nonsingular $(n = 0,1,2,\ldots)$.

Example 3.6 In section 8.7 of Chapter 8, a reaction–diffusion equation in genetics is to be solved. The numerical procedure followed calls for the solution, at some time $t = (n+1)l$, of a nonlinear system of the form

$$\mathbf{f}(\mathbf{U}^{n+1}) = \mathbf{0}\;, \qquad n = 0,1,2,\ldots \tag{3.20}$$

where $\mathbf{f} = [f_1, f_2, \ldots, f_N]^{\mathrm{T}}$ and $\mathbf{U}^n = [U_1^n, U_2^n, \ldots, U_N^n]^{\mathrm{T}}$ (note: in the notation of Chapter 8, the superscript refers to a particular time and not an iteration). Each component of $\mathbf{f}$ is defined by

$$f_m(\mathbf{U}^{n+1}) = \theta r U_{m-1}^{n+1} + (1+2\theta r)U_m^{n+1} - \theta r U_{m+1}^{n+1} - \theta l\phi(U_m^{n+1})$$
$$- (1-\theta)rU_{m-1}^{n} - [1-2(1-\theta)r]U_m^n - (1-\theta)rU_{m+1}^n$$
$$- (1-\theta)l\phi(U_m^n) - b_m \qquad (3.21)$$

where $m = 1,2,\ldots,N$, θ is a parameter in the interval $0 \leqslant \theta \leqslant 1$, r, b and l are constants and $\phi(\mathbf{U})$ is a given function of $\mathbf{U}$.

It follows from (3.21) that the nonzero derivatives of f_m are given by

$$\partial f_m/\partial U_{m-1}^{n+1} = -\theta r,\ \partial f_m/\partial U_m^{n+1} = (1+2\theta r) - \theta l\partial\phi/\partial U_m^{n+1},\ \partial f_m/\partial U_{m+1}^{n+1} = -\theta r \qquad (3.22)$$

and it is these derivatives which form the elements of the Jacobian of $\mathbf{f}(\mathbf{U}^{n+1})$; this matrix is tridiagonal. The Newton–Raphson method for finding $\mathbf{U}^{n+1}$ is thus

$$\mathbf{U}_{(k+1)}^{n+1} = \mathbf{U}_{(k)}^{n+1} - [J_{(k)}^{n+1}]^{-1}\mathbf{f}(\mathbf{U}_{(k)}^{n+1}) \qquad (3.23)$$

in which the bracketed subscript $k = 0,1,2,\ldots$ refers to the iteration number.

PROBLEMS

1. With x_0 sufficiently close to 2, find the order of convergence and the asymptotic error constant of the iteration function

$$g(x) = 1 + \frac{2}{x}$$

which may be used to find one root of the equation $x^2 - x - 2 = 0$.

Starting with $x_0 = \frac{1}{2}$ and carrying four decimal places, perform six iterations of this procedure. Why are the iterates oscillating about the root $\alpha = 2$?

2. Show that the fixed points of the two iterative processes

(i) $$x_{n+1} = \frac{3}{x_n} - 1\ ,$$

(ii) $$x_{n+1} = \frac{x_n^2 + 3}{2x_n + 1}\ ,$$

are roots of the equation $x^2 + x - 3 = 0$ and determine the order of convergence of each method to the positive root.

3. The iterative process

$$x_{n+1} = g(x_n) \ , \qquad n = 0,1,2,\ldots$$

converges to a fixed point ξ, where g is a continuously differentiable function.
If $g'(\xi) \neq 0$, prove that

$$\lim_{n \to \infty} e_{n+1}/e_n = g'(\xi)$$

where $e_n = \xi - x_n$.
State and justify conditions for which the process has quadratic order convergence.
Show that the fixed points of the two iterative processes

(i) $\qquad x_{n+1} = 1 + 1/x_n \ ,$

(ii) $\qquad x_{n+1} = (x_n^2 + 1)/(2x_n - 1) \ ,$

are roots of $x^2 - x - 1 = 0$ and calculate the order of convergence of each method to the positive root.
Verify that (ii) is Newton's method for the equation $x^2 - x - 1 = 0$.

4. With x_0 sufficiently close to $\sqrt[3]{N}$, $N > 0$, find the order of convergence and the asymptotic error constant of the iteration function

$$g(x) = \frac{2x^3 + N}{3x^2}$$

which may be used to find the cube root of the finite, positive number N.

5. Show that there is a root of the equation

$$f(x) \equiv x^3 - x^2 - x - 1 = 0$$

between $x = 1$ and $x = 2$.
Prove that the scheme

$$x_{n+1} = x_n^3 - x_n^2 - 1$$

is not convergent for any x_0 in $1 \leq x_0 \leq 2$.
Show that $1.75 \leq x_0 \leq 2$ is a sufficient condition for the convergence of the scheme

$$x_{n+1} = 1 + \frac{1}{x_n} + \frac{1}{x_n^2};\ n = 0,1,2,\ldots$$

to this zero of $f(x)$.

6. The iterative process

$$x_{n+1} = g(x_n),\ n = 0,1,2,\ldots$$

converges to a fixed point ξ, where g is a twice continuously differentiable function. If $g'(\xi) = 0$ and $g''(\xi) \neq 0$ prove that

$$\lim_{n \to \infty} e_{n+1}/e_n^2 = -g''(\xi)/2$$

where $e_n = \xi - x_n$.

State the corresponding result for the case $g'(\xi) \neq 0$.

Compare the rates of convergence of the two iterative processes

(a) $x_{n+1} = (x_n + 1)/2 + 1/x_n$

(b) $x_{n+1} = (x_n + 4/x_n)/2$

to the fixed point $\xi = 2$ of the equation $x^2 - x - 2 = 0$.

7. Show that, if the iteration formula

$$x_{n+1} = (x_n^k + kQx_n)/(kx_n^{k-1} + Q)\ ;\quad n = 0,1,2,\ldots,\ k \neq 1$$

converges, then the limit α of the sequence $\{x_0, x_1, x_2, \ldots\}$ is the $(k-1)$th root of Q.

Find the values of k for which the order of convergence p is at least 2. Show that only one of these values of k leads to a non-trivial iteration.

8. With x_0 sufficiently close to $\sqrt{A}$, $A > 0$, and $g_1(x)$ and $g_2(x)$ given by

$$g_1(x) = \frac{x^2 + A}{2x} \quad \text{and} \quad g_2(x) = \frac{x(3A - x^2)}{2A},$$

verify that

$$x_{n+1} = g_1(x_n) \quad \text{and} \quad x_{n+1} = g_2(x_n)\ ;\quad n = 0,1,2,\ldots$$

are both second order convergent iteration functions for finding $\sqrt{A}$.

Prove that $g_3(x)$ defined by

$$g_3(x) = \tfrac{3}{4}g_1(x) + \tfrac{1}{4}g_2(x)$$

is at least third order convergent.

9. The equation

$$\text{(B)} \qquad f(x) \equiv x^3 + 4x^2 - 10 = 0$$

has a unique root in [1,2]. Starting with $x_0 = 1.5$, perform three iterations with each of the following one-point iteration functions, giving each iterate to three decimal places:

$$\text{(i)} \qquad x_{n+1} = g_1(x_n) = x_n - x_n^3 - 4x_n^2 + 10 \ ; \quad n = 0,1,2 \ ,$$

$$\text{(ii)} \qquad x_{n+1} = g_2(x_n) = \left(\frac{10}{x_n} - 4x_n\right)^{\frac{1}{2}} \ ; \quad n = 0,1,2 \ ,$$

$$\text{(iii)} \qquad x_{n+1} = g_3(x_n) = \frac{2x_n^2 + 4x_n^2 + 10}{3x_n^2 + 8x_n} \ ; \quad n = 0,1,2 \ .$$

Verify that $g_3(x)$ has satisfied the sufficient condition for convergence, and show that (iii) is, in fact, the Newton–Raphson method for solving (B).

10. (a) The continuous differentiable function g is such that

$$a \leqslant g(x) \leqslant b \quad \text{for all} \quad a \leqslant x \leqslant b.$$

By considering $h(x) \equiv g(x) - x$, show that there exists a fixed point $\xi = g(\xi)$.

(b) If the function g is also such that

$$|g'(x)| \leqslant K < 1 \quad \text{for all} \quad a \leqslant x \leqslant b,$$

prove that the iterative process

$$x_{n+1} = g(x_n) \ ; \quad n = 0,1,2,\ldots$$

converges to ξ, where $x_0 \in [a,b]$.

Hence use (a) and (b) to show that the iterative process

$$x_{n+1} = \frac{x_n}{2} + \frac{1}{x_n} \ ; \quad n = 0,1,2,\ldots$$

with $x_0 \in [1,2]$, converges to a fixed point in $[1,2]$ of the equation $x^2 = 2$.

Explain and justify a condition for quadratic order convergence of $x_{n+1} = g(x_n)$ to a fixed point ξ and show that the given process satisfies this condition.

11. (a) With x_0 sufficiently close to 3, find the order of convergence and the aymptotic error constant of the iteration function

$$g(x) = 2 + \frac{3}{x}$$

which may be used to find one root of the equation $x^2 - 2x - 3 = 0$.

Starting with $x_0 = 0.5$ and carrying four decimal places, perform six iterations of this procedure. Why are the iterates oscillating about the zero $\alpha = 3$ of the function $f(x) = x^2 - 2x - 3$?

(b) Find the Newton–Raphson formulation for $f(x) = x^2 - 2x - 3$.

Starting with $x_0 = 4$ perform two iterations of this method. Why is convergence monotonic to the zero at $\alpha = 3$?

12. Show that the Newton–Raphson method for the equation $f(x) \equiv x^2 - x - 2 = 0$ may be written in the form

$$x_{n+1} = \frac{x_n^2 + 2}{2x_n - 1} \; ; \quad n = 0,1,2, \ldots \quad .$$

Starting with $x_0 = 3$ perform two iterations of this algorithm. Why is convergence monotonic to the root at $\alpha = 2$?

13. Give a geometric description of the derivation of the Newton–Raphson algorithm

$$x_{n+1} = x_n - f(x_n)/f'(x_n) \; ; \quad n = 0,1,2, \ldots$$

used in the numerical solution of $f(x) = 0$.

By writing $f(x) = x^2 - N$ use the Newton–Raphson method to find an algorithm which will converge to the positive square root of N. Use this algorithm to find the positive square root of 17. Take $x_0 = 4$ and calculate each iterate to 3 decimal places giving the final iterate correct to 2 decimal places.

4

Differentiation and integration

4.1 NUMERICAL DIFFERENTIATION

4.1.1 First-order derivatives

Differentiation of functions is an important mathematical operation, and at some stage of pre-university education, many students will have practised techniques to carry out this operation. The student will also have learned that the first derivative of a function with respect to one of its independent variables describes how the function is changing with respect to that variable. Often this is called the *rate of change* of the function, especially when the independent variable is time.

In many practical situations, the function is not known explicitly in terms of its independent variable or variables. Instead, the numerical value of the function is known at a number of discrete values of an independent variable. This results in a set of data pairs, which may have been obtained in a laboratory experiment, and it is the purpose of numerical differentiation to estimate the rate at which the dependent variable is changing, as the independent variable changes, from the set of data pairs.

Suppose, then, that the independent variable is x and that $y = y(x)$ depends on x only. Suppose, further, that y is found to have the numerical value y_n at the point where $x = x_n$ ($n = \ldots, 2, -1, 0, 1, 2, \ldots$) with $x_{n+1} - x_n = h$, a constant.

Example 4.1 Given the three data pairs (x_n, y_n), (x_{n+1}, y_{n+1}) and (x_{n+2}, y_{n+2}) with $x_{n+2} - x_{n+1} = x_{n+1} - x_n = h$, find an approximation y'_n to the derivative $y'(x_n)$ in terms of the three y-values.

Let

$$y'_n = ay_n + by_{n+1} + cy_{n+2} \tag{4.1}$$

where a, b, c are constants, and consider the equation

$$y'(x_n) = ay(x_n) + by(x_n + h) + cy(x_n + 2h). \tag{4.2}$$

Using Taylor's series, equation (4.2) becomes

$$\begin{aligned} y'(x_n) &= ay(x_n) + b\{y(x_n) + hy'(x_n) + \tfrac{1}{2}h^2y''(x_n) + \dots\} \\ &\quad + c\{y(x_n) + 2hy'(x_n) + 2h^2y''(x_n) + \dots\} \\ &= (a+b+c)y(x_n) + (b+2c)hy'(x_n) + (\tfrac{1}{2}b + 2c)h^2y''(x_n) + \dots . \end{aligned} \tag{4.3}$$

Equating coefficients in (4.3) gives

$$a + b + c = 0, \quad (b + 2c)h = 1, \quad \tfrac{1}{2}b + 2c = 0$$

from which if follows that $a = -3/(2h)$, $b = 2/h$, $c = -1/(2h)$. Returning to (4.2), the required value y'_n is given by

$$y'_n = h^{-1}(-\tfrac{3}{2}y_n + 2y_{n+1} - \tfrac{1}{2}y_{n+2}) \tag{4.4}$$

Example 4.2 Show that the error in approximating $y'(x_n)$ by y'_n as given in (4.4) is $O(h^2)$.

Consider the following expresion for the error:

$$y'(x_n) - h^{-1}\{-\tfrac{3}{2}y(x_n) + 2y(x_n + h) - \tfrac{1}{2}y(x_n + 2h)\}. \tag{4.5}$$

Taylor's series gives

$$\begin{aligned} y'(x_n) - h^{-1}[-\tfrac{3}{2}y(x_n) &+ 2\{y(x_n) + hy'(x_n) + \tfrac{1}{2}h^2y''(x_n) + \tfrac{1}{6}h^3y'''(x_n) + \dots\} \\ &- \tfrac{1}{2}\{y(x_n) + 2hy'(x_n) + 2h^2y''(x_n) + \tfrac{4}{3}h^3y'''(x_n) + \dots\} \end{aligned}$$

which simplifies to $-\tfrac{1}{3}h^2y'''(x_n) - \dots$. Hence the error is $O(h^2)$ and (4.4) is a *second-order* differentiation formula.

Equation (4.4) is a *forward difference* approximation to $y'(x_n)$ in that it determines y'_n in terms of values of y at the point x_n and points to the right of it on the real number line. In a similar fashion $y'(x_n)$ can be estimated in terms of the values of y at the point x_n and points to its left. This gives rise to a *backward difference* approximation to $y'(x_n)$. For instance, defining y'_n by the formula $y'_n = ay_n + by_{n-1} + cy_{n-2}$ and proceeding as in Examples 4.1 and 4.2 leads to the backward difference formula

$$y'(x_n) = h^{-1}(\tfrac{1}{2}y_{n-2} - 2y_{n-1} + \tfrac{3}{2}y_n) + O(h^2). \tag{4.6}$$

In the same way, defining y'_n by the formula $y'_n = ay_{n-1} + by_n + cy_{n+1}$, which is a *central difference* formula, and proceeding as in Examples 4.1 and 4.2 gives $c = -a = \tfrac{1}{2}h^{-1}$ and $b = 0$, so that

$$y'(x_n) = \tfrac{1}{2}h^{-1}(-y_{n-1} + y_{n+1}) + O(h^2). \tag{4.7}$$

Using only two data points instead of three leads to an inferior error term in the numerical differentiation formula. Starting with the equations $y'_n = ay_n + by_{n+1}$ and $y'_n = \alpha y_n + \beta y_{n-1}$ and proceeding as in Examples 4.1 and 4.2 gives $a = -b = \alpha = -\beta = h^{-1}$ and, thus,

$$y'(x_n) = h^{-1}(-y_n + y_{n+1}) + O(h) \tag{4.8}$$

and

$$y'(x_n) = h^{-1}(-y_{n-1} + y_{n+1}) + O(h) \ ; \tag{4.9}$$

these are, respectively, forward difference and backward difference numerical differentiation formulas of the first order.

Higher-order formulas may be obtained by involving more data points in the numerical differentiation formula. Consider, by way of example, the formula

$$y'_n = ay_{n-1} - by_n + cy_{n+1} + dy_{n+2} + ey_{n+3} \ . \tag{4.10}$$

Proceeding in the, by now, familiar way shows that $a = -1/(4h)$, $b = -5/(6h)$, $c = 3/(2h)$, $d = -1/(2h)$, $e = 1/(12h)$, so that

$$y'(x_n) = h^{-1}(-\tfrac{1}{4}y_{n-1} - \tfrac{5}{6}y_n + \tfrac{3}{2}y_{n+1} - \tfrac{1}{2}y_{n+2} + \tfrac{1}{12}y_{n+3}) + O(h^4). \tag{4.11}$$

Equation (4.11) is clearly a fourth-order formula; it is not a central difference formula, as it does not involve the same number of points on each side of the 'base point' (x_n, y_n), nor is it a wholly forward nor a wholly backward difference formula. It could perhaps be termed a 'mainly forward' difference formula.

In the same way, the 'mainly backward' formula

$$y'(x_n) = h^{-1}(-\tfrac{1}{12}y_{n-3} - \tfrac{1}{2}y_{n-2} + \tfrac{3}{2}y_{n+1} - \tfrac{5}{6}y_n + \tfrac{1}{4}y_{n+1}) + O(h^4). \tag{4.12}$$

and the central difference formula

$$y'(x_n) = h^{-1}(-\tfrac{1}{12}y_{n-2} - \tfrac{2}{3}y_{n-1} + \tfrac{2}{3}y_{n+1} - \tfrac{1}{12}y_{n+2}) + O(h^4) \tag{4.13}$$

may be derived, both of which are fourth order.

Comparing formulas {(4.4), (4.6), (4.7)} and formulas {(4.11), (4.12), (4.13)} the reader will see that, to achieve a given order, one fewer data point needs to be used if a central difference formula is used to approximate a derivative.

Example 4.3 Creatinine is a component of urine orginally formed from creatinine phosphate, a substance which supplies energy to the muscle (Cullen, 1985). In an experiment carried out by Sapirstein *et al.* (1955), 2 g of creatinine were injected into the bloodstream of dogs, and subsequent concentrations in the bloodstream, estimated by a mathematical model (Cullen, 1985), are given in Table 4.1 where t

Table 4.1

t	2	4	6	8	10
c	0.439	0.383	0.336	0.298	0.266

represents time (in minutes) and c represents concentration (in mg/ml). Estimate the rate at which the concentration is decaying after 6 and 8 minutes using fourth-order numerical differentiation formulas. The data of Table 4.1 could be represented more conveniently as in Table 4.2. Equation (4.13) with $h = 2$ may then be adapted to give

Table 4.2

n	1	2	3	4	5
c_n	0.439	0.383	0.336	0.298	0.266

$$\begin{aligned} c'(6) \simeq c_3' &= \tfrac{1}{2}(\tfrac{1}{12}c_1 - \tfrac{2}{3}c_2 + \tfrac{2}{3}c_4 - \tfrac{1}{12}c_5) \\ &= \tfrac{1}{2}[\tfrac{1}{12}(0.439) - \tfrac{2}{3}(0.383) + \tfrac{2}{3}(0.298) - \tfrac{1}{12}(0.266)] \\ &= -0.021 \text{ mg/ml/min (to three decimal places),} \end{aligned}$$

while equation (4.12) with $h = 2$ may be adapted to give

$$\begin{aligned} c'(8) \simeq c_4' &= \tfrac{1}{2}(-\tfrac{1}{12}c_1 + \tfrac{1}{2}c_2 - \tfrac{3}{2}c_3 + \tfrac{5}{6}c_4 + \tfrac{1}{4}c_5) \\ &= \tfrac{1}{2}[-\tfrac{1}{12}(0.439) + \tfrac{1}{2}(0.383) - \tfrac{3}{2}(0.336) + \tfrac{5}{6}(0.298) + \tfrac{1}{4}(0.266)] \\ &= -0.018 \text{ mg/ml/min (to three decimal places).} \end{aligned}$$

The mathematical model in Cullen (1985, p. 133) reveals these estimates to be accurate to three decimal places.

4.1.2 Second-order derivatives

Numerical differentiation formulas which give approximations to second-order derivatives are derived in the same way as the formulas of subsection 4.1.1.

Example 4.4 Given the three data pairs of Example 4.1, find an appoximation y_n'' to the derivative $y''(x_n)$ in terms of the three y-values.

Let

$$y''_n = ay_n + by_{n+1} + cy_{n+2} \,, \tag{4.14}$$

where a,b,c are constants and consider the equation

$$y''(x_n) = ay(x_n) + by(x_n + h) + cy(x_n + 2h) \;. \tag{4.15}$$

It may be shown, following Example 4.1, that $a = c = 1/h^2$ and $b = -2/h^2$. Thus

$$y''_n = h^{-2}(y_n - 2y_{n+1} + y_{n+2}). \tag{4.16}$$

Example 4.5 Find the order of (4.16).
Considering the error expression

$$y''(x_n) - h^{-2}[y(x_n) - 2y(x_n + h) + y(x_n + 2h)] \tag{4.17}$$

and using Taylor's series gives

$$\begin{aligned} y''(x_n) - h^{-2}[y(x_n) - 2\{y(x_n) + hy'(x_n) + \tfrac{1}{2}h^2y''(x_n) + \tfrac{1}{6}h^3y'''(x_n) + \ldots\} \\ + \{y(x_n) + 2hy'(x_n) + 2h^2y''(x_n) + \tfrac{4}{3}h^3y'''(x_n) + \ldots\}] \end{aligned}$$

which simplifies to $-hy'''(x_n) - \ldots$. Hence the error is $O(h)$ and (4.16) is a first-order forward difference approximation to $y''(x_n)$.

Not that the error of (4.16) is one order lower than (4.4), though each is derived ina similar way and each is based on the same data set. This observation may be made in respect of backward difference and central difference numerical differentiation formulas also. The first-order backward difference formula analogous to (4.16) is

$$y''_n = h^{-2}(y_{n-2} - 2y_{n-1} + y_n) \;. \tag{4.18}$$

Second-order forward difference and backward difference formulas each require four data points: these are given by

$$y''(x_n) = h^{-2}(2y_n - 5y_{n+1} + 4y_{n+2} - y_{n+3}) + O(h^2) \tag{4.19}$$

and

$$y''(x_n) = h^{-2}(-y_{n-3} + 4y_{n-2} - 5y_{n-1} + 2y_n) + O(h^2) \;. \tag{4.20}$$

The second-order central difference approximation to $y''(x_n)$, on the other hand, requires only three data points: it is given by

$$y''(x_n) = h^{-2}(y_{n-1} - 2y_n + y_{n+1}) + O(h^2) \ . \qquad (4.21)$$

A fourth-order central difference formula which uses five data points is given by

$$y''(x_n) = h^{-2}(-\tfrac{1}{12}y_{n-2} + \tfrac{4}{3}y_{n-1} - \tfrac{5}{2}y_n + \tfrac{4}{3}y_{n+1} - \tfrac{1}{12}y_{n+2}) + O(h^4) \ . \qquad (4.22)$$

The reader will have noted the symmetry of the coefficients about the base data point (x_n,y_n) in the central difference formulas (4.21) and (4.22).

Example 4.6 Find the rate at which the concentration is accelerating or decelerating at time $t = 6$ in Example 4.3.

Using Table 4.2 and equation (4.21) with $n = 3$ gives (with an obvious replacement of y by c)

$$\begin{aligned} c''(6) \simeq c_3'' &= \tfrac{1}{4}(c_2 - 2c_3 + c_4) \\ &= +0.002 \text{ mg/ml/min/min (to three decimal places)} \end{aligned}$$

while equation (4.22) gives (also to three decimal places)

$$c''(6) \simeq c_3'' = \tfrac{1}{4}(-\tfrac{1}{12}c_1 + \tfrac{4}{3}c_2 - \tfrac{5}{2}c_3 + \tfrac{4}{3}c_4 - \tfrac{1}{12}c_5) = +0.002 \text{ mg/ml/min/min.}$$

The mathematical model noted in Cullen (1985, p. 133) also gives this figure. The positive sign indicates that the rate of decay, $c'(t)$, is slowing down.

4.1.3 Higher-order derivatives

Numerical differentiation formulas approximating higher-order derivatives are derived in exactly the same way. A few such formulas are as follows:

$$y'''(x_n) = h^{-3}(-y_n + 3y_{n+1} - 3y_{n+2} + y_{n+3}) + O(h) \ , \qquad (4.23)$$

$$y'''(x_n) = h^{-3}(-y_{n-3} + 3y_{n-2} - 3y_{n+1} + y_n) + O(h) \ , \qquad (4.24)$$

$$y'''(x_n) = h^{-3}(-\tfrac{1}{2}y_{n-2} + y_{n-1} - y_{n+1} + \tfrac{1}{2}y_{n+2}) + O(h^2) \ , \qquad (4.25)$$

$$y^{(iv)}(x_n) = h^{-4}(y_n - 4y_{n+1} + 6y_{n+2} - 4y_{n+3} + y_{n+4}) + O(h) \ , \qquad (4.26)$$

$$y^{(iv)}(x_n) = h^{-4}(y_{n-4} - 4y_{n-3} + 6y_{n-2} - 4y_{n-1} + y_n) + O(h) \ , \qquad (4.27)$$

$$y^{(iv)}(x_n) = h^{-4}(y_{n-2} - 4y_{n-1} + 6y_n - 4y_{n+1} + y_{n+2}) + O(h^2) \ . \qquad (4.28)$$

The reader will note the familiar binomial coefficients in equations (4.8), (4.9), (4.16), (4.18), (4.21), (4.23), (4.24), (4.26), (4.27) and (4.28).

4.2 NUMERICAL INTEGRATION

4.2.1 Newton–Cotes formulas

From time to time it is necessary to evaluate the definite integral of a function that has no explicit indefinite integral (called the antiderivative) or an indefinite integral which cannot easily be found. Suppose that

$$I = \int_a^b f(x)\,\mathrm{d}x \tag{4.29}$$

is the definite integral, then *numerical integration*, sometimes called *quadrature* or *numerical quadrature*, uses a sum of the form

$$\sum_{i=0}^{N} a_i f(x_i) \tag{4.30}$$

to approximate I; the resulting approximation will be called I_N.

The $N+1$ point *closed Newton–Cotes* formula uses *nodes* $x_i = x_0 + ih$, for $i = 0,1,\ldots,N$, with $x_0 = a$, $x_N = b$ and $h = (b-a)/N$. The formula assumes that, in (4.30), a_i is given by

$$a_i = \int_{x_0}^{x_N} L_i(x)\,\mathrm{d}x \tag{4.31}$$

where

$$L_i(x) = \prod_{\substack{j=0 \\ j\neq i}}^{N} (x - x_j)/(x_i - x_j) \tag{4.32}$$

is a component of the Lagrange interpolation polynomial given in equation (2.7) of Chapter 2.

To evaluate a_i it is convenient to change the variable from x to t, say, where t is defined by $t = (x - x_0)h$. Then $t - i$ when $x - x_i$ and $\mathrm{d}x = t\mathrm{d}h$ so that

$$a_i = h \int_0^N \prod_{\substack{j=0 \\ j\neq i}}^{N} \frac{t-j}{i-j}\,\mathrm{d}t\,. \tag{4.33}$$

Taking, as an example, the case $N=1$, it follows that

$$a_0 = h\int_0^1 \frac{t-1}{0-1}\,\mathrm{d}t = h[t-\tfrac{1}{2}t^2]_0^1 = \tfrac{1}{2}h$$

and

$$a_1 = h\int_0^1 \frac{t-0}{1-0}\,\mathrm{d}t = h[\tfrac{1}{2}t^2]_0^1 = \tfrac{1}{2}h$$

and the Newton–Cotes formula is

$$\int_{x_0}^{x_1} f(x)\,\mathrm{d}x \simeq \tfrac{1}{2}h[f(x_0)+f(x_1)] \tag{4.34}$$

which is known as the *Trapezoidal Rule*.

Taking, next, $N=2$ the quadrature formula is

$$\int_{x_0}^{x_2} f(x)\,\mathrm{d}x \simeq \frac{h}{2}[f(x_0)+4f(x_1)+f(x_2)] \tag{4.35}$$

which is known as *Simpson's Closed Formula* or *Simpson's Rule*.

Taking $N=3$ leads to the quadrature formula

$$\int_{x_0}^{x_3} f(x)\,\mathrm{d}x = \frac{3h}{8}[f(x_0)+3f(x_1)+3f(x_2)+f(x_3)] \tag{4.36}$$

which is known as *Simpson's Three-Eighths Rule*.

Error formulas for the closed Newton–Cotes quadrature rules can be derived by using the error formula (2.10) of Chapter 2 associated with the Lagrange interpolating polynomial. Writing $N=1$, $X=x$ and $y(x)=f(x)$ in equation (2.10) gives

$$f(x)-p_1(x) = \tfrac{1}{2}(x-x_0)(x-x_1)f''(r_1) \tag{4.37}$$

where $x_0 < r_1 < x_1$. Integrating (4.37) gives

$$e(p_1(x)) = \int_{x_0}^{x_1} f(x)\mathrm{d}x - \int_{x_0}^{x_1} p_1(x)\mathrm{d}x = \int_{x_0}^{x_1} [f(x)-\tfrac{1}{2}h\{f(x_0)+f(x_1)\}]\mathrm{d}x$$

$$
\begin{aligned}
&= \tfrac{1}{2}f''(r_1)\left[\frac{x^3}{3} - \tfrac{1}{2}x_0x^2 - \tfrac{1}{2}x_1x^2 + x_0x_1x\right]_{x_0}^{x_1} \\
&= \tfrac{1}{2}f''(r_1)\cdot[-(x_1-x_0)^3/6] \\
&= -\tfrac{1}{12}h^3f''(r_1)\,. \qquad (4.38)
\end{aligned}
$$

The *local error* associated with the trapezoidal rule is therefore $O(h^3)$. It can be shown in a similar way that the error associated with Simpson's closed rule is $e(p_2(x)) = -h^5f^{(iv)}(r_2)/90$, where $x_0 < r_2 < x_2$, and that the error associated with Simpson's three-eighths rule is $e(p_3(x)) = -3h^5f^{(iv)}(r_3)/80$, where $x_0 < r_3 < x_3$.

Example 4.7 Use the trapezoidal rule with $h = 2$ and Simpson's rule with $h = 1$ to estimate $I = \int_0^2 f(x)\,dx$ when $f(x) = 1, x, x^2, x^3, x^4$.

The results for these computations are contained in Table 4.3. This example

Table 4.3

$f(x)$	1	x	x^2	x^3	x^4
Trapezoidal	2	2	4	8	16
Simpson	2	2	$\frac{8}{3}$	4	6.40
I	2	2	$\frac{8}{3}$	4	$\frac{32}{5}$

highlights the fact that the trapezoidal rule integrates polynomials of the first degree exactly whereas Simpson's rule integrates exactly polynomials of degree at most three.

Definition 4.1 The *degree of precision* of a quadrature formula is the positive integer q such that $e(P_k(x)) = 0$ for *all* polymonials $P_k(x)$ of degree $k \leq q$, but for which $e(P_{q+1}(x)) \neq 0$ for some polynomial of degree $q+1$.

The trapezoidal rule has degree of precision 1 whereas Simpson's rule has degree of precision 3.

Composite formulas are obtained by repeatedly applying the quadrature formulas just derived to integrate over large intervals. To find an approximation to I in (4.29) using the trapezoidal rule for instance, this quadrature formula is applied in each subinterval $[x_i, x_{i+1}]$ for $i = 0, 1, \ldots, N-1$ and the results added to give

$$\int_{x_0}^{x_N} f(x)\,dx \simeq \frac{h}{2}\,[f(x_0) + 2f(x_1) + \ldots + 2f(x_{N-1}) + f(x_N)] \qquad (4.39)$$

which will be called the *composite trapezoidal rule*.

Assuming the second-derivative bound $m < f''(x) < M$, the *global error* associated with (4.39), which is the sum of N errors of the form (4.38), will be between $-\frac{1}{12}h^3Nm$ and $-\frac{1}{12}h^3NM$. Noting that $Nh = b - a$ and assuming that $f''(x)$ is continuous for $x_0 < x < x_N$, the global error associated with (4.39) is equal to $-\frac{1}{12}(b-a)h^2y''(\xi)$, where $x_0 < \xi < x_N$; the global error is thus $O(h^2)$.

Simpson's composite rule requires N to be even, since Simpson's rule (4.36) needs two subintervals for each application. The composite rule is thus applied over each subinterval $[x_i, x_{i+2}]$ for $i = 0, 2, \ldots, N-2$ (N even) and it follows that

$$\int_{x_0}^{x_N} f(x)\,dx \simeq \frac{h}{3}[f(x_0) + 4f(x_1) + 2f(x_2) + \ldots + 2f(x_{N-2}) + 4f(x_{N-1}) + f(x_N)]\,. \tag{4.40}$$

The global error associated with (4.40) is equal to $-\frac{1}{180}h^4y^{(\text{iv})}(\xi) = O(h^4)$ where $x_0 < \xi < x_N$.

It is plain to see from (4.34), (4.35), (4.36), (4.39) and (4.40) that closed Newton–Cotes quadrature formulas use the end points of the interval of integration. *Open Newton–Cotes* quadrature formulas do not use the end points and are thus suitable for use in estimating I in (4.29) whenever $f(x)$ has a singularity at $x = a$ or $x = b$.

The open formulas are derived in a slightly different way to the closed formulas: the nodes $x_i = x_0 + ih$ are used for each $i = 0, 1, \ldots, N$ where now $b - a = (N+2)h$. This means now that $a = x_{-1}$ and $b = x_{N+1}$ and so

$$I = \int_a^b f(x)\,dx = \int_{x_{-1}}^{x_{N+1}} f(x)\,dx \simeq \sum_{i=0}^{N} a_i f(x_i) \tag{4.41}$$

where again the a_i are given by (4.31). Some common open Newton–Cotes formulas may be obtained by writing $N = 0, 1, 2, 3$ in {(4.41), (4.31), (4.32)}. These formulas, together with local errors, are as follows:

$$N = 0: \int_{x_{-1}}^{x_1} f(x)\,dx = 2hf(x_0) + \frac{h^3}{3}f''(r_0),\ x_{-1} < r_0 < x_1 \tag{4.42}$$

(this formula is known as the *Midpoint Rule*);

$$N = 1: \int_{x_{-1}}^{x_2} f(x)\,dx = \frac{3h}{2}[f(x_0) + f(x_1)] + \frac{3h^3}{4}f''(r_1),\ x_{-1} < r_1 < x_2; \tag{4.43}$$

$$N=2: \int_{x_{-1}}^{x_3} f(x)\,dx = \frac{4h}{3}[2f(x_0)-f(x_1)+2f(x_2)] + \tfrac{14}{45}h^5 f^{(iv)}(r_2), \tag{4.44}$$

where $x_{-1}<r_2<x_3$ (this formula is known as *Simpson's Open Formula*);

$$N=3: \int_{x_{-1}}^{x_4} f(x)\,dx = \frac{5h}{24}[11f(x_0)+f(x_1)+f(x_2)+11f(x_3)] + \tfrac{95}{144}h^5 f^{(iv)}(r_3), \tag{4.45}$$

where $x_{-1}<r_3<x_4$.

Comparing the local errors of these four formulas with the closed Newton–Cotes formulas of the same orders, it is evident that open quadrature formulas have inferior errors to closed formulas of the same orders.

Composite open formulas can be formed in the same way as composite closed formulas. The *composite midpoint rule* for instance requires an even number of subintervals and is given by

$$\int_{x_{-1}}^{x_{N+1}} f(x)\,dx \simeq 2h[f(x_0)+f(x_2)+\ldots+f(x_N)] + \frac{(b-a)}{6}h^2 f''(\xi) \tag{4.46}$$

where $x_{-1}<\xi<x_{N+1}$.

Definition 4.2 Suppose that A is the approximation obtained using a numerical integration formula with subintervals each of width h to the definite integral I. The integration formula is said to be *convergent* if $\lim_{h\to 0}(I-A)=0$.

Assuming that the derivatives occurring in their error terms are bounded, all of the Newton–Cotes formulas considered above are covergent.

4.2.2 Romberg integration

It is true that the trapezoidal rule is one of the easiest Newton–Cotes formulas to apply. It is, however, one of the least accurate and thereby loses some of its appeal. Romberg integration is based upon the fact that the error of the composite trapezoidal rule is $O(h^2)$. The method uses the trapezoidal rule to give a preliminary approximation to the definite integral I. Halving h and reapplying the trapezoidal rule reduces the error by a factor of $\frac{1}{4}$ and so, by taking a linear combination of the two approximations to I, improved accuracy is obtained. Continued refinement of the subinterval by systematically halving it leads to further improvements in accuracy by taking linear combinations of approximations to I.

More generally, suppose that the order of a numerical quadrature formula is n. Then two applications of the formula, with intervals h and $\frac{1}{2}h$, involve errors $e^{(1)}=Ch^n+Kh^m+O(h^p)$, $n<m<p$, and $e^{(2)}=C(\frac{1}{2}h)^n+K(\frac{1}{2}h)^m+O(h^p)$, where C and K are constants (note that $e^{(1)}$ and $e^{(2)}$ are both $O(h^n)$). Now,

$$I = A_1 + Ch^n + Kh^m + O(h^p) \tag{4.47}$$

and

$$I = A_2Ch^n/2^n + Kh^m/2^m + O(h^p) \tag{4.48}$$

where A_1 and A_2 are the estimates of I using subintervals h and $\frac{1}{2}h$ respectively. Subtracting equation (4.47) from 2^n times equation (4.48) gives

$$(2^n - 1)I = 2^n A_2 - A_1 + (2^{n-m} - 1)Kh^m + O(h^p)$$

so that

$$I = (2^n A_2 - A_1)/(2^n - 1) + O(h^m) \; . \tag{4.49}$$

The estimate

$$B = (2^n A_2 - A_1)/(2^n - 1) \tag{4.50}$$

is thus of order m and, since $m > n$, is therefore a higher-order approximation than either A_1 or A_2. This process of improving the order uses *Richardson extrapolation* which will be met again in the later chapters on differential equations.

Returning to the trapezoidal rule for which $n = 2$ and it is known that $m = 4$. Clearly $B_1 = \frac{4}{3}A_2 - \frac{1}{3}A_1$ gives an approximation to I which is $O(h^4)$. The process of Romberg integration is implemented by applying the trapezoidal rule several times, continually halving h. Call the results $A_1, A_2, A_3, \ldots$ and apply the formula $\frac{4}{3}A_{i+1} - \frac{1}{3}A_i$ to each pair of consecutive A_i. Call these results $B_1, B_2, B_3, \ldots$; then each B_i is an $O(h^4)$ approximation to I. Now apply the formula $\frac{16}{15}B_{i+1} - \frac{1}{15}B_i$ to each pair of consecutive B_i. Call this set of results $C_1, C_2, C_3, \ldots$; then each C_i may be shown to be $O(h^6)$. The process may be continued if desired to improve the order even further. The reader should note, however, that although the order is increased by Romberg integration, the increase in accuracy might not be what is hoped for, because the A_i computed using very small subintervals may have been subjected to relatively large round-off errors.

4.2.3 Gaussian quadrature

Newton–Cotes and Romberg formulas all use equally spaced x-values which are specified, so that with a formula of three terms (for example, equation (4.35)) there were three parameters or *weights* applied to the three functional values.

Gauss noted that if the x-values are not known, a three-point formula contains six parameters, the three weights and the three unknown x-values, and should correspond to an interpolating polynomial of degree five. Such integration formulas are known as *Gaussian Quadrature Formulas* and are only applicable when $f(x)$ is known explicitly so that $f(\alpha)$ can be determined for any $x = \alpha$.

Consider the two-term formula

$$\int_{-1}^{1} f(x)\,\mathrm{d}x \simeq w_0 f(x_0) + w_1 f(x_1)\ . \tag{4.51}$$

There are four unknowns in (4.51), so the integration formula will be made exact for the integration of all polynomials of degrees 0,1,2 and 3. In particular, let $f(x)$ be the monomial functions $1, x, x^2$ and x^3, so that

$$\int_{-1}^{1} \mathrm{d}x = 2,\ \Rightarrow\ w_0 + w_1 = 2 \tag{4.52}$$

$$\int_{-1}^{1} x\mathrm{d}x = 0,\ \Rightarrow\ w_0 x_0 + w_1 x_1 = 0, \tag{4.53}$$

$$\int_{-1}^{1} x^2\mathrm{d}x = \tfrac{2}{3},\ \Rightarrow\ w_0 x_0^2 + w_1 x_1^2 = \tfrac{2}{3}, \tag{4.54}$$

$$\int_{-1}^{1} x^3\mathrm{d}x = 0,\ \Rightarrow\ w_0 x_0^3 + w_1 x_1^3 = 0. \tag{4.55}$$

Multiplying (4.53) by x_0^2 and subtracting from (4.55) gives

$$0 = w_1(x_1^3 - x_0^2 x_1) = w_1 x_1 (x_1 - x_0)(x_1 + x_0)\ . \tag{4.56}$$

Equation (4.56) is satisfied by either $w_1 = 0$, $x_1 = 0$, $x_0 = x_1$ or $x_0 = -x_1$. The first three of these cause degeneracy of (4.51), so $x_0 = -x_1$ is chosen, this choice leading to

$$w_0 = w_1 = 1 \text{ and } x_1 = -x_0 = 3^{-\frac{1}{2}} \simeq 0.5773503\ .$$

Equation (4.51) then becomes

$$\int_{-1}^{1} f(x)\,\mathrm{d}x \simeq f(-3^{-\frac{1}{2}}) + f(3^{-\frac{1}{2}}). \tag{4.57}$$

Suppose now that the interval of integration is $[a,b]$ instead of $[-1,1]$. Let $t = \frac{1}{2}[(b-a)x + b + a]$, then $\mathrm{d}t = \frac{1}{2}(b-a)\mathrm{d}x$ and

$$\int_a^b f(x)\,\mathrm{d}x = \tfrac{1}{2}(b-a)\int_{-1}^{1} f(\tfrac{1}{2}[\{b-a\}x+b+a])\mathrm{d}t \qquad (4.58)$$

which may be used in association with (4.57) to integrate $f(x)$ over $[a,b]$.

The parameters for Gaussian formulas of higher order lead to a set of simultaneous nonlinear equations in the same way. Some of the results are given in Table 4.4

Table 4.4

n	x-values	Weights	q
2	0.5773503	1.0000000	3
	−0.5773503	1.0000000	
3	0.7745967	0.5555556	5
	0.0000000	0.8888889	
	−0.7745967	0.5555556	
4	0.8611363	0.3478548	7
	0.3399810	0.6521452	
	−0.3399810	0.6521452	
	−0.8611363	0.3478548	
5	0.9061798	0.2369269	9
	0.5384693	0.4786287	
	0.0000000	0.5688889	
	−0.5384693	0.4786287	
	−0.9061798	0.2369269	

where n is the number of weights and the number of x-values (nodes) and q is the highest degree polynomial which may be integrated exactly. The information in Table 4.4 refers to the integration of an arbitrary function $f(x)$ over $[-1,1]$.

PROBLEMS

1. The mid-point qudrature rule is defined by

$$\int_0^h f(x)\,\mathrm{d}x \simeq h\,f(h/2).$$

Verify that this rule is exact for, at most, linear polynomial functions. Hence derive an expression for the error by expanding $f(x)$ in a Taylor expansion about $x=0$.

2. Simpson's quadrature rule on $[-h,h]$ is given by

$$\int_{-h}^{h} f(x)\,\mathrm{d}x \simeq \frac{h}{3}\,[f(-h) + 4f(0) + f(h)].$$

Verify that this rule is exact for polynomials of degree $\leqslant 3$.

Hence, by expanding $f(x)$ in a Taylor expansion about $x=0$, show that the error is of order h^5.

3. (i) Derive the Gauss quadrature rule

$$\int_{-1}^{1} |x| f(x)\,\mathrm{d}x \simeq w_0 f(x_0) + w_1 f(x_1)$$

which is exact when $f(x)$ is a polynomial of degree $\leqslant 3$.

(ii) The corrected trapezoidal rule on $[0,h]$ is given by

$$\int_{0}^{h} f(x)\,\mathrm{d}x \simeq \frac{h}{2}[f(0) + f(h)] + \frac{h^2}{12}\,[f'(0) - f'(h)].$$

Use a Taylor expansion analysis to show that the error is of order h^5.

4. Define the composite corrected trapezoidal rule for the integral

$$\int_{0}^{1} f(x)\,\mathrm{d}x$$

which uses N intervals of length $h = 1/N$.

5. Verify that

$$\int_{-1}^{1} x^2(x + \sqrt{3/5})\,(x - \sqrt{3/5})\,q(x)\,\mathrm{d}x = 0,$$

where $q(x)$ is any linear polynomial.

Hence derive the Gauss quadrature rule

$$\int_{-1}^{1} x^2 f(x)\,\mathrm{d}x \simeq w_0 f(x_0) + w_1 f(w_1)$$

which is exact when $f(x)$ is any polynomial of degree $\leqslant 3$.

6. Verify that

$$\int_{-1}^{1} q(x)\,(x+1/\sqrt{3})\,(x-1/\sqrt{3})\,\mathrm{d}x = 0$$

where $q(x)$ is any linear polynomial.

Hence derive the Gauss quadrature rule

$$\int_{-1}^{1} f(x)\,\mathrm{d}x \simeq w_0 f(x_0) + w_1 f(x_1)$$

which is exact when $f(x)$ is a polynomial of degree $\leqslant 3$.

5

First-order ordinary differential equations

5.1 INTRODUCTION

First-order ordinary differential equations arise in the mathematical modelling of many biological, medical, chemical and physical systems such as predator–prey problems, chemical reactions, drug absorption problems, the transmission of electrochemical pulses in nerves, heart beat cycles, blood flow and pressure, and the flow of heat in the body.

Many of these ordinary equations arise in system form following the replacement of space derivatives in hyperbolic or parabolic partial differential equations by finite difference approximants. Such systems will be discussed in Chapters 7 and 8. Other systems of first-order differential equations, as well as single equations, arise naturally from the modelling procedure, and it is these differential equations which will be solved in the present chapter.

Many of the numerical methods which will be developed and analysed will be for single differential equations but their developments and analyses carry over in a natural way to systems of equations.

Given an independent variable t and a function $y(t)$, a typical (single) *first-order initial value problem* encountered in the biomedical sciences takes the general form

$$y'(t) = f(t,y),\ t > t_0,\ y(t_0) = g. \tag{5.1}$$

A first-order initial value problem always consists of two parts: the differential equation $y'(t) = f(t,y)$ and the initial condition $y(t_0) = g$.

The following theorem, which is proved in Henrici (1962), gives conditions on $f(t,y)$ which guarantee the existence of a unique solution to the initial value problem (5.1).

Theorem 5.1 Let $f(t,y)$ be defined and continuous for all points (t,y) in the region R

defined by $t_0 \leqslant t \leqslant T$, $-\infty < y < \infty$, t_0 and T finite, and let there exist a constant L^* such that, for every t, y, y^* such that (t,y) and (t,y^*) are both in R,

$$|f(t,y) - f(t,y^*)| \leqslant L^*|y - y^*|. \tag{5.2}$$

Then, if t_0 is any given number, there exists a unique solution $y(t)$ of the initial value problem (5.1), where $y(t)$ is continuous and differentiable for all (t,y) in R.

The requirement (5.2) is known as a *Lipschitz condition* and the constant L^* as a *Lipschitz constant*. The Lipschitz condition is satisfied if the Lipschitz constant is chosen to be

$$L^* = \sup_{(t,y)\in R} \left| \frac{\partial f(t,y)}{\partial y} \right| . \tag{5.3}$$

In system form, the first-order initial value problem is written as

$$\mathbf{y}'(t) = \mathbf{f}(t,\mathbf{y}),\ t > t_0;\ \mathbf{y}(t_0) = \mathbf{g}, \tag{5.4}$$

where $\mathbf{y}(t) = [y_1(t),\ y_2(t), \ldots,\ y_N(t)]^{\mathrm{T}}$, $\mathbf{f}(t,\mathbf{y}) = [f_1(t,\mathbf{y}),\ f_2(t,\mathbf{y}), \ldots,\ f_N(t,\mathbf{y})]^{\mathrm{T}}$ and $\mathbf{g} = [g_1,\ g_2, \ldots,\ g_N]^{\mathrm{T}}$ are N-dimensional vectors. In the interests of brevity, the notations $y_i = y_i(t)$ and $f_i = f_i(t,\mathbf{y})$ will be used often $(i = 1,2,\ldots,N)$.

Theorem 5.1 generalizes immediately to give necessary conditions for the existence of a unique solution to (5.4). The region R is now defined to be $t_0 \leqslant t \leqslant T, -\infty < y_i < \infty$ $(i = 1,2,\ldots,N)$ and (5.2) is replaced by

$$\|\mathbf{f}(t,\mathbf{y}) - \mathbf{f}(t,\mathbf{y}^*)\| \leqslant L^*\|\mathbf{y} - \mathbf{y}^*\|, \tag{5.5}$$

where $(t,\mathbf{y})$ and $(t,\mathbf{y}^*)$ are in R and $\|.\|$ denotes a vector norm. Assuming that each of the $f_i (i = 1,2,\ldots,N)$ possesses a continuous derivative with respect to each of the $y_j (j = 1,2,\ldots,N)$, the Lipschitz constant in (5.3) may be replaced by

$$L^* = \sup_{(t,\mathbf{y})\in R} \left\| \frac{\partial \mathbf{f}}{\partial \mathbf{y}} \right\| , \tag{5.6}$$

where

$$\frac{\partial \mathbf{f}}{\partial \mathbf{y}} = \begin{bmatrix} \partial f_1/\partial y_1 & \partial f_1/\partial y_2 & \ldots & \partial f_1/\partial y_N \\ \partial f_2/\partial y_1 & \partial f_2/\partial y_2 & \ldots & \partial f_2/\partial y_N \\ \ldots & \ldots & & \ldots \\ \partial f_N/\partial y_1 & \partial f_N/\partial y_2 & \ldots & \partial f_N/\partial y_N \end{bmatrix} \tag{5.7}$$

is the *Jacobian* of **f** with respect to **y**.

The first-order initial value problem (5.4) is said to be *nonlinear* if **f** depends nonlinearly on **y**. If **f** depends linearly on **y** the system is said to be *linear* and may be written in the form

$$\mathbf{y}'(t) = A(t)\mathbf{y}(t) + \boldsymbol{\phi}(t), \; t > t_0; \; \mathbf{y}(t_0) = \mathbf{g}, \tag{5.8}$$

where the matrix $A(t)$ and the vector $\boldsymbol{\phi}(t)$ are each of order N. This classification into linear and nonlinear types applies also, of course, to single equations.

In (5.1), (5.4) and (5.8) the derivatives $y'(t)$ and $\mathbf{y}'(t)$ are given explicitly in terms of various functions. These differential equations are therefore *explicit differential equations*; *implicit differential equations* of the forms

$$F(t,y,y') = 0 \; \textit{or} \; \mathbf{F}(t,\mathbf{y},\mathbf{y}') = \mathbf{0}, \tag{5.9}$$

where $\mathbf{0}$ is the zero-vector of order N, are also possible, but very few numerical methods exist for their solution and only explicit problems will be discussed in this book.

In the following sections, numerical methods will be developed and analysed for the solution of single and systems of first-order initial value problems. Analyses in much greater depth and other methods of solution are given in texts such as Henrici (1962) and Lambert (1973).

5.2 TAYLOR SERIES METHODS

5.2.1 Notations

Suppose that the independent variable t is discretized at the points $t_0, t_1, t_2, \ldots$ *with* $l = t_{n+1} - t_n (n = 0,1,2,\ldots)$ being a fixed *increment* in t. In Chapters 7 and 8, where t will represent time, l will be called a (fixed) *time step*. The solution of the initial value problem (5.1) at t_n is clearly $y(t_n)$. In using a numerical method to solve (5.1), the value of y at $t = t_n$ is not $y(t_n)$ but some other numerical value which will be denoted by y_n. Clearly it is desirable that y_n should be as close as possible to $y(t_n)$: how close depends on a number of factors which will be analysed later in the chapter. Using this notation, it follows that $y_0 = g$, the initial condition.

5.2.2 The Taylor method of order p

It is clear from equation (A.6) of Appendix A that Taylor's series may be written in the form

$$y(t + l) = y(t) + \frac{l}{1!}\, y''(t) + \frac{l^2}{2!}\, y''(t) + \ldots . \tag{5.10}$$

Writing $t = t_n$ and truncating the right-hand side of (5.10) after the term in l^p gives

$$y_{n+1} = y_n + \frac{l}{1!}\, y'_n + \frac{l^2}{2!}\, y''_n + \ldots + \frac{l^p}{p!}\, y_n^{(p)}, \tag{5.11}$$

where $y'_n, y'_n, \ldots$ denote the first, second, ... derivatives of y evaluated at t_n, the subscript notation having been introduced because (5.11) is an approximating numercial method whereas (5.10) is not. Equation (5.11) is known as the *Taylor series method of order p* or as the *Taylor algorithm of order p*.

Taylor series methods are useful in providing accurate starting values for linear multistep methods to be discussed in section 5.4. As numerical methods in their own right, however, they are not often used because of the need to determine the higher-order derivatives. These derivatives are obtained by successively differentiating the differential equation $y' = f(t,y)$; this can be a very tedious chore even for uncomplicated functions $f(t,y)$ as in the following example.

Example 5.1 Use the Taylor series method of order 5 with a step size $l = 0.1$ to compute an approximation to $y(1.1)$ given the initial value problem

$$y' = ty^2 - \frac{3}{t^3},\ t > 1;\ y(1) = 1. \tag{5.12}$$

Here $t_0 = 1$ and $g = y_0 = 1$ so that $y'_0 = -2$. Differentiating the differential equation four times, and substituting for $t_0, y_0, y'_0, y'_0, y'''_0, y_0^{(iv)}$ as they are required, gives

$$y''_0 = 6, \quad y'''_0 = -24, \quad y_0^{(iv)} = 120,\ y_0^{(v)} = -720$$

so that, from (5.11),

$$y_1 = 1 - 2l + 3l^2 - 4l^3 + 5l^4 - 6l^5. \tag{5.13}$$

Putting $l = 0.1$ gives $y_1 = 0.82644$. The theoretical solution of (5.12) is $y(t) = t^{-2}$ so that $y(1.1) = 0.82645$ to five decimal places. The Taylor series method of order 5 thus gives an accutate answer, but calculating the higher-order derivatives for use in (5.11) was a little tricky.

Substituting $l = 0.2$ in (5.13) gives $y_1 = 0.69408$ which is an approximation to $y(1.2)$. The theoretical solution of (5.12) at $t = 1.2$ is 0.69444 to five decimal places; the reader can thus see that the error is much greater than before. One way of overcoming this growth in error is to use $y_0 = 0.82644$ as the initial condition and to derive a new Taylor series method. This, of course, is very time-consuming and the user may be well advised to use another method.

In some problems which arise in mathematical biology the function f is a function of the dependent variable only, making it easier to determine the higher derivatives.

One such example arises in predator–prey interactions. In the absence of predators, the prey population will either settle down to some steady value, fluctuate between various levels, decline, or grow without bound; the parameters built into the mathematical model will determine which growth pattern is followed, as in the following example.

Example 5.2 Use the Taylor algorithm of order 3 with a step size $l = \frac{1}{12}$ to compute an approximate to $x(\frac{1}{12})$ given the initial value problem

$$x' = ax - bx^2, \quad t > 0; \quad x(0) = 1100$$

with $a = 0.5$ and $b = 0.0005$. (The parameter a is called the *intrinsic rate of increase* and the ratio a/b is called the *carrying capacity*.)

In this example, x is the dependent variable, $t_0 = 0$ and $x_0 = 1100$ so that $x_0' = -55$. Differentiating the differential equation as in Example 5.1 gives

(1) $x'' = 0.5x' - 0.001xx'$ so that $x_0'' = 33$,
(2) $x''' = 0.5x'' - 0.001xx' - 0.001(x')^2$ so that $x_0''' = -55.925$.

The Taylor algorithm of order 3 is thus

$$x = 1100 + \frac{l}{1}(-55) + \frac{l^2}{2}(33) + \frac{l^3}{6}(-55.925).$$

Substituting $l = \frac{1}{12}$ gives $x_1 = 1095.5313$. The theoretical solution of this initial value problem is

$$x = \frac{ax_0}{bx_0 + (a - bx_0)e^{-at}}. \tag{5.14}$$

Letting $t \to \infty$ shows that $x \to a/b = 1000$ monotonically, irrespective of x_0, so that the Taylor method has correctly predicted a population decline. Substituting $t = \frac{1}{12}$ in (5.14) gives $x(\frac{1}{12}) = 1095.5291$ so that the third-over Taylor method has given the correct value $x = 1095$ at time $t = \frac{1}{12}$.

5.3 RUNGE–KUTTA METHODS

Returning to the initial value problem (5.1), integration from t_n to t_{n+1} may be carried out using one of the family of explicit *Runge–Kutta* methods

$$y_{n+1} = y_n + l\phi(t_n, y_n, l). \tag{5.15}$$

In the general *r-stage Runge–Kutta method*, ϕ is given by

$$\phi(t, y, l) = \sum_{i=1}^{r} c_i k_i \tag{5.16}$$

with

$$k_1 = f(t,y)$$

$$k_i = f(t + la_i,\ y + l \sum_{s=1}^{i-1} b_{is} k_s), \quad i = 2,3, \ldots r \tag{5.17}$$

$$a_i = \sum_{s=1}^{i-1} b_{is}\,,\ i = 2,3, \ldots r.$$

Note that an r-stage Runge–Kutta method involves r function evaluations to integrate from t_n to t_{n+1}.

Definition 5.1 The Runge–Kutta method (5.15) is said to have *order p* if p is the largest integer for which

$$y(t + l) - y(t) - l\phi(t,y(t),l) = O(l^{p+1})$$

holds, where $y(t)$ is the theoretical solution of (5.1).

Definition 5.2 The method (5.15) is said to be *consistent* with (5.1) if

$$\phi(t,y,0) = f(t,y)$$

Definition 5.3 The method (5.15) is consistent with (5.1) if the order of (5.15) is at least unity.

Each of the functions $k_i(t,y,l)$, $i = 1,2, \ldots, r$, may be interpreted as an approximation to the derivative $y'(t)$ and the function $\phi(t,y,l)$ as a weighted mean of these approximations. It follows therefore from Definition 5.3 that $\sum_{i=1}^{r} c_i = 1$. The constants c_i, a_i, b_{is} must be chosen such that the expansion of $\phi(t,y,l)$, defined by (5.16) and (5.17), in powers of l differs from the Taylor expansion of $\phi(t,y,l)$ about $\phi(t,y,0) = f(t,y)$ only in the term in l^p and in higher powers of l. The Runge–Kutta method then has order p (note $\phi(t,y,l)$ is multiplied by l in (5.15) and it is assumed that $y(t)$ possesses p continuous derivatives for $t \in [t_n,t_{n+1}]$).

There is a great deal of tedious manipulation involved in deriving Runge–Kutta methods of high orders, particularly implicit methods where the k_i are defined implicitly. Accordingly, only a second-order method will be derived; others will be quoted without derivation.

Define the quantities

$$k_1 = f(t_n, y_n),$$
$$k_2 = f(t_n + \alpha l,\ y_n + \beta l k_1)$$

and form the equation

$$y_{n+1} = y_n + c_1 l k_1 + c_2 l k_2 \tag{5.18}$$

where c_1 and c_2 are the weights to be determined.

Equation (5.18) is formed so that the expansion of $y(t_n + l)$ given by

$$y(t_n + l) = y(t_n) + c_1 l k_1 + c_2 l k_2 \tag{5.19}$$

about $y(t_n)$ in powers of l agrees with the solution of (5.1) at $t = t_{n+1}$ to three terms of the Taylor series. Now

$$y(t_n + l) = y(t_n) + l y'(t_n) + \tfrac{1}{2} l^2 y''(t_n) + O(l^3) \ ,$$

that is

$$y(t_n + l) = y_n + l f_n + \tfrac{1}{2} l^2 \left(\frac{\partial f_n}{\partial t} + f_n \frac{\partial f_n}{\partial y} \right) + O(l^3) \tag{5.20}$$

where $f_n \equiv f(t_n, y_n)$.

Expanding k_2 about (t_n, y_n) gives

$$k_2 = f_n + \alpha l \frac{\partial f_n}{\partial t} + \beta l k_1 \frac{\partial f_n}{\partial y} + O(l^2)$$

and then substituting for k_1 and k_2 in (5.18) gives

$$y_{n+1} = y_n + c_1 l f_n + c_2 l f_n + c_2 \alpha l^2 \frac{\partial f_n}{\partial t} + c_2 \beta l^2 f_n \frac{\partial f_n}{\partial y} + O(l^3). \tag{5.21}$$

Comparing powers of l in (5.20) and (5.21) it follows that

$$c_1 + c_2 = 1, \quad c_2 \alpha = \tfrac{1}{2}, \quad c_2 \beta = \tfrac{1}{2}. \tag{5.22}$$

Choosing $\alpha = \beta = 1$ and $c_1 = c_2 = \frac{1}{2}$ gives

$$k_1 = f(t_n, y_n),$$

$$k_2 = f(t_n + l,\ y_n + lk_1)$$

$$y_{n+1} = y_n + \tfrac{1}{2}l\{f(t_n, y_n) + f(t_n + l,\ y_n + lk_1)\} + O(l^3) \tag{5.23}$$

which is the best known second-order explicit Runge–Kutta method.

In (5.22) there are three equations in the four unknowns c_1, c_2, α and β and there are thus infinitely many second-order two-stage Runge–Kutta methods. One other second-order method is found by writing $c_1 = 0$, $c_2 = 1$, $\alpha = \beta = \frac{1}{2}$ to give

$$k_1 = f(t_n, y_n),$$

$$k_2 = f(t_n + \tfrac{1}{2}l,\ y_n + \tfrac{1}{2}lk_1),$$

$$y_{n+1} = y_n + lf(t_n + \tfrac{1}{2}l,\ y_n + \tfrac{1}{2}lk_1) + O(l^3). \tag{5.24}$$

There are no two-stage explicit Runge–Kutta methods of order higher than two.

In a similar way it may be shown that there are infinitely many three-stage explicit Runge–Kutta methods of order three. Two of these are

$$k_1 = f(t_n, y_n),$$

$$k_2 = f(t_n + \tfrac{1}{3}l,\ y_n + \tfrac{1}{3}lk_1),$$

$$k_3 = f(t_n + \tfrac{2}{3}l,\ y_n + \tfrac{2}{3}lk_2),$$

$$y_{n+1} = y_n + \tfrac{1}{4}l(k_1 + 3k_3) + O(l^4) \tag{5.25}$$

and

$$k_1 = f(t_n, y_n),$$

$$k_2 = f(t_n + \tfrac{1}{2}l,\ y_n + \tfrac{1}{2}lk_1),$$

$$k_3 = f(t_n + l, y_n - lk_1 + 2lk_2),$$

$$y_{n+1} = y_n + \tfrac{1}{6}l(k_1 + 4k_2 + k_3) + O(l^4), \tag{5.26}$$

the former being known as *Heun's third-order rule*, the latter as *Kutta's third-order method*. There are no three-stage methods of order higher than three.

The derivation of the parameters of each four-stage explicit Runge–Kutta method involves a lot of tedious manipulation. There are no four-stage methods of order higher than four. Arguably the most popular fourstage methods are

$$k_1 = f(t_n, y_n),$$

$$k_2 = f(t_n + \tfrac{1}{3}l,\ y_n + \tfrac{1}{3}lk_1),$$

$$k_3 = f(t_n + \tfrac{2}{3}l,\ y_n - \tfrac{1}{3}lk_1 + lk_2),$$

$$k_4 = f(t_n + l, y_n + lk_1 - lk_2 + lk_3),$$

$$y_{n+1} = y_n + \tfrac{1}{8}l(k_1 + 3k_2 + 3k_3 + k_4) + O(l^5) \tag{5.27}$$

and

$$k_1 = f(t_n, y_n),$$

$$k_2 = f(t_n + \tfrac{1}{2}l,\ y_n + \tfrac{1}{2}lk_1),$$

$$k_3 = f(t_n + \tfrac{1}{2}l,\ y_n + \tfrac{1}{2}lk_2),$$

$$k_4 = f(t_n + l, y_n + lk_3),$$

$$y_{n+1} = y_n + \tfrac{1}{6}l(k_1 + 2k_2 + 2k_3 + k_4). \tag{5.28}$$

Often, (5.28), because of its great popularity, is referred to as *the* fourth-order Runge–Kutta method and sometimes as *the* Runge-Kutta method, both statements being inaccurate. The reader should note that there are no five-stage Runge–Kutta methods of order five and should consult a specialist text, such as Lambert (1973), for analyses of Runge–Kutta methods.

Example 5.3 Use the fourth-order Runge–Kutta method of order four given by (5.28) to find $x(\tfrac{1}{12})$, given the initial value problem of Example 5.2.

Taking $a = 0.5$, $b = 0.0005$ and $l = \tfrac{1}{12}$ gives

$$k_1 = f(0, 1100) = 0.5(1100) - 0.0005(1100)^2 = -55,$$

$$k_2 = f(\tfrac{1}{24}, 1100 + \tfrac{1}{2}(\tfrac{1}{12})k_1) = -53.62764,$$

$$k_3 = f(\tfrac{1}{24}, 1100 + \tfrac{1}{2}(\tfrac{1}{12})k_2) = -53.66181,$$

$$k_4 = f(\tfrac{1}{12}, 1100 + \tfrac{1}{12}k_3) = -52.32691,$$

$$y_1 = 1100 + \tfrac{1}{6}(\tfrac{1}{12})\{-55 + 2(-53.62765) + 2(-53.66181) - 52.32691\}$$

that is $y(\tfrac{1}{12}) = 1095.5291$ which agrees with the theoretical solution given by (5.14) to four decimal places.

5.4 LINEAR MULTISTEP METHODS

5.4.1 Introduction

A computational method which describes the numerical solution y_{n+k} of (5.1) at the point t_{n+k} in terms of a linear relationship involving y_{n+j}, f_{n+j} $(j=0,1,\ldots,k)$, is called a *linear multistep method* (*LMM*) *with stepnumber k* or a *linear k-step method*. The general linear multistep method may thus be written

$$\sum_{j=0}^{k} \alpha_j y_{n+j} = l \sum_{j=0}^{k} \beta_j f_{n+j}, \tag{5.29}$$

where the α_j and β_j are constants; it will be assumed that $\alpha_k \neq 0$ and that not both α_0 and β_0 are zero. Equation (5.29) can be multiplied throughout by any constant without altering the relationship so that the coefficients α_j and β_j are arbitrary to the extent of a constant multiplier. This arbitrariness is removed by stipulating that $\alpha_k = 1$.

The linear multistep method (5.29) is said to be *explicit* if $\beta_k = 0$ and *implicit* if $\beta_k \neq 0$. For an explicit method, (5.29) yields y_{n+k} directly in terms of y_{n+j}, f_{n+j} $(j=0,1,\ldots,k-1)$ which have already been computed. An implicit LMM, however, seeks the solution of the equation

$$y_{n+k} = l\beta_k f(t_{n+k}, y_{n+k}) - \sum_{j=0}^{k-1} (\alpha_j y_{n+j} - l\beta_j f_{n+j}) \tag{5.30}$$

(note: $\alpha_k = 1$). When the function $f(t,y)$ is linear in y, equation (5.30) is linear in y also and y_{n+k} can easily be found. When $f(t,y)$ is nonlinear in y, however, then, from Chapter 3, there exists a unique solution for y_{n+k} which can be approached arbitrarily closely by the iteration

$$y_{n+k}^{[s+1]} = l\beta_k f(t_{n+k}, y_{n+k}^{[s]}) - \sum_{j=0}^{k-1} (\alpha_j y_{n+j} - l\beta_j f_{n+j}), \tag{5.31}$$

where $s = 0,1,2,\ldots$ is the iteration number and $y^{[0]}{}_{n+k}$ is arbitrary, provided $0 \leqslant M < 1$, M being the Lipschitz constant with respect to y_{n+k} of the right-hand side of (5.31). If the Lipschitz constant of f with respect to y is L, then M has the value $Ll|\beta_k|$, a unique solution for y_{n+k} exists and (5.31) converges to it provided

$$l < 1/(L|\beta_k|). \tag{5.32}$$

Clearly implicit methods require a great deal more computational effort than do explicit methods. On the other hand, as later analyses will reveal, for a given stepnumber k they are usually more accurate and have other more favourable properties.

When using a linear k-step method to determine y_{n+k}, the *starting values* $y_n, y_{n+1}, \ldots, y_{n+k-1}$ are required. These values must be calculated to an accuracy at least as high as that of the LMM to be used to determine y_{n+k}. Suitable methods for determining these starting values are the one-step Taylor and Runge–Kutta methods which were discussed in sections 5.2 and 5.3.

5.4.2 Derivation of some well-known methods

Consider the Taylor method of order $p = 1$; it is given by

$$y_{n+1} = y_n + lf_n; \quad n = 0,1,2, \ldots \tag{5.33}$$

and is in fact an explicit *linear one-step method*. It is the simplest linear multistep method of all and, regarded as an LMM, it is known as *Euler's method* or *Euler's rule* or the *Euler predictor formula* (see section 5.4.8).

Consider, next, the identity

$$y(t_{n+1}) - y(t_n) \equiv \int_{t_n}^{t_{n+1}} y'(t)\mathrm{d}t = \int_{t_n}^{t_{n+1}} f(t,y)\mathrm{d}t\ . \tag{5.34}$$

Using the trapezium rule for numerical integration (see Chapter 4), (5.34) becomes

$$y_{n+1} = y_n + \tfrac{1}{2}l(f_n + f_{n+1}); \quad n = 0,1,2, \ldots \tag{5.35}$$

which is known as *Euler's modified rule* or the *Euler corrector formula* (see section 5.4.8). Equation (5.35) is an implicit one-step method.

Numerical integration techniques can be used to derive other linear multistep methods. Consider, for instance, the identity

$$y(t_{n+4}) - y(t_n) \equiv \int_{t_n}^{t_{n+4}} y'(t)\mathrm{d}t = \int_{t_n}^{t_{n+4}} f(t,y)\mathrm{d}t\ . \tag{5.36}$$

Using Simpson's Open Formula (see Chapter 4), (5.36) becomes

$$y_{n+4} = y_n + \frac{4l}{3}(2f_{n+1} - f_{n+2} + 2f_{n+3}) \tag{5.37}$$

which is an explicit, four-step method and which is known as *Simpson's predictor formula*. Using Simpson's Closed Formula (also known as Simpson's Rule, see Chapter 4) in the identity

$$y(t_{n+2}) - y(t_n) \equiv \int_{t_n}^{t_{n+2}} y'(t)\mathrm{d}t = \int_{t_n}^{t_{n+2}} f(t,y)\mathrm{d}t\ . \tag{5.38}$$

gives the implicit two-step method

$$y_{n+2} = y_n + \frac{l}{6}(f_n + 4f_{n+1} + f_{n+2}) \tag{5.39}$$

which is known as *Simpson's corrector formula*.

Equation (5.39) can also be derived by interpolation. Suppose that $y(t)$, the solution of (5.1), can be approximated locally in the interval $t_n \leqslant t \leqslant t_{n+2}$ by a polynomial $P(t)$ which satisfies the six conditions

$$P(t_{n+j}) = y_{n+j}; \quad P'(t_{n+j}) = f_{n+j}, \quad j = 0,1,2\ . \tag{5.40}$$

Let $P(t)$ have five free parameters; it can therefore be described by the fourth-degree polynomial

$$P(t) = at^4 + bt^3 + ct^2 + dt + e$$

with

$$P'(t) = 4at^3 + 3bt^2 + 2ct + d.$$

Eliminating the five undetermined coefficients between the six conditions in (5.40) yields (5.39).

This interpolation technique is an inefficient way of determining the α_j and β_j of an LMM. Also, it pays no heed to accuracy. Based as it is on the assumption that $y(t)$ can be approximated in some interval by a polynomial (possibly of high degree), the approach is inappropriate for use when the initial value problem is stiff (see section 5.5.6), for then the solution is a rapidly decaying exponential and any LMM, with high stepnumber, so derived will not give good results.

In addition to the four common linear multistep methods quoted in this subsection of the chapter, numerous others exist and the reader is referred to the specialist texts by Henrici (1962) and Lambert (1973) for details of some of them.

5.4.3 Convergence

The concept of *convergence* is concerned with ensuring that the solution $\{y_n\}$ generated by an LMM coincides with the theoretical solution $y(t)$ as the steplength l tends to zero. In general, l should be considered as tending continuously to zero, but in leading to a formal definition of convergence, it is usual to employ the idea of *fixed station convergence*.

Consider, for example, a fixed point $t = T$ and let the initial choice of steplength l_0 be such that $T = t_0 + 2l_0$. In the special case when the steplength is successively halved, the situation depicted in Fig. 5.1 arises. Using the notation $y_n(l)$ to denote the

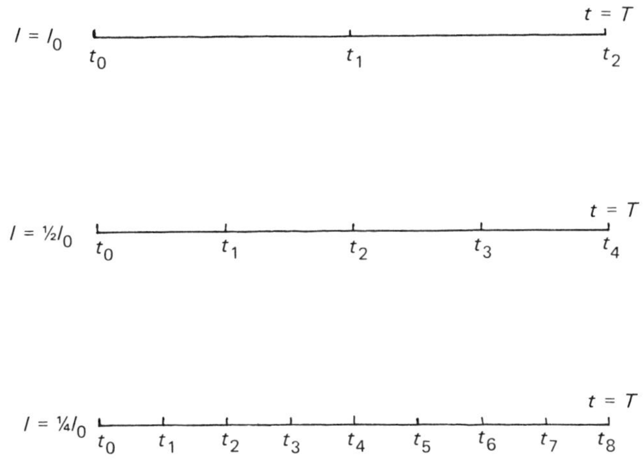

Fig. 5.1 — Convergence.

value y_n given by the LMM (5.29) when the steplength is l, it is observed that the sequence

$$y_2(l_0), \quad y_2(\tfrac{1}{2}l_0), \quad y_2(\tfrac{1}{4}l_0), \ldots$$

converges to the value $y(t_0)$. It is therefore inappropriate to consider n as remaining fixed while $l \to 0$. On the other hand, the sequence

$$y_2(l_0), \quad y_4(\tfrac{1}{2}l_0), \quad y_8(\tfrac{1}{4}l_0). \ldots$$

converges to the value $y(T)$ and fixed station convergence is concerned with convergence in the limit as $l \to 0$, $n \to \infty$, $nl = t - t_0$ remaining fixed. Such a limit is denoted by

$$\lim_{\substack{l \to 0 \\ nl = t - t_0}} \quad .$$

A formal definition of convergence must take account of the starting values which must be supplied when the stepnumber k exceeds unity. These need not be exact solutions of the initial value problem (5.1) at $t_1, t_2, \ldots, t_{k-1}$ but, regarded as functions of l, they should all tend to y_0 as $l \to 0$. A formal definition of convergence must also ensure that the convergence property holds for *all* initial value problems of the form (5.1) which are subject to the hypotheses of Theorem 5.1. Such a formal definition is as follows.

Definition 5.4 The linear multistep method is said to be *convergent* if, for all initial value problems of the form (5.1) subject to the hypotheses of Theorem 5.1,

$$\lim_{\substack{l \to 0 \\ nl = t - t_0}} y_n = y(t_n)$$

holds for all t in some interval $t_0 \leqslant t \leqslant T$ and for all solutions $\{y_n\}$ of (5.29) satisfying starting conditions $y_r = y_{0,r}(l)$ for which $\lim_{l \to 0} y_{0,r}(l) = y_0$, $r = 0,1,\ldots,k-1$.

5.4.4 Order, consistency and truncation errors

Consider once more the LMM (5.29) and associate with it the linear difference operator $\mathscr{L}$ defined by

$$\mathscr{L}[y(t);l] = \sum_{j=0}^{k} [\alpha_j y(t+jl) - l\beta_j y'(t+jl)], \tag{5.41}$$

where $y(t)$ is some function which is continuously differentiable on some interval $t_0 \leqslant t \leqslant T$. Expanding the terms $y(t+jl)$ and $y'(t+jl)$ $(j = 1,2,\ldots,k)$ in (5.41) about $y(t)$ and $y'(t)$, respectively, using Taylor's series, then adding together like powers of l, it is clear that

$$\mathscr{L}[y(t);l] = C_0 y(t) + C_1 l y'(t) + C_2 l^2 y''(t) + \ldots \tag{5.42}$$

where the $C_0, C_1, C_2, \ldots$ are constants.

Definition 5.5 The linear multistep method (5.29) is said to be of *order p* if, in (5.42), $C_0 = C_1 = \ldots = C_p = 0$ and $C_{p+1} \neq 0$.

Definition 5.6 The linear multistep method (5.29) is said to be *consistent* with the initial value problem (5.1) if it has order $p \geqslant 1$.

Clearly, the difference operator $\mathscr{L}$ associated with an LMM of order $p \geqslant 1$ is given by

$$\mathcal{L}[y(t);l] = C_{p+1}l^{(p+1)}y^{(p+1)}(t) + C_{p+2}l^{p+2}y^{(p+2)}(t) + \ldots . \tag{5.43}$$

The expression on the right-hand side of (5.43) is known as the *local truncation error* of the linear multistep method; the term $C_{p+1}l^{p+1}y^{(p+1)}(t)$ is known as its *principal part* and the constant C_{p+1} as its *error constant.*

Example 5.4 Show that Euler's modified formula (5.35) is consistent with (5.1) and find its error constant.

Here, $\mathcal{L}[y(t),l] = y(t+l) - y(t) - \frac{1}{2}ly'(t) - \frac{1}{2}ly'(t+l)$, so that Taylor's series gives

$$\begin{aligned}\mathcal{L}[y(t);l] &= y(t) + ly'(t) + \tfrac{1}{2}l^2y''(t) + \tfrac{1}{6}l^3y'''(t) + \tfrac{1}{24}l^4y^{(\mathrm{iv})}(t) + \ldots \\ &\quad - y(t) - \tfrac{1}{2}ly'(t) \\ &\quad - \tfrac{1}{2}l[y'(t) + ly''(t) + \tfrac{1}{2}l^2y'''(t) + \tfrac{1}{6}l^3y^{\mathrm{iv}}(t) + \ldots] \\ &= -\tfrac{1}{12}l^3y'''(t) - \tfrac{1}{24}l^4y^{(\mathrm{iv})}(t) - \ldots .\end{aligned}$$

Since $C_0 = C_1 = C_2 = 0$ the method has order $p = 2$ and is therefore consistent with (5.1). Its error constant is $C_3 = \frac{1}{12}$.

The simple calculation involved in deriving (5.42) from (5.41) reveals that

$$C_0 = \alpha_0 + \alpha_1 + \alpha_2 + \ldots + \alpha_k,$$

$$C_1 = \alpha_1 + 2\alpha_2 + \ldots + k\alpha_k - (\beta_0 + \beta_1 + \beta_2 + \ldots + \beta_k),$$

$$C_q = \frac{1}{q!}(\alpha_1 + 2^q\alpha_2 + \ldots + k^q\alpha_k)$$

$$-\frac{1}{(q-1!)}(\beta_1 + 2^{q-1}\beta_2 + \ldots + k^{q-1}\beta_k); \quad q = 2,3,\ldots .$$

For consistency, $p \geqslant 1$; it follows therefore that $C_0 = C_1 = 0$ is a minimum requirement and this leads to the following theorem.

Theorem 5.2 The linear multistep method (5.29) is consistent with the initial value problem (5.1) if and only if

$$\sum_{j=0}^{k} \alpha_j = 0 \quad \text{and} \quad \sum_{j=0}^{k} j\alpha_k = \sum_{j=0}^{k} \beta_j . \tag{5.44}$$

Introduce, now, the *influence function* $G(s)$ of the LMM (5.29) defined by

$$G(s)=\sum_{j=0}^{k}\left[\alpha_j(j-s)_+^p - p\beta_j(j-s)_+^{p-1}\right]$$

where

$$(j-s)_+ = \begin{cases} j-s, & \text{if } j \geqslant s \\ 0, & \text{if } j < s \end{cases}.$$

Then, if $G(s)$ does not change sign in the interval $[0,k]$, the local truncation error is bounded by

$$|\mathscr{L}\,[y(t);\,l]| \leqslant l^{p+1}GY \tag{5.45}$$

where

$$G=|C_{p+1}|=\frac{1}{p!}\left|\int_0^k G(s)\mathrm{d}s\right|$$

and

$$Y=\max_{t_0\leqslant t\leqslant T}|y^{(p+1)}(t)|\ ,$$

while, if $G(s)$ does change sign in this interval, $\mathscr{L}[y(t);l]$ is still bounded by (5.45) but now

$$G=\frac{1}{p!}\int_0^k |G(s)|\mathrm{d}s.$$

Defining $e_{n+k}=y(t_{n+k})-y_{n+k}$ ty be the *accumulated* or *global truncation error*, it may be shown that

$$|e_{n+k}|\leqslant \delta\,\exp[LB(t_{n+k}-t_0)]+\frac{l^pGY}{LB}\,\{\exp[\mathrm{LB}(t_{n+k}-t_0)]-1\}\ ,$$

where

$$\delta = \max_{r=0,1,\ldots,k-1} |e_r| \quad \text{and} \quad B = \sum_{j=0}^{k} |\beta_j|$$

from which it follows that $e_{n+k} = O(l^p)$. The global truncation error is one order fewer than the local truncation error. A full treatment of bounding local and global truncation errors is given in, for example, Lambert (1973).

5.4.5 Zero-stability

Introduce, now, the *first* and *second characteristic polynomials* of the linear multistep method (5.29) defined by

$$\rho(r) = \sum_{j=0}^{k} \alpha_j r^j \quad \text{and} \quad \sigma(r) = \sum_{j=0}^{k} \beta_j r^j \tag{5.46}$$

respectively, Putting $r = 1$ in (5.46) and using Theorem 5.2 gives the following theorem.

Theorem 5.3 The LMM (5.29) is consistent with the initial value problem (5.1) if and only if

$$\rho(1) = 0 \quad \text{and} \quad \rho'(1) = \sigma(1). \tag{5.47}$$

Corollary The first characteristic polynomial of a consistent LMM always has a zero at $r = +1$.

This zero is called the *principal zero* of $\rho(r)$ and is usually named r_1; the other zeros of $\rho(r)$, named $r_2, r_3, \ldots, r_k$, are called *spurious zeros*. The spurious zeros of $\rho(r)$ need not be distinct and they may be complex. It is shown in Lambert (1973), page 32, that the linear multistep method (5.29) will not be convergent if any zero of $\rho(r)$ has modulus greater than unity.

Definition 5.7 The linear multistep method (5.29) is said to be *zero-stable* if no zero of the first characteristic polynomial $\rho(r)$ has modulus greater than unity and if every zero with modulus unity is simple.

It follows that any or all of $+1$, -1, $e^{\pm i\theta}$ (θ real) may be zeros of $\rho(r)$ provided that each is simple. An LMM which is zero-stable is often referred to as *D-stable* or *Dahlquist-stable*, the theory having been published by Dahlquist (1956). The word 'zero' is chosen because the phenomenon under consideration is concerned with the convergence of (5.29) in the limit as $l \to 0$.

For a one-step method, the first characteristic polynomial $\rho(r)$ has degree unity and if the method is consistent, the only zero is $r = +1$. Thus a consistent one-step method is necessarily zero-stable. It is also easy to see that a consistent zero-stable LMM cannot have $\sigma(1)=0$ for, if it did, then, by Theorem 5.3, $\rho(1)=0=\rho'(1)$ which implies that $\rho(r)$ has a double zero at $r=+1$, contrary to Definition 5.7.

The phenomena of convergence, consistency and zero-stability are connected by the following theorem, the proof of which appears in Henrici (1962).

Theorem 5.4 (*The Fundamental Theorem of Dahlquist*) The necessary and sufficient conditions for a linear multistep method to be convergent are that it be consistent and zero-stable.

The reader will deduce that using a linear multistep method which is not consistent or which is not zero-stable will give numerical results which are meaningless, irrespective of the steplength l being used. It is also known that zero-stable methods can only attain a certain order as the following theorem indicates (see Henrici (1962) for a proof).

Theorem 5.5 No zero-stable linear multistep method of stepnumber k can have order exceeding $k+1$ when k is odd, or exceeding $k+2$ when k is even. A zero-stable linear k-step method which has order $k+2$ is said to be *optimal.*

5.4.6 Weak stability theory: stability intervals

Consider again, the general linear multistep method (5.29) and assume that it is consistent and zero-stable and, therefore, convergent. The role of consistency is to limit the size of local inaccuracies while that of zero-stablity is to ensure that these inaccuracies are not propagated in a disastrous manner; as a result, convergence is concerned with the situation in the limit as $l\to 0$, $n\to\infty$ with $t=t_0+nl$ fixed.

The role of *weak stability* is to provide some criterion under which the global error $\tilde{e}_n=y(t_n)-\tilde{y}_n$, where $\tilde{y}_n$ is the solution of (5.29) when a round-off error is committed at the nth application of the method, will be damped out as the integration of the initial value problem proceeds from $t=t_0$. The criterion involves l, which is now assumed to be fixed and arises from a study of $\tilde{y}_n$ as $n\to\infty$. The criterion gives an indication of the maximum value of l which may be used to integrate from $t=t_n$.

Introduce now, the *stability polynomial* $\Pi(r,\bar{l})$ of a numerical method defined by

$$\Pi(r,\bar{l})=\rho(r)-\bar{l}\sigma(r) \tag{5.48}$$

where $\bar{l}=l(\partial f/\partial y)$ and $\rho(r)$ and $\sigma(r)$ are defined by (5.46). The global error will not grow as $n\to\infty$ if the roots $r_1,r_2,\ldots,r_k$ of the *stability equation*

$$\Pi(r,\bar{l})=0 \tag{5.49}$$

satisfy the condition $|r_j|<1$ for $j=1,2,\ldots,$k.

Definition 5.8 The linear multistep method (5.29) is said to be *absolutely stable* for a given $\bar{l}$ if, for that $\bar{l}$, all the roots r_j of (5.49) satisfy $|r_j|<1, j=1,2,\ldots,k$, and to be *absolutely unstable* otherwise.

Definition 5.9 An interval (a,b) of the real number line is said to be an *interval of absolute stability* if the method is absolutely stable for all $\bar{l}\in(a,b)$.

If the method is absolutely unstable for all $\bar{l}$ it is said to have *no interval of absolute stability*.

The interval of absolute stability is determined only by the coefficients α_j, β_j $(j=1,2,\ldots,k)$ of the LMM (5.29). However, the largest value of l for which the global error will not increase depends on $\partial f/\partial y$, since $\bar{l}=l(\partial f/\partial y)$, and hence on the differential equation being integrated. It is clear that, when $f(t,y)$ is linear in y, the maximum value of l which may be used is the same for each step of the computation; when $f(t,y)$ is nonlinear in y, however, $\partial f/\partial y$ is not constant and the maximum value of l changes at each step. In practice, a bound or typical value of $\partial f/\partial y$ is used to estimate the maximum value of l which is then used throughout the computation.

The reader is referred to a specialist text, such as Lambert (1973), for the derivation of (5.49) and for a proof that

$$r_1=\exp(\bar{l})+O(\bar{l}^{p+1}),\quad \text{as } \bar{l}\to 0, \tag{5.50}$$

where r_1 is the zero of $\Pi(r,\bar{l})$ which tends to the principal zero of $\rho(r)$ as $\bar{l}\to 0$. It follows from (5.50) that for a sufficiently small $\bar{l}$, $r_1<1$ whenever $l<0$, that is, whenever $\partial f/\partial y>0$. This gives rise to the following theorem.

Theorem 5.6 Every consistent zero-stable linear multistep method is absolutely unstable for small positive $\bar{l}$.

Consistent, zero-stable methods which have no interval of absolute stability are often referred to as *weakly stable* methods (causing confusion with the title of the whole theory) and those with nonvanishing interval are termed *strongly stable*.

Definition 5.10 The linear multistep method (5.29) is said to be *relatively stable* for a given $\bar{l}$ if, for that $\bar{l}$, the roots r_j of (5.49) satisfy $|r_j|<|r_1|, j=2,3,\ldots,k$, and to be *relatively unstable* otherwise.

Definition 5.11 An interval (α,β) of the real line is said to be an *interval of relative stability* if the method is relatively stable for all $l\in(\alpha,\beta)$.

Linear multistep methods which have no interval of absolute stability may have nonvanishing intervals of relative stability. Arguably the commonest example of this is the Simpson corrector formula given by (5.39).

Example 5.5 Find the interval of absolute stability of the Euler predictor formula $y_{n+1}=y_n+lf_n$.

Writing this formula in the form of (5.29) as $-y_n+y_{n+1}=lf_n$ gives $\rho(r)=r-1$ and $\sigma(r)=1$ so that $\Pi(r,\bar{l})=r-(1+\bar{l})$. The stability equation $\Pi(r,\bar{l})=0$ therefore has only one root given by $r=r_1=1+\bar{l}$ and the interval of absolute stability clearly consists of those values of $\bar{l}$ for which $-1<1+\bar{l}<1$, that is, those values of $\bar{l}$ for which $-1<1+\bar{l}$ and $1+\bar{l}<1$. This gives $-2<\bar{l}$ and $\bar{l}<0$ and the stability interval is the intersection of these two intervals giving $\bar{l}\in(-2,0)$.

Example 5.6 In percutaneous drug absorption, the initial value problem

$$y'(t)=-400.01y(t),\ t>0;\ y(0)=1$$

may be associated with the removal of ointment from the skin after the steady state has been reached. Show that the maximum time step which may be used when solving this problem by the Euler predictor formula is less than 0.005. Tabulate the errors in the Euler approximations to $y(t)$ for $t=0.0(0.05)0.4$ using time steps of $l=0.004$, 0.005 and 0.008. Discuss the numerical results. (This problem is discussed in greater detail in subsection 8.5.2 of Chapter 8. In the notation of that subsection, $\omega=0.001$, $h=0.1$ and $\phi(x)\equiv 1$ so that the eigenvalue of the matrix A given by equation (8.73) which has greatest modulus is -400.01).

Here, $f(t,y)=-400.01y(t)$ so that $\partial f/\partial y=-400.01$. It was seen in Example 5.5 that, for this LMM, $-2<\bar{l}<0$ and so $-2<-400.01\ l<0$. This gives $l<2/400.01<0.005$. The theoretical solution of this initial value problem is $y(t)=\exp(-400.01t)$ and the differences between $y(t)$ and the computed values for $t=0.0(0.05)0.4$ are given in Table 5.1.

The time step $l=0.004$ is less than the maximum permitted for stability and the errors are clearly decreasing as t increases (Table 5.1) though they are still bigger in magnitude than the theoretical solution which decays *very* rapidly. The time step $l=0.005$ is very slightly bigger than the maximum permitted and Table 5.1 reveals that, to two significant figures, the error does not grow for t in the interval $0.0<t<0.4$, though the computed solution $y\simeq 1.0$ is unacceptable. The time step $l=0.008$ is much bigger than the maximum permitted and it is obvious from Table 5.1 that the error is growing disastrously in magnitude and is completely swamping the theoretical solution.

This example clearly illustrates the good behaviour of a stable numerical method and the bad behaviour of an unstable method.

Example 5.7 Find the interval of absolute stability of the Euler corrector formula $y_{n+1}=y_n+\frac{1}{2}l(f_n+f_{n+1})$.

It is easy to see that, here, $\rho(r)=r-1$ and that $\sigma(r)=\frac{1}{2}(1+r)$ so that $\Pi(r,\bar{l})=(1-\frac{1}{2}\bar{l})r-(1+\frac{1}{2}\bar{l})$, giving $r=r_1=(1+\frac{1}{2}\bar{l})/(1-\frac{1}{2}\bar{l})$ as the only zero of the stability

Table 5.1

t	Theoretical solution	Errors $\bar{l}=0.004$	$\bar{l}=0.005$	$\bar{l}=0.008$
0.00	0.10E + 01	0	0	0
0.04	0.11E − 06	− 0.61E − 02	− 0.10E + 01	+ 0.52E + 02
0.08	0.13E − 13	− 0.37E − 04	− 0.10E + 01	− 0.27E + 04
0.12	0.14E − 20	− 0.22E − 06	− 0.10E + 01	+ 0.14E + 05
0.16	0.16E − 27	− 0.13E − 08	− 0.10E + 01	− 0.71E + 07
0.20	0.18E − 34	− 0.81E − 11	− 0.10E + 01	+ 0.36E + 09
0.24	0.00E + 00	− 0.49E − 13	− 0.10E + 01	− 0.19E + 11
0.28	0.00E + 00	− 0.30E − 15	− 0.10E + 01	+ 0.97E + 12
0.32	0.00E + 00	− 0.18E − 17	− 0.10E + 01	− 0.50E + 14
0.36	0.00E + 00	− 0.11E − 19	− 0.10E + 01	+ 0.26E + 16
0.40	0.00E + 00	− 0.66E − 22	− 0.10E + 01	− 0.13E + 18

equation. The stability criterion then requires $-(1-\frac{1}{2}\bar{l})<1+\frac{1}{2}\bar{l}<1-\frac{1}{2}\bar{l}$. The left side of this three-way inequality gives $-1+\frac{1}{2}\bar{l}<1+\frac{1}{2}\bar{l}$ or $-1<1$, which is trivially true for $-\infty<\bar{l}<\infty$, while the right side gives $\bar{l}<0$. The stability interval is thus $\bar{l}\in(-\infty,0)$.

Definition 5.12 An LMM for which the interval of absolute stability includes the whole of the negative real number line is said to be *A_0-stable*.

5.4.7 Weak stability theory: stability regions

It often happens that, for some reason, $\bar{l}$ is complex and it is more appropriate to talk of *stability regions* (in the complex $\bar{l}$-plane) than stability intervals.

Definition 5.13 A *region of absolute stability* is a region of the complex $\bar{l}$-plane such that the roots of $\Pi(r,\bar{l})=0$ lie within the unit circle whenever $\bar{l}$ lies in the interior of the region.

A similar definition may be made for a region of relative stability.

Let the region of absolute stability of an LMM be R and its boundary be ∂R. The roots of $\Pi(r,\bar{l})=0$ are continuous functions of $\bar{l}$ so that $\bar{l}$ will lie on ∂R when one of the roots of stability equation lies on the boundary of the unit circle $r=e^{i\theta}$, $-\pi<\theta\leqslant\pi$. That is when $\Pi(e^{i\theta},\bar{l})=\rho(e^{i\theta})-\bar{l}\sigma(e^{i\theta})=0$.

Thus the locus ∂R is found by solving $\bar{l}=\bar{l}(\theta)=\rho(e^{i\theta})/\sigma(e^{i\theta})$. This boundary is plotted in the complex $\bar{l}$-plane by plotting the real and imaginary parts of $\bar{l}$ for a number of values of θ, the interior of the locus being the region of absolute stability (the word 'interior' is perhaps inappropriate, as the region may, in fact, be open).

This is known as the *boundary locus method* for finding the region of absolute stability. When $\bar{l}$ is real, the end points of the interval of absolute stability coincide with the points at which ∂R cuts the real axis in the $\bar{l}$-plane. This provides the user with the means to find the stability interval of an LMM when the approach followed in Example 5.6 may be much more difficult.

Example 5.8 Use the boundary locus method to find the interval of absolute stability of the LMM $y_{n+2}=y_n+\frac{1}{2}l(3f_n+f_{n+1})$.

Here $\rho(r)=r-1$ and $\sigma(r)=\frac{1}{2}(3+r)$ so that

$$\bar{l}=\frac{\rho(e^{i\theta})}{\sigma(e^{i\theta})}=\frac{e^{2i\theta}-1}{\frac{1}{2}(e^{i\theta}+3)}=\frac{3(\cos 2\theta-1)}{5+3\cos\theta}+i\frac{3\sin 2\theta+2\sin\theta}{5+3\cos\theta}.$$

It is noted that $\bar{l}$ is of the form

$$\bar{l}=\bar{l}(\theta)=\frac{u(\theta)}{d(\theta)}+i\frac{\upsilon(\theta)}{d(\theta)},$$

where $u(\theta)=3(\cos 2\theta-1)$, $\upsilon(\theta)=3\sin 2\theta+2\sin\theta$ and $d(\theta)=5+3\cos\theta$, and the interval of absolute stability is found by first finding the values of θ which cause $\upsilon(\theta)$ to vanish and then by substituting these values into $u(\theta)/d(\theta)$. It is easy to show that $\upsilon(\theta)=0$ when $\sin\theta=0$ or $\cos\theta=-\frac{1}{3}$, that is, when $\theta=0$, π, $\pi+\cos^{-1}\frac{1}{3}$ or $\pi-\cos^{-1}\frac{1}{3}$. Substituting these values into the expressions for $u(\theta)$ and $d(u)$ gives $\bar{l}(\theta)=0, 0, -\frac{4}{3}, -\frac{4}{3}$, respectively from which it is obvious that the stability interval is $\bar{l}\in(-\frac{4}{3},0)$.

Example 5.9 Plot the boundary of the stability region of the LMM of Example 5.8.

It is easy to see that $u(\theta)=u(-\theta)$, $d(\theta)=d(-\theta)$, $\upsilon(\theta)=-\upsilon(-\theta)$ so that the stability region is symmetric about the real $\bar{l}$-axis. Taking $\theta=0°(10°)110°$ gives Table 5.2.

Plotting these 12 points and using the symmetry property shows that the region of absolute stability is the circle on the interval $(-\frac{4}{3},0)$ as diameter (see Fig. 5.2).

Example 5.10 Find the maximum steplength which may be used to solve the initial value problem $y'(t)=(-10+20i)y(t)$, $y(t_0)=y_0$, using the LMM of Examples 5.8 and 5.9.

In this problem, $\partial f/\partial y=-10+20i$ so that $\bar{l}=(-10+20i)l$ and it follows that $\text{Im}(\bar{l})=-2\text{Re}(\bar{l})$. Drawing the straight line $\text{Im}(\bar{l})=-\text{Re}(\bar{l})$ in a more accurate version of Fig. 5.2 reveals that the line cuts the circle where $\text{Re}(\bar{l})=-0.26$ approximately and $\text{Im}(\bar{l})=0.53$ approximately. Thus

Table 5.2

θ	u/d	v/d
0	0.00	0.00
10	−0.02	0.17
20	−0.09	0.33
30	−0.20	0.47
40	−0.34	0.58
50	−0.51	0.65
60	−0.69	0.67
70	−0.88	0.63
80	−1.05	0.57
90	−1.20	0.40
100	−1.30	0.21
110	−1.33	−0.01

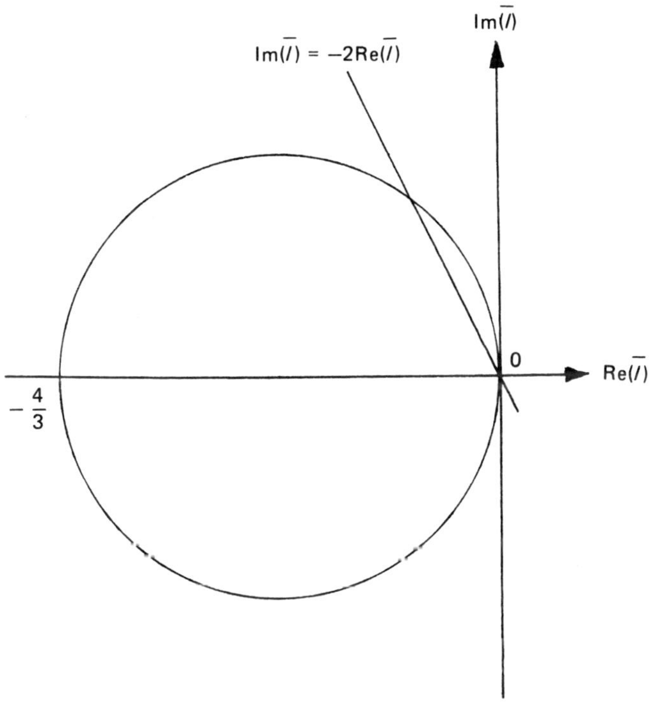

Fig. 5.2 Stability region of the LMM $y_{n+1} = y_n + \frac{1}{4}l(3f_n + f_{n+1})$.

$$l_{\max}=\min\left(\frac{-0.26}{-10},\frac{0.53}{20}\right)=0.026 \text{ (approximately).}$$

An accurate calculation reveals that $l_{\max}=\frac{4}{150}=0.0266\ldots$.

Example 5.11 Find the stability region of the Euler corrector formula $y_{n+1}=y_n+\frac{1}{2}l(f_n+f_{n+1})$.

Following the procedure of Example 5.8 gives $u(\theta)=0$, $v(\theta)=2\sin\theta$ and $d(\theta)=1+\cos\theta$ and clearly the region of absolute stability is the whole of the left half plane.

Definition 5.14 A numerical method is said to be *A-stable* if its region of absolute stability contains the whole of the left half plane $\mathrm{Re}(\bar{l})<0$.

5.4.8 Predictor–corrector combinations

Explicit linear multistep methods suffer from poor stability properties, and implicit methods can be difficult to implement for nonlinear problems, though their stability properties are usually good. To use the best properties of these two types of method they are often used in *predictor–corrector* combination, the explicit *predictor* providing a starting value for the implicit *corrector* which may be used iteratively as indicated by the discussion associated with equation (5.31).

Let P indicate an application of the predictor, E an evaluation of $f(t,y)$ in terms of known values of t and y, and C a single application of the corrector. The value of $y_{n+k}^{[0]}$ is computed using the predictor, $f_{n+k}^{[0]}\equiv f(t_{n+k},y_{n+k}^{[0]})$ is evaluted from the differential equation, then $y_{n+k}^{[1]}$ is computed by using the corrector once. This value of y is said to have been determined by applying the predictor–corrector pair in *PEC mode*. A further evaluation of f_{n+k}, named $f_{n+k}^{[1]}$, indicates that *PECE* mode has been used, and yet one more application of the corrector to yield $y_{n+k}^{[2]}$ indicates that $P(EC)^2$ mode has been used. Applying the evaluation and correction stages m times, the mode so used is indicated by $P(EC)^m$.

Suppose, now, that m is fixed, then $y_{n+k}^{[m]}$ is accepted as the numerical solution of (5.1) at t_{n+k}. The last available computed value for f_{n+k} is $f_{n+k}^{[m-1]}\equiv f(t_{n+k},y_{n+k}^{[m-1]})$ and the user must decide whether or not to evaluate $f_{n+k}^{[m]}\equiv f(t_{n+k},y_{n+k}^{[m]})$. If this final evaluation is not made, the mode used is indeed $P(EC)^m$, but if it is made the mode used is $P(EC)^mE$. The decision made by the user will affect the value of y_{n+k+1} since the predicted value $y_{n+k+1}^{[0]}$ and subsequent corrected values $y_{n+k+1}^{[s]}$, $s=1,2,\ldots,m$, will depend on whether f_{n+k} is taken to be $f_{n+k}^{[m-1]}$ or $f_{n+k}^{[m]}$.

Suppose next that m is not fixed, then steps E and C may be used sufficiently often to obtain convergence of the sequence of corrected values $y_{n+k}^{[s]}$, $s=1,2\ldots$. Each iterate $y_{n+k}^{[s]}$, $s=1,2,\ldots$, corresponds to one application of the corrector and this mode of operation is known as *correcting to convergence*. In practice, steps E and C are used continually until some criterion such as $|y_{n+k}^{[s+1]}-y_{n+k}^{[s]}|<\varepsilon$, where ε is a predetermined tolerance, is satisfied. The value $y_{n+k}^{[s+1]}$ is then taken to be the numerical solution of (5.1) at the point t_{n+k}.

Example 5.12 Use the Euler predictor formula (5.33) and the Euler corrector formula (5.35) to obtain the value of x at time $t=\frac{1}{12}$ to two decimal places in the predator–prey model of Example 5.2. Use a step length $l=\frac{1}{12}$.

In the notation used in this section (with an obvious change from y to x) the information available from Example 5.2 is $x_0=1100$ and $f_0\equiv f(t_0,x_0)=-55$. An approximation to $x(\frac{1}{12})$ is sought to two decimal places so, carrying three decimal places, the computation proceeds as follows:

$$P: x_1^{[0]}=x_0+lf_0=1100+\tfrac{1}{12}(-55)=1095.417;$$

$$\begin{aligned}E: f_1^{[0]}&\equiv f(t_1,x_1^{[0]})=f(\tfrac{1}{12},\ 1095.417)\\ &=0.5(x_1^{[0]})-0.0005(x_1^{[0]})^2=-52.260;\end{aligned}$$

$$C: x_1^{[1]}=x_0+\tfrac{1}{2}l(f_0+f_1^{[0]})=1095.531.$$

Comparing $x_1^{[1]}$ to $x_1^{[0]}$, it is easy to see that convergence to two decimal places has not been attained and so at least one more application of the corrector is called for (note: *PEC* mode has been used so far). Continuing,

$$E: f_1^{[1]}\equiv f(t_1,x_1^{[1]})=-52.328;$$

$$C: x_1^{[2]}=x_0+\tfrac{1}{2}l(f_0+f_1^{[1]})=1095.528.$$

Comparing $x_1^{[2]}$ to $x_1^{[1]}$ shows that convergence to two decimal places has been obtained following the use of $P(EC)^2$ mode giving $x_1=1095.53$. This approximation to $x(\frac{1}{12})$ agrees with those of Examples 5.2 and 5.3 to two decimal places (realistically, of course, only integer values of x are meaningful in this particular problem).

An in-depth study of local truncation errors of linear multistep methods used in predictor–corrector combinations is well beyond the scope of this book and the reader is referred to the excellent treatment of this topic in Lambert (1973). However, a summary of certain important points to note has a place in this book, and to avoid confusion an asterisk will be added to terms relating to the predictor formula.

Suppose, for instance, that a predictor of order p^* with error constant $C^*_{p^*+1}$ and a corrector of order p with error constant C_{p+1} are used in $P(EC)^m$ or $P(EC)^mE$ mode where p^*, p, m are integers with $p^*\geqslant 0$, $p\geqslant 1$, $m\geqslant 1$. Then, if $p^*\geqslant p$, the principal local truncation error of the predictor–corrector combination at $t=t_{n+k}$ is that of the corrector alone, namely $C_{p+1}l^{p+1}y^{(p+1)}(t_n)$. If $p^*=p-r, 0<r\leqslant p$, then the principal local truncation error of the combination is

(a) that of the corrector along when $m\leqslant r+1$,
(b) of the same order as that of the corrector, but not identical with it, when $m=r$,
(c) of the form $Kl^{p-r+m+1}+O(l^{p-r+m+2})$ when $m\leqslant r-1$.

In the mode of correcting to convergence, the principal part of the local truncation error is equal to that of the corrector alone, irrespective of the order of the predictor. No matter which mode is chosen, there is little to be gained by choosing a predictor formula of order higher than that of the corrector. Such an LMM would probably have a large stepnumber k, thus requiring a sequence of starting values, and the higher accuracy would not be carried over to the overall predictor–corrector combination.

When $p^* = p$, the principal local tuncation errors of the predictor and corrector may be written as

$$C^*_{p+1} l^{p+1} y^{(p+1)}(t_n) = y(t_{n+k}) - y^{[0]}_{n+k} + O(l^{p+2}) \tag{5.51}$$

and

$$C_{p+1} l^{p+1} y^{(p+1)}(t_n) = y(t_{n+k}) - y^{[m]}_{n+k} + O(l^{p+2}), \tag{5.52}$$

respectively. Subtracting (5.52) from (5.51), dividing by $C^*_{p+1} - C_{p+1}$, then multiplying by C_{p+1} gives

$$C_{p+1} l^{p+1} y^{(p+1)}(t_n) = \frac{C_{p+1}}{C^*_{p+1} - C_{p+1}} (y^{[m]}_{n+k} - y^{[0]}_{n+k}) \tag{5.53}$$

after dropping $O(l^{p+2})$ terms. The expression on the right-hand side of (5.53), which assumes that $C^*_{p+1} \neq C_{p+1}$, gives an estimate of the principal local tuncation error in $P(EC)^m$ or $P(EC)^m E$ mode, and is known as *Milne's device*.

Retaining the notation of adding an asterisk to terms relating to a predictor formula, the first and second characteristic polynomials of the predictor and corrector are named $\rho^*(r)$, $\sigma^*(r)$, $\rho(r)$, $\sigma(r)$, respectively. The stability polynomial of the predictor–corrector algorithm in $P(EC)^m E$ mode is known to be

$$\Pi_{P(EC)^m E}(r,\bar{l}) = \rho(r) - \bar{l}\sigma(r) + M_m(\bar{l})[\rho^*(r) - \bar{l}\sigma^*(r)] \tag{5.54}$$

where

$$M_m(\bar{l}) = (\bar{l}\beta_k)^m (1 - \bar{l}\beta_k)/[1 - (\bar{l}\beta_k)^m] \tag{5.55}$$

with $m = 1,2\ldots$ (see Lambert (1973) for full details), and in $P(EC)^m$ mode to be

$$\Pi_{P(EC)^m}(r,\bar{l}) = \beta_k r^k\,[\rho(r) - \bar{l}\sigma(r)] + M_m(\bar{l})\,[\rho^*(r)\sigma(r) - \rho(r)\sigma^*(r)]. \tag{5.56}$$

Taking $L = \partial f/\partial y$, (5.32) becomes

$$|\bar{l}\beta_k| < 1. \tag{5.57}$$

Then, provided (5.57) is satisfied, $M_m(\bar{l}) \to 0$ and $m \to \infty$. It follows that the stability polynomials given by (5.54) and (5.56) then tend to that of the corrector alone given by (5.48). Thus, correcting to convergence ensures that the weak stability property of the predictor–corrector algorithm is the same as that of the corrector.

In practice, algorithms in *PECE* mode prove attractive to the user. Here, $m = 1$ and, provided (5.57) is satisfied, the right-hand side of (5.54) becomes $M_1(\bar{l}) = \bar{l}\beta_k$ and (5.54) gives

$$\Pi_{PECE} = \rho(r) - \bar{l}\sigma(r) + \bar{l}\beta_k[\rho^*(r) - \bar{l}\sigma^*(r)]. \tag{5.58}$$

It can be concluded from this brief discussion of weak stability that if a mode other than correcting to convergence is employed, the predictor, as well as the corrector, influences the weak stability of the predictor–corrector algorithm.

Example 5.13 Find the stability interval of the Euler-modified Euler algorithm in *PECE* mode.

Here, $\rho^*(r) = \rho(r) = r - 1$, $\sigma^*(r) = 1$, $\sigma(r) = \frac{1}{2}(r + 1)$, $\beta_k = \frac{1}{2}$ and so, from (5.58),

$$\Pi_{PECE}\,(r,\bar{l}) = r - (1 + \bar{l} + \tfrac{1}{2}\bar{l}^2).$$

it is then easy to show that $|r| < 1$ provided $\bar{l} \in (-2,0)$. Unusually, the stability interval in *PECE* mode is thus the same as the predictor alone, but the combination is a second-order method whereas the predictor is first order.

5.5 SYSTEMS OF EQUATIONS

5.5.1 Applicability of Runge–Kutta methods

First-order systems of linear ordinary differential equations have the form (5.8) while nonlinear systems have the form (5.4). The family of explicit Runge–Kutta methods for a system of differential equation takes the form

$$\mathbf{y}_{n+1} = \mathbf{y}_n + l\,\boldsymbol{\phi}\,(t,\mathbf{y}_n,l) \tag{5.59}$$

and, from Definition 5.1, a Runge–Kutta method is of order p if p is the largest integer for which

$$\mathbf{y}(t + l) - \mathbf{y}(t) - l\,\boldsymbol{\phi}\,(t,\mathbf{y}(t),l) = \boldsymbol{O}(l^{p+1}), \tag{5.60}$$

where $y(t)$ is the theoretical solution of (5.4); the method is consistent with (5.4) if $\boldsymbol{\phi}(t,y,0) = \mathbf{f}(t,\mathbf{y})$ and $p \geqslant 1$.

The Runge–Kutta methods developed in section 5.3 are applicable to a system if the scalars y, f, ϕ and the terms $k_r (r = 1,2, \ldots ,p)$ are replaced by vectors $\mathbf{y}, \mathbf{f}, \boldsymbol{\phi}$ and $\mathbf{k}_r$ respectively. In general, a Runge–Kutta method has order p for a single equation has order p for a system if $p \leqslant 4$; for $p > 4$, the order for a system may be less than p.

5.5.2 Applicability of linear multistep methods

All the linear multistep methods occurring in section 5.4 are applicable to first-order differential systems, either individually or in predictor–corrector combinations. The definitions of order, consistency, error constant and zero-stability, which depend only on the coefficients $\alpha_j, \beta_j (j = 0,1,2, \ldots ,k)$, are all valid for a system of equations. The local truncation error (5.43) is now a vector, the linear difference operation (5.41) leading to it being written as

$$\mathscr{L}[\mathbf{y}(t);l] = \sum_{j=0}^{k} [\alpha_j \mathbf{y}(t + jl) - l\beta_j \mathbf{y}'(t + jl)]. \tag{5.61}$$

Definition 5.4 which defines convergence is modified to the the following.

Definition 5.15 The vector form of the LMM (5.29) is said to be convergent if, for all initial value problems (5.4) satisfying the hypotheses of the vector form of Theorem 5.1, the limit

$$\lim_{\substack{l \to 0 \\ nl = t - t_0}} \|\mathbf{y}_n - \mathbf{y}(t_n)\| \to 0$$

holds for all t in some interval $t_0 \leqslant t \leqslant T$ and for all solutions $\{\mathbf{y}_n\}$ of the vector form of (5.29) satisfying starting conditions $\mathbf{y}_r = \mathbf{y}_{0,r}(l)$ for which $\lim_{l \to 0} \|\mathbf{y}_{0,r}(l) - \mathbf{y}_0\| = 0$, $r = 0,1, \ldots ,k - 1$.

Theorem 5.4 which connects consistency, zero-stability and convergence is valid for a system of differential equations.

The theory of weak stability outlined in subsections 5.4.6 and 5.4.8 is also valid, though $\bar{l}$ now takes the form $\bar{l} = l\lambda$ where λ is *any* of the eigenvalues of the Jacobian $\partial\mathbf{f}/\partial\mathbf{y}$. Thus any condition imposed on $\bar{l}$ must be satisfied for *all* of the eigenvalues of the Jacobian.

5.5.3 Linear systems

The general linear initial value problem in system form may be written as in equation (5.8). A particular form of (5.8) which arises frequently in the mathematical modelling of biomedical systems is

$$\mathbf{y}'(t) = A\mathbf{y}(t) + \boldsymbol{\psi}(t),\ t > 0;\ \mathbf{y}(0) = \mathbf{y}_0, \tag{5.62}$$

where A is a constant square matrix of order N and each element of $\boldsymbol{\psi}(t)$ is a continuous function of t; $\boldsymbol{\psi}(t)$ represents a continuous *source term*. The differential system of (5.62) is said to be nonhomogeneous; when $\boldsymbol{\psi}(t) \equiv \mathbf{0}$, the system is said to be *homogeneous*.

It is easy to verify that the solution of the homogeneous problem

$$\mathbf{y}'(t) = A\mathbf{y}(t), \quad t > 0; \ \mathbf{y}(0) = \mathbf{y}_0 \tag{5.63}$$

is

$$\mathbf{y}(t) = \exp(tA)\mathbf{y}_0 \tag{5.64}$$

and that this solution satisfies the recurrence relation

$$\mathbf{y}(t+l) = \exp(lA)\mathbf{y}(t), \tag{5.65}$$

where $t = 0, l, 2l, \ldots$ .Suppose that this solution is $\mathbf{y}_{\mathrm{H}}(t)$ and that $\mathbf{y}_{\mathrm{P}}(t)$ is a particular solution of (5.62). Then the general solution of (5.62) is established by the following theorem.

Theorem 5.7 Suppose that $\mathbf{y}_{\mathrm{P}}(t)$ is a particular solution to the nonhomogeneous system $\mathbf{y}'(t) = A\mathbf{y}(t) + \boldsymbol{\psi}(t)$. Then the general solution is

$$\mathbf{y}(t) = \mathbf{y}_{\mathrm{H}}(t) + \mathbf{y}_{\mathrm{P}}(t)$$

where $\mathbf{y}_{\mathrm{H}}(t)$ is a solution to $\mathbf{y}'(t) = A\mathbf{y}(t)$.

Proof of Theorem 5.7 If $\mathbf{y}_{\mathrm{S}}(t)$ is a second solution, let $\mathbf{z}(t) = \mathbf{y}_{\mathrm{S}}(t) - \mathbf{y}_{\mathrm{P}}(t)$. Then

$$\begin{aligned} \mathbf{z}'(t) = \mathbf{y}_{\mathrm{S}}'(t) - \mathbf{y}_{\mathrm{P}}'(t) &= [\mathrm{A}\mathbf{y}_{\mathrm{S}}(t) + \boldsymbol{\psi}(t)] - [A\mathbf{y}_{\mathrm{P}}(t) - \boldsymbol{\psi}(t)] \\ &= A[\mathbf{y}_{\mathrm{S}}(t) - \mathbf{y}_{\mathrm{P}}(t)] = A\mathbf{z}(t). \end{aligned}$$

Hence, $\mathbf{y}_{\mathrm{S}}(t) = \mathbf{z}(t) + \mathbf{y}_{\mathrm{P}}(t)$ where $\mathbf{z}(t)$ is clearly equal to $\mathbf{y}_{\mathrm{H}}(t)$.

The next step is to determine a particular solution: this is achieved in the following theorem.

Theorem 5.8 A particular solution to $\mathbf{y}'(t) = \mathrm{A}\mathbf{y}(t) + \boldsymbol{\psi}(t)$ satisfying the initial vector $\mathbf{y}_{\mathrm{P}}(0) = \mathbf{0}$ is given by

$$\mathbf{y}_{\mathrm{P}}(t) = \int_0^t \exp[(t-s)A] \ \boldsymbol{\psi}(s)\mathrm{d}s.$$

Proof of Theorem 5.8 If $\mathbf{y}'_P(t) - A\mathbf{y}_P(t) = \boldsymbol{\psi}(t)$ then $\exp(-tA)\ [\mathbf{y}'(t) - A\mathbf{y}_P(t)] = \exp(-tA)\ \psi(t)$. By the product rule

$$\frac{\mathrm{d}}{\mathrm{d}t}[\exp(-tA)\mathbf{y}_P(t)] = \exp(-tA)\mathbf{y}'(t) - \exp(-tA)A\mathbf{y}_P(t)$$

$$= \exp(-tA)\ [\mathbf{y}'_P(t) - A\mathbf{y}_P(t)] = \exp(-tA)\ \boldsymbol{\psi}(t).$$

Integration then gives

$$\exp(-tA)\mathbf{y}_P(t) = \int_0^t \exp(-sA)\ \boldsymbol{\psi}(s)\mathrm{d}s$$

and multiplying both sides by $\exp(tA)$ gives the result sought.

Combining Theorems 5.7 and 5.8 gives Theorem 5.9, the proof of which is immediate, and leads to Theorem 5.10.

Theorem 5.9 The solution of the nonhomogeneous initial value problem $\mathbf{y}'(t) = A\ \mathbf{y}(t),\ t > 0,\ \mathbf{y}(0) = \mathbf{y}_0$ is given by

$$\mathbf{y}(t) = \exp(tA)\mathbf{y}_0 + \int_0^t \exp[(t-s)A]\ \psi(s)\mathrm{d}s. \tag{5.66}$$

Theorem 5.10 The solution (5.66) of the nonhomogenous initial value problem (5.62) satisfies the recurrence relation

$$\mathbf{y}(t+l) = \exp(lA)\mathbf{y}(t) + \int_t^{t+l} \exp[(t+l-s)A]\ \boldsymbol{\psi}(s)\mathrm{d}s. \tag{5.67}$$

where $t = 0, l, 2l, \ldots$.

The proof of this theorem is mechanical and is left as an exercise to the reader.

Systems of the form (5.63) arise in the numerical solution of second-order parabolic equations which are discussed in Chapter 8 while systems of the form (5.62), with $\boldsymbol{\psi}$ equal to a constant vector, arise in the numerical solution of first-order hyperbolic equations which are treated in Chapter 7. Systems of the form (5.62), with the time-dependent source term $\boldsymbol{\psi}(t)$, arise in the mathematical modelling of ecosystems, cardiac output, and pharmacokinetics, and equation (5.67) provides a useful recurrence relation for determining a numerical solution to such systems.

Using the trapezoidal rule to approximate the integral in (5.67) gives

$$\mathbf{y}(t+l) = \exp(lA)\mathbf{y}(t) + \tfrac{1}{2}l[\boldsymbol{\psi}(t+l) + \exp(lA)\boldsymbol{\psi}(t)] + O(l^3). \tag{5.68}$$

Replacing, now, the matrix exponential terms in (5.68) by the rational approximant

$$\exp(lA) \simeq (I - \theta lA)^{-1}[I + (1-\theta)lA],$$

where $0 \leqslant \theta \leqslant 1$ is a parameter, leads to

$$\begin{aligned}(I - \theta lA)\mathbf{y}_{n+1} = [I + (1-\theta)lA]\mathbf{y}_n + \tfrac{1}{2}l\{(I - \theta lA)\boldsymbol{\psi}_{n+1} \\ + [I + (1-\theta)lA]\boldsymbol{\psi}_n\} + O(l^r)\end{aligned} \tag{5.69}$$

in which $r = 2$ if $\theta \neq \frac{1}{2}$. Choosing $\theta = 0$ gives the first-order explicit method

$$\mathbf{y}_{n+1} = (I - lA)\mathbf{y}_n + \tfrac{1}{2}l[\boldsymbol{\psi}_{n+1} + (I + lA)\boldsymbol{\psi}_n] + O(l^2) \tag{5.70}$$

while choosing $\theta = \frac{1}{2}$ gives the second-order implicit method

$$(I - \tfrac{1}{2}lA)\mathbf{y}_{n+1} = (I + \tfrac{1}{2}lA)\mathbf{y}_n + \tfrac{1}{2}l[(I - \tfrac{1}{2}lA)\boldsymbol{\psi}_{n+1} + (I + \tfrac{1}{2}lA)\boldsymbol{\psi}_n] + O(l^3) \tag{5.71}$$

which is more expensive to implement but has better stability properties.

As has been noted already, linear differential systems arise frequently in the modelling of ecological, physiological and pharmaceutical systems. An appropriate example is in the compartmental modelling of tracer kinetics. In most cases, it is clinically possible to obtain tracer concentration measurements from one compartment only (usually a urine or blood sample), so that a mathematical model with more than two compartments (usually tissue and the blood stream) is unlikely to be useful. Such a two-compartment model is depicted in Fig. 5.3 where a tracer $Q = Q(t)$ is injected continuously into the blood stream giving, in the terminology used in (5.62), $\mathbf{y}(0) = [y_{1,0}, y_{2,0}]^{\mathrm{T}}$ and $\boldsymbol{\psi}(t) = [Q(t), 0]^{\mathrm{T}}$. In Fig. 5.3, F_{ij} represents the flow rate of the tracer into compartment i from compartment j, the volume in compartment j being V_j (a subscript 0 represents an external reservoir).

It is wise in compartmental modelling to monitor carefully the units of the variables. Typically, $Q(t)$ (and consequently the elements of $\boldsymbol{\psi}$) is measured in mg/s, $\mathbf{y}_0$ and the elements of $\mathbf{y}(t)$ in mg, the elements of $\mathbf{y}'(t)$ in mg/s, F_{ij} in litres/s, and V_j in litres. Denoting by $\mathbf{c}(t) = [c_1(t), c_2(t)]^{\mathrm{T}}$ the concentration of tracer (measured in mg/litre) in the two compartments of the model at time t, the rates of increase of tracer in the two compartments at time t are described by the linear differential equations

$$y_1'(t) = -(F_{01} + F_{21})c_1(t) + F_{12}c_2(t) + Q(t), \tag{5.72}$$

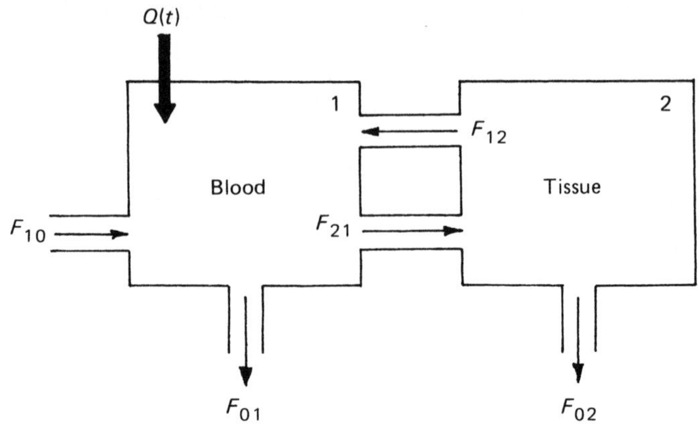

Fig. 5.3 — General two-compartmental model.

$$y_2'(t) = F_{21}c_1(t) - (F_{02} + F_{12})c_2(t), \tag{5.73}$$

the negative signs signifying flow out of a compartment.

Noting that $y_i = c_iV_i$ $(i = 1,2)$, equations (5.72) and (5.73) may be rewritten as

$$\begin{bmatrix} y_1'(t) \\ y_2'(t) \end{bmatrix} = \begin{bmatrix} a_{11} & a_{12} \\ a_{21} & a_{22} \end{bmatrix} \begin{bmatrix} y_1(t) \\ y_2(t) \end{bmatrix} + \begin{bmatrix} Q(t) \\ 0 \end{bmatrix}, \quad \begin{bmatrix} y_1(0) \\ y_2(0) \end{bmatrix} = \begin{bmatrix} y_{1,0} \\ y_{2,0} \end{bmatrix} \tag{5.74}$$

which is of the form (5.62), with

$$a_{11} = -(F_{01} + F_{21})/V_1,\ a_{12} = F_{12}/V_2,$$

$$a_{21} = F_{21}/V_1,\ a_{22} = -(F_{02} + F_{12})/V_2. \tag{5.75}$$

Example 5.14 Use (5.70) and (5.71) to obtain solutions to (5.74) at time $t = (n+1)l$ for $n = 0,1,2,\ldots$.

The explicit method (5.70) gives

$$y_{1,n+1} = (1 + la_{11})y_{1,n} + la_{12}y_{2,n} + \tfrac{1}{2}lQ_{n+1} + \tfrac{1}{2}l(1 + la_{11})Q_n, \tag{5.76}$$

$$y_{2,n+1} = la_{21}y_{1,n} + (1 + la_{22})y_{2,n} + \tfrac{1}{2}l^2a_{21}Q_n \tag{5.77}$$

where $y_{i,n}$ $(i = 1,2;\ n = 0,1,2,\ldots)$ denotes the computed value of $y_i(nl)$ and Q_n denotes the value of Q at time $t = nl$, while the implicit method (5.71) gives

$$(1 - \tfrac{1}{2}la_{11})y_{1,n+1} - \tfrac{1}{2}la_{12}y_{2,n+1} = (1 + \tfrac{1}{2}la_{11})y_{1,n} + \tfrac{1}{2}la_{12}y_{2,n}$$

$$+\tfrac{1}{2}l(1-\tfrac{1}{2}la_{11})Q_{n+1}+\tfrac{1}{2}l(1+\tfrac{1}{2}la_{11})Q_n, \tag{5.78}$$

$$\begin{aligned}-\tfrac{1}{2}la_{21}y_{1,n+1}+(1-\tfrac{1}{2}la_{22})y_{2,n+1}&=\tfrac{1}{2}la_{21}y_{1,n}+(1+\tfrac{1}{2}la_{22})y_{2,n}\\ &\quad-\tfrac{1}{4}l^2a_{21}Q_{n+1}+\tfrac{1}{4}l^2a_{21}Q_n.\end{aligned} \tag{5.79}$$

Equations (5.78) and (5.79) can easily be solved to obtain $\mathbf{y}_{n+1}$.

5.5.4 Nonlinear systems

The general nonlinear first-order initial value problem in system form is given by (5.4). A family of numerical methods for the solution of (5.4) is based on the Taylor formula

$$\mathbf{y}(t+l)=\exp(lD)\mathbf{y}(t) \tag{5.80}$$

where D is given by

$$D=\operatorname{diag}\{\mathrm{d}/\mathrm{d}t\}=\begin{bmatrix}\mathrm{d}/\mathrm{d}t & & & & \\ & \mathrm{d}/\mathrm{d}t & & & 0\\ & & \cdot & & \\ & & & \cdot & \\ & & & & \cdot \\ 0 & & & & \mathrm{d}/\mathrm{d}t\end{bmatrix}$$

Replacing $\exp(lD)$ in (5.80) by the approximation

$$\exp(lD)\simeq(I-\theta lD)^{-1}\,[I+(1-\theta)lD],$$

where $0\leqslant\theta\leqslant1$ is a parameter, leads to

$$(I-\theta lD)\mathbf{y}(t+l)=[I+(1-\theta)lD]\mathbf{y}(t)+O(l^r) \tag{5.81}$$

in which $r=2$ if $\theta\neq\frac{1}{2}$ and $r=3$ if $\theta=\frac{1}{2}$. $D\mathbf{y}(t)$ and $D\mathbf{y}(t+l)$ are obtained from (5.4) which may be written as

$$D\mathbf{y}(t)\equiv\mathbf{y}'(t)=\mathbf{f}(t,\mathbf{y}(t)),\quad t>t_0;\quad \mathbf{y}(t_0)=\mathbf{y}_0. \tag{5.82}$$

Taking $\theta=0$ give the first-order explicit method

$$\mathbf{y}_{n+1}-\mathbf{y}_n+l\mathbf{f}_n \tag{5.83}$$

while $\theta=\frac{1}{2}$ give the second-order implicit method

$$\mathbf{y}_{n+1} - \tfrac{1}{2}l\mathbf{f}_{n+1} = \mathbf{y}_n + \tfrac{1}{2}l\mathbf{f}_n. \tag{5.84}$$

In equation (5.83) and (5.84), which are the multidimensional equivalents of (5.33) and (5.35), $\mathbf{f}_n = \mathbf{f}(t,\mathbf{y}_n)$ and $\mathbf{f}_{n+1} = \mathbf{f}(t+l,\mathbf{y}_{n+1})$ with $t = nl$. This pair of equations can be used in predictor–corrector combination as the following example illustrates.

Example 5.15 The dynamics of a continously fed biochemical reaction tank, in which a biological organism promotes chemical reaction, may be modelled by the nonlinear differential system

$$P' = \mu(P)P + \beta(P)G \tag{5.85}$$

$$G' = [-\mu(P) + \beta(P) - \alpha]G, \tag{5.86}$$

where $P = P(t)$ represents the total population at time t, $\mu(P)$ is the mortality function, $\beta(P)$ is the fertility function, α is a parameter, and $G(t) = B(t)/\beta(P)$ with $B(t)$ representing the birth rate. Use methods (5.83) and (5.84) in *PECE* mode to obtain the solution at time $=(n+1)l$ $(n = 0,1,2,\ldots)$ assuming that the initial vector $[P(0), G(0)]^{\mathrm{T}}$ is given. (The reader may obtain full details of the model described by (5.85) and (5.86) in the paper by Swart (1986).)

Comparing (5.82) with (5.85) it is clear that $\mathbf{y}(t) = [y_1(t),y_2(t)]^{\mathrm{T}} = [P(t), G(t)]^{\mathrm{T}}$ and $\mathbf{f}(t,\mathbf{y}(t)) = [f_1,f_2]^{\mathrm{T}}$ with

$$f_1(t,P,G) = -\mu(P)P + \beta(P)G, \tag{5.87}$$

$$f_2(t,P,G) = [-\mu(P) + \beta(P) - \alpha]G. \tag{5.88}$$

Using the familiar subscript and superscript notations (section 5.4.8) on P and G, the predictor–corrector combination in *PECE* mode gives

$$n := 0;\ f_{1,n} = -\mu(P_n)P_n + \beta(P_n)G_n,\ f_{2,n} = [-\mu(P_n) + \beta(P_n) - \alpha]G_n;$$

$$\text{P (predicting): } \begin{aligned} P_{n+1}^{[0]} &= P_n + lf_{1,n}, \\ G_{n+1}^{[0]} &= G_n + lf_{2,n}; \end{aligned}$$

$$E:\ \begin{aligned} f_{1,n+1}^{[0]} &= -\mu(P_{n+1}^{[0]})\,P_{n+1}^{[0]} + \beta(P_{n+1}^{[0]})\,G_{n+1}^{[0]}, \\ f_{2,n+1}^{[0]} &= [-\mu(P_{n+1}^{[0]}) + \beta(P_{n+1}^{[0]}) + \beta(P_{n+1}^{[0]}) - \alpha]\,G_{n+1}^{[0]}; \end{aligned}$$

$$\text{C: } \begin{aligned} P_{n+1}^{[1]} &= P_n + \tfrac{1}{2}l(f_{1,n} + f_{1,n+1}^{[0]}), \\ G_{n+1}^{[1]} &= G_n + \tfrac{1}{2}l(f_{2,n} + f_{2,n+1}^{[0]}); \end{aligned}$$

$$E:\ \begin{aligned} f_{1,n+1}^{[1]} &= -\mu(P_{n+1}^{[1]})\,P_{n+1}^{[1]} + \beta(P_{n+1}^{[1]})\,G_{n+1}^{[1]}, \\ f_{2,n+1}^{[1]} &= [-\mu(P_{n+1}^{[1]}) + \beta(P_{n+1}^{[1]}) + \beta(P_{n+1}^{[1]}) - \alpha]\,G_{n+1}^{[1]}; \end{aligned}$$

$$n:=n+1,\quad f_{1,n}:=f_{1,n+1}^{[1]},\quad f_{2,n}:=f_{2,n+1}^{[1]}\,.$$

This predictor–corrector combination in *PECE* mode has a stability restriction, as noted in Example 5.13 of section 5.4.8. No such restriction is encountered when using the fully implicit method obtained by writing $\theta=1$ in (5.69). This method is given by

$$\mathbf{y}_{n+1}-l\mathbf{f}_{n+1}=\mathbf{y}_n \tag{5.89}$$

and, in conjunction with a method such as the Newton–Raphson method for a nonlinear algebraic system, may be used with a large time step to compute the solution at time $t=(n+1)l$, $n=0,1,\ldots$, as in the following example.

Example 5.16 One method of therapy of autoimmune diseases is to use substances capable of suppressing immunity. Smirnova and Stepanova (1975) proposed an optimal control model of autoimmunity which was described by the nonlinear differential system

$$x'=x-x^2-xz \tag{5.90}$$

$$z'=a(xz^2-bxz-cz), \tag{5.91}$$

in which $x=x(t)$ denotes the weighted concentration of cells in a certain body tissue and $z=z(t)$ denotes the weighted concentration of aggressive T-lymphocytes which cause the death of cells in a given tissue. Letting $\mathbf{y}(t)=[x(t),z(t)]^{\mathrm{T}}$ and assuming that the initial vector $\mathbf{y}(0)=[x(0),z(0)]^{\mathrm{T}}=[x_0,z_0]^{\mathrm{T}}$ is given, use (5.89) to find the solution vector $\mathbf{y}_{n+1}(n=0,1,2\ldots)$. (Swan (1984) notes that typical values of a,b,c are $a=1.2$, $b=0.17$, $c=0.1$).

In the notation of this subsection of the chapter

$$f_1(t,x,z)=x-x^2-xz$$
$$f_2(t,x,z)=a(xz^2-bxz-cz)$$

so that, from (5.89) (with $n=0,1,2,\ldots$),

$$x_{n+1}-l(x_{n+1}-x_{n+1}^2-x_{n+1}z_{n+1})=x_n, \tag{5.92}$$

$$z_{n+1}-la(x_{n+1}z_{n+1}^2-bx_{n+1}z_{n+1}-cz_{n+1})=z_n, \tag{5.93}$$

which must be solved simultaneously to find $\mathbf{y}_{n+1}=[x_{n+1},z_{n+1}]^{\mathrm{T}}$.
Writing

$$F_{1,n+1}=x_{n+1}-l(x_{n+1}-x_{n+1}^2-x_{n+1}z_{n+1})-x_n, \tag{5.94}$$

$$F_{2,n+1} = z_{n+1} - la(x_{n+1}z_{n+1}^2 - bx_{n+1}z_{n+1} - cz_{n+1}) - z_n, \tag{5.95}$$

$\mathbf{y}_{n+1}$ is computed by solving the nonlinear algebraic system

$$\mathbf{F}_{n+1} = [F_{1,n+1}, F_{2,n+1}]^{\mathrm{T}} = \mathbf{O}. \tag{5.96}$$

In the notation of Chapter 3

$$J_{11} = \partial F_{1,n+1}/\partial x_{n+1} = 1 - l(1 - 2x_{n-1} - z_{n+1}), \quad J_{12} = \partial F_{1,n+1}/\partial z_{n+1} = lx_{n+1},$$

$$J_{21} = \partial F_{2,n+1}/\partial x_{n+1} = -la(z_{n+1} - bz_{n+1}), \quad J_{22} = \partial F_{2,n+1}/\partial z_{n+1} = 1 - la(2x_{n+1}z_{n+1} - bx_{n+1} - c),$$

giving the Jacobian $J = [J_{rs}]$ $(r,s = 1,2)$. The Newton–Raphson method is given by

$$\mathbf{y}_{n+1}^{[k+1]} = \mathbf{y}_{n+1}^{[k]} - (J^{[k]})^{-1}\mathbf{F}_{n+1}^{[k]}, \tag{5.97}$$

where the superscript $k = 0,1,2,\ldots$ relates to the kth iterate. The vector $\mathbf{y}_{n+1}^{[k+1]}$ in (5.97) is found from

$$\mathbf{y}_{n+1}^{[k+1]} = \mathbf{y}_{n+1}^{[k]} + e^{[k]},$$

where $\mathbf{e}^{[k]}$ is the correction vector found by solving the *linear* system

$$\mathbf{J}^{[k]}\,\mathbf{e}^{[k]} = -\mathbf{F}_{n+1}^{[k]}. \tag{5.98}$$

The iteration procedure ceases when the components x_{n+1}, z_{n+1} have both converged to a specified tolerance.

5.5.5 Stiffness

The mathematical modelling of many biomedical systems yields initial value problems involving differential systems which exhibit a phenomenon known as *stiffness*. Under the assumption that the matrix A has N distinct eigenvalues with negative real parts, it is known that the solutions of (5.62) at time t may be written as

$$\mathbf{y}(t) = \sum_{r=1}^{n} k_r \exp(\lambda_r t)\mathbf{c}_r + \mathbf{y}_{\mathrm{P}}(t), \tag{5.99}$$

where $\mathbf{c}_r$ is the eigenvector corresponding to the eigenvalue $\lambda_r (r = 1,2,\ldots,N)$, the k_r are constants and $\mathbf{y}_{\mathrm{P}}(t)$ is a particular solution of (5.62).

Since $\mathrm{Re}(\lambda_r) < 0$, the term

$$\sum_{r=1}^{N} k_r \exp(\lambda_r t)\mathbf{c}_r \to \mathbf{0} \text{ as } t \to \infty \ ;$$

this term is called the *transient solution* while $\mathbf{y}_P(t)$ is the *steady-state solution*, so called because it is the limiting solution vector as $t \to \infty$. When the (negative) real parts of the eigenvalues are widely varying in magnitude, some components of the transient solution tend to zero much more rapidly than others and attempts to use some of the familiar numerical methods to solve such 'stiff' problems encounter substantial difficulties.

The phenomenon of stiffness is formalized in the following definitions.

Definition 5.16 The linear system $\mathbf{y}'(t) = A(t)\mathbf{y}(t) + \boldsymbol{\psi}(t)$ is said to be *stiff* in the interval $t_0 \leqslant t \leqslant T$ if, in this interval, (i) $\text{Re}(\lambda_r(t)) < 0$, $r = 1,2,\ldots,N$, and (ii) $\max_r|\text{Re}(\lambda_r(t))| \gg \min_r|\text{Re}(\lambda_r(t))|$ where $\lambda_r(t)$, $r = 1,2,\ldots,N$, are the eigenvalues of $A(t)$.

Definition 5.17 The nonlinear system $\mathbf{y}'(t) = \mathbf{f}(t,\mathbf{y}(t))$ is said to be stiff in the interval $t_0 \leqslant t \leqslant T$ if, in this interval, (i) $\text{Re}(\lambda_r) < 0$, $r = 1,2,\ldots,N$, and (ii) $\max_r|\text{Re}(\lambda_r)| \gg \min_r|\text{Re}(\lambda_r)|$ where, now, $\lambda_r (r = 1,2,\ldots,N)$ are the eignevalues of the Jacobian $\partial\mathbf{f}/\partial\mathbf{y}$.

Definition 5.18 The ratio

$$\max_r|\text{Re}(\lambda_r)| : \min_r|\text{Re}(\lambda_r)|, \ r = 1,2,\ldots,N,$$

is called the *stiffness ratio* of the differential system.

Definition 5.18 applies to linear and nonlinear systems, $\lambda_r (r = 1,2,\ldots,N)$ having the meanings assigned in Definitions 5.16 and 5.17 respectively.

5.5.6 A further note on stability

Consider the model single initial value problem

$$y'(t) = \lambda y(t), \ -\infty < \lambda < 0, \ t > t_0; \ y(t_0) = y_0 \tag{5.100}$$

and the Euler corrector formula $y_{n+1} = y_n + \frac{1}{2}l(f_n + f_{n+1})$. Applied to (5.100), this formula becomes $y_{n+1} - \frac{1}{2}l\lambda y_{n+1} = y_n + \frac{1}{2}l\lambda y_n$, which gives

$$y_{n+1} = \left(\frac{1+\frac{1}{2}l\lambda}{1-\frac{1}{2}l\lambda}\right) y_n = \left(\frac{1+\frac{1}{2}\bar{l}}{1-\frac{1}{2}\bar{l}}\right) y_n \tag{5.101}$$

where $\bar{l} = l\lambda$. The theoretical solution of (5.100) is $y(t) = e^{\lambda(t-t_0)}y_0$ which decays to zero as $t \to \infty$ since $\lambda < 0$ and $l > 0$. Therefore, the quantity $(1+\frac{1}{2}\bar{l})/(1-\frac{1}{2}\bar{l})$, which is known as the *amplification factor* or *amplfication symbol* or *symbol* of this numerical method, must not exceed unity in modulus. This condition is met by the amplification factor in (5.101) since $\bar{l} \in (-\infty, 0)$ and the reader will note that this condition placed on the amplification factor coincides with the requirement for A_0-stability in section 5.4.6.

Writing the amplification symbol as $[(\bar{l})^{-1}+\frac{1}{2}]/[(\bar{l})^{-1}-\frac{1}{2}]$, it is easy to see that it tends to -1 as $\bar{l} \to -\infty$ so that the sequence $y_0, y_1, y_2, \ldots$ generated by the Euler corrector formula does not decay to zero sufficiently quickly to model accurately the theoretical solution of (5.100) which decays to zero rapidly for large negative values of λ.

In the case of (5.100) and, therefore, in the case of stiff systems, a numerical method should be used, the symbol of which never exceeds unity in modulus (the method is therefore A_0-stable or A-stable), and which tends to zero as $\bar{l} \to -\infty$. Such methods are classified as follows:

Definition 5.19 A numerical method is said to be *L_0-stable* or *L-stable* if it is A_0-stable or A-stable and if, in addition, its amplification factor tends to zero as $\bar{l} \to -\infty$.

The reader should note that λ is interpreted as $\partial f/\partial y$ in the case of (5.1), as the eigenvalues of $\partial \mathbf{f}/\partial \mathbf{y}$ in the case of (5.4), as the eigenvalues of $A(t)$ in (5.8) and as the eigenvalues of A in (5.62). The concepts of A_0-stability and L_0-stability have important applications in the numerical solution of second-order parabolic partial differential equations which are discussed in Chapter 8.

Example 5.17 Show that the fully implicit method $y_{n+1} = y_n + lf_{n+1}$ is L_0-stable. Application of this method to the model equation (5.100) gives $y_{n+1} - l\lambda y_{n+1} = y_n$ so that the amplification factor is $1/(1-\bar{l})$. It is easy to show that the magnitude of the amplification symbol never exceeds unity; the method is thus A_0-stable. Furthermore the amplification factor tends to zero as $\bar{l} = l\lambda \to -\infty$ and the method is L_0-stable.

A further characteristic of A_0-stable numerical methods is that they may induce oscillations in the computed solution when these are not present in the theoretical solution of the initial value problem. Such oscillations occur when a time step is used which makes the amplification symbol negative, and they become increasingly accentuated as the time step is increased.

Example 5.18 Tabulate the errors in the solutions generated by the second order

A_0-stable method $y_{n+1} = y_n + \frac{1}{2}l(y'_n + y'_{n+1})$ and by the first order L_0-stable method $y_{n+1} = y_n + ly'_{n+1}$,, in solving the initial value problem

$$y'(t) = -400.01\ y(t),\ t > 0;\ y(0) = 1$$

for $t = 0.0(0.05)0.4$ using a time step of $l = 0.005$ (see Example 5.6).

The errors are given in Table 5.3 where it is seen that both methods generate

Table 5.3

t	Theoretical solution	Errors: L_0-stable method	Errors: A_0-stable method
0.00	0.01E + 01	0	0
0.04	0.11E − 06	− 0.76E − 03	+ 0.65E − 03
0.08	0.13E − 13	− 0.59E − 06	− 0.43E − 06
0.12	0.14E − 20	− 0.49E − 09	+ 0.28E − 09
0.16	0.16E − 27	− 0.34E − 13	− 0.18E − 12
0.20	0.18E − 34	− 0.26E − 15	+ 0.12E − 15
0.24	0.00E + 00	− 0.20E − 18	− 0.79E − 19
0.28	0.00E + 00	− 0.15E − 21	+ 0.52E − 22
0.32	0.00E + 00	− 0.12E − 24	− 0.34E − 25
0.36	0.00E + 00	− 0.90E − 29	+ 0.22E − 28
0.40	0.00E + 00	− 0.69E − 31	− 0.14E − 31

solution sequences with decaying errors, the errors of the second order method being smaller in magnitude than those of the first order method. The errors of the L_0-stable method are decaying monotonically whilst those of the A_0-stable method, though decaying, are alternating in sign (there is an even number of time steps (ten between each tabulated value of t in Table 5.3). The solution curve relating to the L_0-stable method will thus resemble quite well the shape of the solution curve of the initial value problem. On the other hand, the solution curve relating to the A_0-stable method will oscillate about the curve of the theoretical solution, the oscillations being gradually damped as t increases. The oscillations are induced because the symbol of the A_0-stable method has a small negative value ($-0.000\,012\,4$, approximately) for $l = 0.005$, whereas the corresponding symbol for the L_0-stable method is $+0.333\,327\,7$ (approximately). This example shows that an L_0-stable (or L-stable) numerical method should be used to solve an initial value problem with a rapidly decaying solution.

5.6 LOCAL EXTRAPOLATION PROCEDURES

The local extrapolation of one-step methods has enjoyed its greatest popularity in the numerical solution of first-order hyperbolic and second order parabolic partial differential equations which are treated in Chapters 7 and 8 respectively.

The procedure called *local extrapolation* involves, for instance, integrating from time $t_n = t_0 + nl$ to t_{n+1} and then from t_{n+1} to t_{n+2} to give a first approximation $y^{(1)}_{n+2}$ to $y(t_{n+2})$: two single time steps of size l have been used in this part of the procedure. Next a single integration from time t_n to time t_{n+2} is carried out using a double time step $2l$ to obtain a second approximation $y^{(2)}_{n+2}$ to $y(t_{n+2})$. The local extrapolation ends by finding a linear combination $y^{(E)} = \alpha y^{(2)} + (1-\alpha)y^{(1)}$, where α is some parameter, which increases the order of the numerical method by at least one power of l. The reader can show as an exercise that this local extrapolation procedure with $\alpha = 2$ increases the orders of the first-order methods $y_{n+1} = y_n + lf_n$ and $y_{n+1} = y_n + lf_{n+1}$ to two and, with $\alpha = \frac{4}{3}$, the order of the second-order method $y_{n+1} = y_n + \frac{1}{2}l(f_n + f_{n+1})$ to four.

Other local extrapolation methods increase the orders of such low-order methods even higher (see, in the case of parabolic equations for instance, Gourlay and Morris (1980)). However, local extrapolation procedures often interfere with the stability properties of the method. The procedure described above with $\alpha = \frac{4}{3}$, for instance, imposes the stability restriction $\bar{l} \in (-6 - 4\sqrt{3}, 0)$ on the Euler corrector formula $y_{n+1} = y_n + \frac{1}{2}l(f_n + f_{n+1})$ which, before extrapolation, is unconditionally stable. On the other hand, the L-stable fully implicit method $y_{n+1} = y_n + \frac{1}{2}lf_{n+1}$ remains L-stable following local extrapolation over $[t_n, t_{n+2}]$ with $\alpha = 2$. The stability properties of all such one-step methods are left invariant by global extrapolation, which is described in the following section and which carries greater recommendation.

5.7 GLOBAL EXTRAPOLATION PROCEDURES

5.7.1 Global extrapolation on two grids

Consider again the first-order initial value problem (5.1) and suppose that the solution is sought at time $t = T < \infty$. The interval of integration is divided, first of all, into Q subintervals each of width l so that $Ql = T - t_0$, giving a discretization or time grid G_1 consisting of the $Q+1$ points $t_{n,1} = t_0 + n\,l$ $(n = 0, 1, \ldots, Q)$. The theoretical solution of (5.1) at $t = t_{n,1}$ is clearly $y(t_{n,1})$ and the notation $y_{n,1}$ will be used to denote the solution of an approximating method at the same point $t_{n,1}$ of G_1 $(n = 0, 1, \ldots, Q)$.

The application of a convergent numerical method M to find the solution results in a global error at the point $T = t_{Q,1}$ which was seen in section 5.4.4 to be $O(l^p)$, where p is the order of M as defined in Definition 5.5. Suppose that this error is denoted by $e_{Q,1}$, then $e_{Q,1}$ has the form

$$e_{Q,1} = l^p A_0 + l^{p+1} A_1 + l^{p+2} A_2 + l^{p+3} A_3 + \ldots \tag{5.102}$$

where the $A_i (i = 0, 1, 2, \ldots)$ are independent of l.

Suppose now that the interval of integration is divided into $2Q$ subintervals, each of width $\frac{1}{2}l$ giving a grid G_2 consiting of the $2Q+1$ points $t_{i,2} = t_0 + \frac{1}{2}il$ $(i = 0, 1, 2, \ldots, 2Q)$. Clearly the points $t_{r,2}$ $(r = 0, 2, 4, \ldots, 2Q)$ of G_2 are coincident with the points $t_{n,1}=t_0+nl$ ($n=0,1,\ldots,Q$) of G_1. The notation $y_{i,2}$ will be used to denote the solution of (5.1), determined using the same numerical method M, at the point $t_{i,2}$ ($i=0,1,\ldots,2Q$) of grid G_2. This time, the global error, $e_{2Q,2}$, at the point $T=t_{2Q,2}$ is given by

$$e_{2Q,2}=2^{-p}l^pA_0+2^{-p-1}l^{p+1}A_1+2^{-p-2}l^{p+2}A_2+\ldots\,. \tag{5.103}$$

Consider, now, the combined error function

$$e^{(\mathrm{E})}=\alpha e_{2Q,2}+(1-\alpha)e_{Q,1}\,. \tag{5.104}$$

Then

$$e^{(\mathrm{E})}=(\alpha 2^{-p}+1-\alpha)l^pA_0+(\alpha 2^{-p-1}+1-\alpha)l^{p+1}A_1+\ldots\,, \tag{5.105}$$

and it is clear that the term in l^p in (5.105) vanishes provided

$$\alpha=2^p/(2^p-1) \tag{5.106}$$

that is, $e^{(\mathrm{E})}=O(l^{p+1})$. Recalling, now, that $e_{Q,1}=y(t_{Q,1})-y_{Q,1}=y(T)-y_{Q,1}$ and therefore that $e_{2Q,2}=y(T)-y_{2Q,2}$, it follows from (5.104) that

$$\begin{aligned}e^{(\mathrm{E})}&=\alpha e_{2Q,2}+(1-\alpha)e_{Q,1}\\&=\alpha[y(T)-y_{2Q,2}]+(1-\alpha)[y(T)-y_{Q,1}]\\&=y(T)-[\alpha y_{2Q,2}+(1-\alpha)y_{Q,1}]\end{aligned}$$

and, thus, that the linear combination

$$y_2^{(\mathrm{E})}=\alpha y_{2Q,2}+(1-\alpha)y_{Q,1} \tag{5.107}$$

is an approximation to $y(T)$ which is of order $p+1$ where p is the order of M, the numerical method being used, provided α takes the value given by (5.106).

5.7.2 Global extrapolation on three grids

Suppose, now, that the interval of integration is divided into $3Q$ subintervals each of width $\frac{1}{3}l$ giving a discretization G_3 consisting of the $3Q+1$ points $t_{j,3}=t_0+\frac{1}{3}jl$ ($j=0,1,\ldots,3Q$). It is clear that the points $t_{s,3}$ ($s=0,3,6,\ldots,3Q$) of G_3 are coincident with the points $t_{n,1}$ ($n=0,1,\ldots,Q$) of the original grid G_1. The notation $y_{j,3}$ will be used to denote the solution of (5.1) obtained using the numerical method M at the points $t_{j,3}$ ($j=0,1,\ldots,3Q$) of G_3.

The application of M to find the solution at the point $T=t_{3Q,3}$ of G_3 gives a third approximation to $y(T)$ and generates the global error

$$e_{3Q,3}=3^{-p}l^pA_0+3^{-p-1}l^{p+1}A_1+3^{-p-2}l^{p+2}A_2+\ldots \tag{5.108}$$

which, like $e_{Q,1}$ and $e_{2Q,2}$, is $O(l^p)$, so that the third approximation to $y(T)$, given by $y_{3Q,3}$, is also of order p.

Consider the combined error function

$$e^{(E)} = \beta e_{3Q,3} + \gamma e_{2Q,2} + (1 - \beta - \gamma)e_{Q,1}, \tag{5.109}$$

it is easy to show that

$$\begin{aligned} e^{(E)} = {} & (\beta 3^{-p} + \gamma 2^{-p} + 1 - \beta - \gamma)l^p A_0 + (\beta 3^{-p-1} + \gamma 2^{-p-1} + 1 - \beta - \gamma)l^{p+1} A_1 \\ & + (\beta 3^{-p-2} + \gamma 2^{-p-2} + 1 - \beta - \gamma)l^{p+2} A_2 + \ldots \end{aligned} \tag{5.110}$$

and, further, that the terms in l^p and l^{p+1} vanish provided

$$\beta = 3^{p+1}/(1 + 3^{p+1} - 2^{p+2}) \quad \text{and} \quad \gamma = -2^{p+2}/(1 + 3^{p+1} - 2^{p+2}) \tag{5.111}$$

so that $1 - \beta - \gamma = 1/(1 + 3^{p+1} - 2^{p+2})$.

Noting that $e_{3Q,3} = y(T) - y_{3Q,3}$ it follows from (5.108) that

$$\begin{aligned} e^{(E)} &= \beta e_{3Q,3} + \gamma e_{2Q,2} + (1 - \beta - \gamma)e_{Q,1} \\ &= \beta[y(T) - y_{3Q,3}] + \gamma[y(T) - y_{2Q,2}] + (1 - \beta - \gamma)[y(T) - y_{Q,1}] \\ &= y(T) - [\beta y_{3Q,3} + \gamma y_{2Q,2} + (1 - \beta - \gamma)y_{Q,1}] \end{aligned}$$

and, therefore, that the linear combination

$$y_3^{(E)} = \beta y_{3Q,3} + \gamma y_{2Q,2} + (1 - \beta - \gamma)y_{Q,1} \tag{5.112}$$

is an approximation to $y(T)$ which is of order $p + 2$, where p is the order of M, provided β and γ take the values given by (5.111).

Global extrapolation procedures are easy to implement. Having written the code to generate $\{y_{i,1}\}$, $i = 1,2, \ldots ,Q$, it is a trivial task to generate, then, the sequence $y_{j,2}$, $j = 1,2, \ldots ,2Q$, and to compute $y_2^{(E)}$ using (5.107). Following this, the sequence $\{y_{S,3}\}$, $s = 1,2, \ldots ,3Q$, is equally easy to compute followed by the determination of $y_3^{(E)}$ from (5.112). It will be clear to the reader that even higher orders may be obtained by using more, even finer, grids in the same way.

PROBLEMS

1. Approximate the solution of the ordinary differential equation

$$\frac{dy}{dt} = y + \frac{t}{y}, \; y(1) = 1$$

at $t_1 = 1.1$ and $t_{-1} = 0.9$ using the third order Taylor's series method.

2. Approximate the solution of the ordinary differential equation

$$\frac{dy}{dt} = \frac{t^2}{y},\ y(1) = 1$$

at $t_1 = 1.1$ and $t_{-1} = 0.9$ using the third order Taylor's series method.

3. Approximate the solution of the ordinary differential equation

$$\frac{dy}{dt} = t^2 - y^2;\ y(1) = 0$$

at $t_1 = 1.1$ and $t_{-1} = 0.9$ using the fourth order Taylor's series method.

4. Given the first order initial value problem

$$y' \equiv f(t,y)) = t + y;\ t > 0,\ y(0) = 1$$

(i) use the fourth order Runge–Kutta method (5.28) with $l = 0.1$ to find $y(0.1)$ to four decimal places (work to five decimal places);

(ii) given that $y(-0.1) = 0.9997$, use the initial condition, the solution to (i) and the formulas

$$\text{P: } y_{n+2} = y_n + 2l\,(\tfrac{7}{6}y'_{n+1} - \tfrac{1}{3}y'_n + \tfrac{1}{6}y'_{n-1})$$

$$\text{C: } \tfrac{11}{6}y_{n+2} = 3y_{n+1} - \tfrac{3}{2}y_n + \tfrac{1}{3}y_{n-1} + ly'_{n+2}$$

in *PEC* mode with $l = 0.1$ to find $y(0.2)$ to three decimal places (work to four decimal places);

(iii) find the order of the corrector formula C and give the principal part of its local truncation error.

5. Given the linear multistep method

$$\text{(A)} \qquad y_{n+2} = y_{n+1} + \tfrac{1}{2}l(3y'_{n+1} - y'_n)$$

for solving the first order initial value problem

$$\text{(B)} \qquad y' = f(x,y);\ t > t_0,\ y(t_0) = y_0\ ,$$

find the error constant of the method.

Define what is meant by the *order* and *consistency* of a linear multistep method. Deduce the order of method (A) and deduce further that it is consistent with (B). Write down the first and second characteristic polynomials of (A) and deduce that it is zero-stable. Find the range of values of $\bar{l}$ for which (A) is absolutely stable.

Given the initial value problem

$$y' = t - \tfrac{2}{15} y^2;\ t>0,\ y(0)=1$$

(i) use Taylor's expansion with $l=0.1$ to find $y(0.1)$ correct to three decimal places;
(ii) use method (A) above with $l=0.1$ to find $y(0.2)$. Take care not to give too many decimal places in the answer!

6. Given the first order initial value problem

$$y' \equiv f(t,y) = t - \frac{2y^2}{15},\ y(0)=1,$$

(i) use the third order method (5.25) with $l=0.1$ to find $y(0.1)$ to three decimal places (work to four decimal places);
(ii) use the initial condition and the Taylor series with $l=0.2$ to find $y(0.2)$ to three decimal places;
(iii) use the solution to (ii) and the formulas

$$\text{P: } y_{n+1}^{(P)} = y_n + l y_n' + 0(l^2)$$

$$E\text{: } y_{n+1}' = f(t_{n+1}, y_{n+1}^{(P)})$$

$$\text{C: } y_{n+1} = y_n + l y_{n+1}' + 0(l^2)$$

in *PEC* mode with $l=0.1$ to find $y(0.3)$. Take care not to quote this answer to too many decimal places.

7. Using the fourth order Taylor's series method calculate the solutions of the ordinary differential equation

$$y' = t^2 - y^2;\ y(0)=1,$$

at the points $t=-0.1$, $t=0.1$ and $t=0.2$.

Use the following pair of equations to extend the solution to the point $t=0.3$:

$$y_{n+1} = y_{n-3} + \frac{4l}{3}[2y_{n-2}' - y_{n-1}' + 2y_n']$$

$$y_{n+1} = y_{n-1} + \frac{l}{3}[y_{n-1}' + 4y_n' + y_{n+1}'] \ .$$

8. Given the first order initial value problem

$$y' \equiv f(t,y) = t - \frac{2y^2}{15} \text{ ; } t>0,\ y(0)=1$$

(i) use the third order Runge–Kutta method (5.26) with $l=0.1$ to find $y(0.1)$ to three decimal places;
(ii) use the initial condition and the Taylor series with $l=0.2$ to find $y(0.2)$ to three decimal places;
(iii) use the initial condition, the solutions to (i) and (ii) and the formulas

$$\text{P: } y_{n+3}^{(0)} - y_{n+1} = 2l\ (\tfrac{7}{6}f_{n+2} - \tfrac{1}{3}f_{n+1} + \tfrac{1}{6}f_n)$$

$$C\text{: } \tfrac{11}{6}y_{n+3}^{(1)} - 3y_{n+2} + \tfrac{2}{3}y_{n+1} - \tfrac{1}{3}y_n = lf_{n+3}$$

in PEC mode with $l=0.1$, to find $y(0.3)$ to three decimal places.

Find the error constants C_{p+1}^* and C_{p+1}, respectively, of the predictor and corrector formulas, and use Milne's expression $T = \frac{C_{p+1}}{C_{p+1}^* - C_{p+1}}\,[y_{n+3}^{(1)} - y_{n+3}^{(0)}]$ to estimate the principal local truncation error T of the predictor–corrector combination.

9. The linear multistep method

$$y_{n+2} = y_{n+1} + \tfrac{1}{2}l(3y'_{n+1} - y'_n)$$

is suitable for use as a predictor in the numerical solution of the first order initial value problem

$$y' = f(x,y);\ t>t_0,\ y(t_0) = y_0\ .$$

Use the boundary locus method to find the stability region of the method, computing the boundary points for $\theta = 0°\ (45°)\ 180°$ and verifying that the region is symmetric about the real axis.

10. The linear multistep method

$$y_{n+2} = y_{n+1} + \frac{l}{12}\ (5f_{n+2} + 8f_{n+1} - f_n)$$

is to be used to solve the initial value problem

$$y' = f(t,y);\ t>t_0;\ y(t_0) = y_0.$$

Find the principal part of the local truncation error of the method. Show that the method is consistent, zero-stable and convergent.

Use the boundary locus method to show that the stability region of the method may be determined by plotting

$$\text{(A)} \quad \bar{l}(\theta) = -\frac{6(3-4\cos\theta+\cos 2\theta)}{45+32\cos\theta-5\cos 2\theta} + \mathrm{i}\,\frac{6(14\sin\theta-\sin 2\theta)}{45+32\cos\theta-5\cos 2\theta}$$

in the complex $\bar{l}$ plane. (Do not plot the boundary.)

Deduce the interval of absolute stability of the method from the expression for $\bar{l}$ given by (A).

11. Find the initial stiffness ratio of the following initial value problem which arises following the mathematical modelling of the thermal decomposition of ozone:

$$y_1' = -y_1 - y_1 y_2 + 294 y_2;\ t>0,\ y_1(0)=1,$$

(B)

$$y_2' = \frac{y_1}{98} - \frac{y_1 y_2}{98} - 3y_2;\ t>0,\ y_2(0)=0.$$

Use the boundary locus method to show that the stability region of the linear multistep method

$$\text{(C)} \qquad y_{n+2} - \tfrac{3}{2}y_{n+1} + \tfrac{1}{2}y_n = \tfrac{1}{4}l(5y_{n+1}' - 3y_n'),$$

in which l is the increment in t, for solving a single equation of the form (A) may be determined by plotting

$$\text{(D)} \qquad \bar{l}(\theta) = \frac{24\cos\theta - 6\cos 2\theta - 18}{17–15\cos\theta} + \mathrm{i}\,\frac{14\sin\theta - 6\sin 2\theta}{17–15\cos\theta},$$

in the complex $\bar{l}$ plane. (Do not plot the boundary.)

Deduce the interval of absolute stability from the expression for $\bar{l}$ given by (D) and find the maximum initial steplength which may be used to solve (B) using (C), giving the answer to three decimal places.

12. The problem of predicting the populations of two species $y_1(t)$ and $y_2(t)$, that compete for the same food supply, at time t, may be modelled mathematically by the non-linear system

$$y_1' = y_1(4 - 0.0003y_1 - 0.0004y_2),\ y_1(0) = 10\,000\,,$$

$$y_2' = y_2(2 - 0.0002y_1 - 0.0001y_2),\ y_2(0) = 10\,000\,.$$

Use the Euler-modified Euler predictor–corrector combination with a step of 0.1

to find, to the nearest thousand, the populations of the two species at time $t = 0.1$ (correct to convergence).

The mathematical criteria that must be satisfied in order for the populations to be at equilibrium are that, simultaneously,

$$y_1'(t) = 0 \text{ and } y_2'(t) = 0.$$

Use one application of the Newton–Raphson method (Chapter 3) for a system of non-linear algebraic equations to find, to the nearest hundred, estimates $y_1^{(1)}$ and $y_2^{(1)}$ of the equilibrium populations taking as starting values

$$y_1^{(0)} = 8000 \text{ and } y_2^{(0)} = 5000.$$

Use an elementary direct method to find the values to which successive iterates will converge.

13. One method of therapy of autoimmune diseases is to use substances capable of suppressing immunity. This may be modelled mathematically by the nonlinear initial value problem

$$x' = x - x^2 - xz;\ t > 0,\ x(0) = 0.9\ ,$$

$$z' = 1.2(xz^2 - 0.17xz - 0.1z);\ t > 0,\ z(0) = 0.1\ ,$$

in which $x = x(t)$ denotes the weighted concentration of cells in a certain body tissue and $z = z(t)$ denotes the weighted concentration of aggressive T-lymphoctyes which cause the death of cells in the tissue.

Use the predictor–corrector combination

$$\text{P: } \mathbf{y}(t + h) = \mathbf{y}(t) + l\mathbf{y}'(\mathrm{t}) + 0(l^2),$$

$$\text{C: } \mathbf{y}(t + h) = \mathbf{y}(t) + \tfrac{1}{2}l[\mathbf{y}'(t) + \mathbf{y}'(t + l)] + 0(l^3),$$

where $\mathbf{y}(t) = [x(t), z(t)]^{\mathrm{T}}$, T denoting transpose, with a step of $l = 0.1$, to estimate to three decimal places,the two concentrations at time $t = 0.1$.

The mathematical criteria that must be satisfied in order that the concentrations should be critical, are that $x'(t)$ and $z'(t)$ vanish simultaneously. Use one application of the Newton–Raphson method for a system of non-linear algebraic equations to find, to three decimal places, estimates $x^{(1)}$ and $z^{(1)}$ of the critical concentrations taking as starting values

$$x^{(0)} = 0.150 \text{ and } z^{(0)} = 0.850.$$

14. Given the initial value problem

$$y' = f(t,y);\ t > t_0,\ y(t_0) = y_0$$

find expressions for y'', y''' in terms of f and its derivatives and hence determine the third order Taylor method for finding y_{n+1} in terms of y_n and f and its derivatives, where $y_n \equiv y(nl)$ for $n = 0,1,2, \ldots$ and l is an increment in t.

Given also the Runge–Kutta method (5.23) show, by finding an expansion for k_1 in terms of f and its derivatives, that the Runge–Kutta method, when applied to the test equation $y' = \lambda y$ (λ some constant), gives

$$\frac{y_{n+1}}{y_n} = \exp(l\lambda) + 0(l\lambda)^3 .$$

Use this Runge–Kutta method with $l = 0.1$ to find the solution of the initial value problem

$$y' = \frac{1}{t+y}\ ;\ t > 0;\ y(0) = 2$$

when $t = 0.1$.

Use the initial condition and the third order Taylor method with $l = 0.1$ to find another approximation to the solution when $t = 0.1$. Explain the closeness of the two answers.

15. Show that the initial stiffness ratio of the initial value problem

$$(B) \quad \begin{aligned} y_1' &= 0.2(y_2 - y_1) && ;\ t > 0,\ y_1(0) = 0,\\ y_2' &= 10y_1 - (60 - 0.125y_3)y_3 + 0.125y_3 && ;\ t > 0,\ y_2(0) = 0,\\ y_3' &= 1 && ;\ t > 0,\ y_3(0) = 0, \end{aligned}$$

which arises in nuclear reactor theory, is infinite.

The linear one-step methods

$$\begin{aligned} &\text{P: } y_{n+1} = y_n + l y_n' + 0(l^2)\ ,\\ &\text{C: } y_{n+1} = y_n + l y_{n+1}' + 0(l^2), \end{aligned}$$

where l is the increment in t, are to be used in PECE mode to solve problems of the form (B). Determine the interval of absolute stability in *PECE* mode.

Deduce the maximum step length which may be used initially when this predictor–corrector combination is used to solve (B), giving the answer to 5 decimal places.

16. Show that the initial stiffness ratio of the initial value problem

$$(A) \quad \begin{aligned} y_1' &= -0.04y_1 + 0.01y_2y_3) && ;\ t > 0,\ y_1(0) = 1,\\ y_2' &= 400y_1 - 100y_2y_3 - 3000y_2^2 && ;\ t > 0,\ y_2(0) = 0, \end{aligned}$$

$$y_3' = 30y_2^2 \qquad ; \; t>0,\; y_3(0)=0,$$

in which y_1,y_2 and y_3 are functions of t, is infinite.

The predictor–corrector combination

$$\begin{aligned} &P\colon y_{n+1} = y_n + \alpha l y_n',\; \alpha > 0\,,\\ &C\colon y_{n+1} = y_n + (1-\beta) l y_n' + \beta l y_{n+1}',\; 0 \leqslant \beta \leqslant 1, \end{aligned}$$

where l is the increment in t, is to be used in *PECE* mode to solve initial value problems of the form (A). Find the values of the parameters α and β which maximize the individual orders of the predictor and of the corrector, and determine the interval of absolute stability in PECE mode associated with these parameter-values.

Find the maximum initial steplength which may be used when this predictor-corrector combination (with the values of α and β just determined) is used in PECE mode to solve system (A).

17. Certain aspects of the dynamics of a continuously fed biochemical reaction tank, in which a biological organism promotes chemical reaction, may be analysed with recourse to the initial value problem

$$y'(t) = \lambda y(t),\; t > t_0\; y(t_0) = y_0,$$

where λ is a negative real constant. Show that the solution $y(t)$ satisfies the recurrence relation

$$\text{(A)} \qquad y(t+l) = \exp(l\lambda) y(t)$$

where $l > 0$ is an increment in t.

The rational approximation $R_a(l\lambda) = (1 - al\lambda)^{-1}[1 + (1-a)l\lambda]$ to the exponential function, in which $0 < a \leqslant 1$, is to be used in (A). Show that this leads to an implicit one-step method, involving the parameter a, for solving the initial value problem.

$$\text{(B)} \qquad \mathbf{y}'(t) = \mathbf{f}(t,\mathbf{y}),\; t > t_0,\; \mathbf{y}(t_0) = \mathbf{y}_0\; ;\; \mathbf{y},\mathbf{y}',\mathbf{f},\; \mathbf{y}_0 \in E^2,$$

where the components of $\mathbf{y}$ represent the total population, and the ratio of the birth rate to a fertility function, of the organism in the tank. Find the values of a for which the method is

(i) A_0-stable,
(ii) L_0-stable.

18. The linear multistep method

$$\text{(A)} \qquad y_{n+1} = y_n + \tfrac{1}{4} l (3f_n + f_{n+1});\; n = 0,1,2,\ldots$$

with $y_n = y(t_n)$ and $f_n = f(t_n, y_n)$, may be used to solve the general initial value problem

(B) $\quad y'(t) = f(t,y),\ t > t_0;\ y(t_0) = y_0$.

In (A), $l > 0$ is an increment in t and $t_n = t_0 + nl$.

(i) Determine the order and local truncation error of (A) and *hence* show that it is consistent with (B).

(ii) Write down
  (a) the first characteristic polynomial $\rho(r)$ of (A),
  (b) the second characteristic polynomial $\sigma(r)$ of (A),
  (c) the stability polynomial $\Pi(r, \bar{l})$, where $\bar{l} = l\lambda$ with $\lambda = \partial f/\partial y < 0$.

(iii) Find the range of values of $\bar{l}$ for which (A) is absolutely stable.

(iv) find $l_{\max}$, the maximum value of l which may be used with (A) to solve

$$y'(t) = -40y + 100t,\ t > 0,\ y(0) = 1.$$

(v) Describe what happens when increments l_1, l_2, l_3, which satisfy
  (a) $l_1 < l_{\max}$,
  (b) $l_2 = 1.000\,001\ l_{\max}$,
  (c) $l_3 \gg l_{\max}$,
  are used in (A) to solve (B).

19. Given the model first order initial value problem

$$y'(t) = \lambda y(t);\ t > t_0,\ y(t_0) = y_0 \ ,$$

where λ is a complex constant with negative real part, show that the solution satisfies the recurrence relation

$$y(t + l) = \exp(l\lambda)y(t) \ ,$$

where $l > 0$ is an increment in t.

Suppose now that λ is real and negative; the rational approximation $R_\theta(l\lambda) = (1 + \theta l\lambda)/[1 - (1 - \theta)l\lambda]$ to the exponential function is to be used in the recurrence relation, where $0 < \theta < 1$. Show that this leads to the linear one-step method

$$y_{n+1} = y_n + l[\theta y'_n + (1 - \theta)y'_{n+1}] \ ,$$

where $y_n \equiv y(t_n) = y(t_0 + nl)$. Find the local truncation error of this method and determine θ^*, the value of θ which makes the method second order; calculate the error constant when $\theta = \theta^*$. Determine whether the method based on $R_{\theta^*}(l\lambda)$ is L_0-stable or A_0-stable.

20. Define what is meant by the *order* and *consistency* of the linear multistep method

$$\sum_{j=0}^{k} \alpha_j y_{n+j} = l \sum_{j=0}^{k} \beta_j y'_{n+j}$$

for solving the initial value problem

$$y'(t) = f(t,y);\ t > t_0, y(t_0) = y_0.$$

The linear multistep method

$$y_{n+3} = y_{n+1} + 2l(\tfrac{7}{6}y'_{n-2} - \tfrac{1}{3}y'_{n+1} + Ay'_n)$$

is used to solve this problem, where A is some parameter and l is a constant increment in t. Find the value of A if the method is to be consistent and determine the order of the method and its error constant with this value of A.

Determine the stability polynomial of the method with this value of A.

6

Second-order ordinary differential equations

6.1 GENERAL INITIAL VALUE PROBLEMS

Second-order ordinary differential equations arise less frequently following the mathematical modelling of biomedical systems than do first-order differential equation. Nevertheless, this text would be incomplete without reference to second-order systems, and the reader will observe in section 6.2 many similarities between the theories relating to the numerical solutions of first- and second-order initial value problems.

There are four different types of second-order problem to be considered: these are general and special initial value problems, and general and special boundary value problems. The first is summarily dealt with in the next paragraphs of this section of the chapter, the others in greater detail in sections 6.2., 6.3 and 6.4.

The *general nonlinear second-order initial value problem* is given by

$$\mathbf{y}''(t) = \mathbf{f}(t,\mathbf{y}(t),\mathbf{y}'(t)),\ t>t_0; \qquad \mathbf{y}(t_0) = \mathbf{y}_0,\ \mathbf{y}'(t_0) = \mathbf{z}_0 \tag{6.1}$$

where, as in Chapter 5, all vectors have N components. Writing $\mathbf{u}(t)=\mathbf{y}(t)$ and $\mathbf{v}(t)=\mathbf{y}'(t)$, so that $\mathbf{u}'(t)=\mathbf{y}'(t)=\mathbf{v}(t)$ and $\mathbf{v}'(t)=\mathbf{y}''(t)=\mathbf{f}(t,\mathbf{u},\mathbf{v})$, the general nonlinear second-order problem (6.1) may be written as the *first*-order initial value problem

$$\begin{bmatrix}\mathbf{u}'(t)\\ \mathbf{v}'(t)\end{bmatrix} = \begin{bmatrix}0 & I\\ 0 & 0\end{bmatrix}\begin{bmatrix}\mathbf{u}(t)\\ \mathbf{v}(t)\end{bmatrix} + \begin{bmatrix}\mathbf{0}\\ \mathbf{f}(t,\mathbf{u},\mathbf{v})\end{bmatrix},\quad t>t_0\,; \quad \begin{bmatrix}\mathbf{u}(t_0)\\ \mathbf{v}(t_0)\end{bmatrix} = \begin{bmatrix}\mathbf{y}_0\\ \mathbf{z}_0\end{bmatrix} \tag{6.2}$$

where 0 and I are the zero and identity matrices of order N and $\mathbf{0}$ is the zero-vector of order N. Note, the system in (6.2) has $2N$ equations.

The linear form of (6.1) is given by

$$\mathbf{y}''(t) = A(t)\mathbf{y}(t) + B(t)\mathbf{y}'(t) + \mathbf{C}(t),\ t > t_0; \quad \mathbf{y}(t_0) = \mathbf{y}_0, \mathbf{y}'(t_0) = \mathbf{z}_0 \tag{6.3}$$

in which $A(t)$, $B(t)$ are matrices of order N. Making the same substitions for $\mathbf{y}(t)$ and $\mathbf{y}'(t)$, the initial value problem (6.3) becomes

$$\begin{bmatrix} \mathbf{u}'(t) \\ \mathbf{v}'(t) \end{bmatrix} = \begin{bmatrix} 0 & I \\ A(t) & B(t) \end{bmatrix} \begin{bmatrix} \mathbf{u}(t) \\ \mathbf{v}(t) \end{bmatrix} + \begin{bmatrix} \mathbf{0} \\ \mathbf{C}(t) \end{bmatrix}, \quad t > t_0; \quad \begin{bmatrix} \mathbf{u}(t_0) \\ \mathbf{v}(t_0) \end{bmatrix} = \begin{bmatrix} \mathbf{y}_0 \\ \mathbf{z}_0 \end{bmatrix}. \tag{6.4}$$

Writing (6.1) and (6.3) as first-order systems means that the numerical methods discussed in Chapter 5 may be used to compute their solutions.

Example 6.1 The effect of an external oscillatory disturbance on a nonlinear system is typified by the nonhomogeneous van der Pol equation

$$x'' = \varepsilon(1 - x^2)x' - x + E\sin\Omega t, \quad t > 0; \quad x(0) = x_0, \quad x'(0) = z_0 \tag{6.5}$$

in which $x = x(t)$, $\varepsilon > 0$ is a small constant, and E, Ω, x_0, z_0 are constants. Express (6.5) as a first-order initial value problem.

The differential equation in (6.5) is of the form $x''(t) = f(t,x,x')$ where $f(t,x,x') = \varepsilon(1 - x^2)x' - x + E\sin\Omega t$. Writing $u(t) = x(t)$ and $\upsilon(t) = x'(t)$ it follows that $u'(t) = x'(t) = \upsilon(t)$ and $\upsilon'(t) = x''(t) = \varepsilon(1 - u^2)\upsilon - u + E\sin\Omega t$, and that $u(0) = x_0$, $\upsilon(0) = z_0$. The equivalent first-order initial value problem is thus

$$\begin{bmatrix} u'(t) \\ v'(t) \end{bmatrix} = \begin{bmatrix} 0 & 1 \\ 0 & 0 \end{bmatrix} \begin{bmatrix} u(t) \\ v(t) \end{bmatrix} + \begin{bmatrix} 0 \\ \varepsilon(1 - u^2)v - u + E\sin\Omega t \end{bmatrix}, \quad t > 0; \quad \begin{bmatrix} u(0) \\ v(0) \end{bmatrix} = \begin{bmatrix} x_0 \\ z_0 \end{bmatrix}$$

Example 6.2 In the mathematical modelling of the physiological control of blood glucose levels, a function $g(t)$ satisfies the second-order initial value problem

$$g''(t) + 2ag'(t) + \omega^2 g(t) = B\delta(t), \quad t > 0; \quad g(0) = 0, \quad g'(0) = B. \tag{6.6}$$

This is known as the Ackerman model (Ackerman *et al.*, 1965). In (6.6), $g(t) = G(t) - G$ where $G(t)$ denotes the level of blood glucose at time t with G representing its constant fasting value, $a > 0$, ω and B are constants and $\delta(t)$ is the Dirac delta function with dimension $(\text{time})^{-1}$. Express this model as a first-order initial value problem

Writing $u(t) = g(t)$ and $\upsilon(t) = g'(t)$, it follows that $u'(t) = g'(t) = \upsilon(t)$, $\upsilon'(t) = g''(t) = -2ag'(t) - \omega^2 g(t) + B\delta(t) = -2a\upsilon(t) - \omega^2 u(t) + B\delta(t)$ and that $u(0) = g(0) =$

$0, v(0) = g'(0) = B$. The equivalent first-order system is therefore

$$\begin{bmatrix} u'(t) \\ v'(t) \end{bmatrix} = \begin{bmatrix} 0 & 1 \\ -\omega^2 & -2a \end{bmatrix} \begin{bmatrix} u(t) \\ v(t) \end{bmatrix} + \begin{bmatrix} 0 \\ B\delta(t) \end{bmatrix}, \quad t>0\,; \quad \begin{bmatrix} u(0) \\ v(0) \end{bmatrix} = \begin{bmatrix} 0 \\ B \end{bmatrix}$$

The reader should note that the general nonlinear initial value problem of order q given by

$$y^{(q)}(t) = f(t, y(t), y'(t), \ldots, y^{(q-1)}(t)), \quad t > t_0; \quad y^{(r)}(t_0) = z_r, \quad r = 0,1,\ldots,q-1$$

may be transformed into a first-order system by the substitutions $u_1(t) = y(t)$, $u_2(t) = y'(t)$, $u_3(t) = y''(t), \ldots, u_q(t) = y^{(q-1)}(t)$ so that $u_1'(t) = y'(t) = u_2(t)$ $u_2'(t) = y''(t) = u_3(t), \ldots, u_q'(t) = y^{(q)}(t) = f(t, u_1, u_2, \ldots, u_q)$. The initial conditions become $u_1(t_0) = y(t_0) = z_0$, $u_2(t_0) = y'(t_0) = z_1, \ldots, u_q(t_0) = y^{(q-1)}(t_0) = z_{q-1}$ and the equivalent first-order system is

$$\begin{bmatrix} u_1'(t) \\ u_2'(t) \\ \cdots \\ u_{q-1}'(t) \\ u_q'(t) \end{bmatrix} = \begin{bmatrix} 0 & 1 & & & \\ & 0 & 1 & & 0 \\ & & \cdots & & \\ 0 & & & 0 & 1 \\ & & & & 0 \end{bmatrix} \begin{bmatrix} u_1(t) \\ u_2(t) \\ \cdots \\ u_{q-1}(t) \\ u_q(2) \end{bmatrix} + \begin{bmatrix} 0 \\ 0 \\ \cdots \\ 0 \\ f(t, u_1, \ldots, u_q) \end{bmatrix}, \quad t>t_0$$

with initial vector $[u_1(t_0), u_2(t_0), \ldots, u_q(t_0)]^{\mathrm{T}} = [z_0, z_1, \ldots, z_{q-1}]^{\mathrm{T}}$. In the case of a system of N nonlinear initial value problems, $y(t)$ and its first $q-1$ derivatives, f, $y^{(r)}(t_0)$, $z_r (r = 0,1,\ldots,q-1)$ and u_i $(i = 1,2,\ldots,q)$ are all N-vectors.

In a similar way, the general linear initial value problem of order q given by

$$y^{(q)}(t) = a_1(t)y(t) + a_2(t)y'(t) + a_3(t)y''(t) + \ldots + a_q(t)y^{(q-1)}(t) + b(t), \quad t > t_0; \quad y^{(r)}(t_0) = z_r (r = 0,1,\ldots,q-1),$$

is transformed into the equivalent first-order system

$$\begin{bmatrix} u_1'(t) \\ u_2'(t) \\ \cdots \\ u_{q-1}'(t) \\ u_q'(t) \end{bmatrix} = \begin{bmatrix} 0 & 1 & & & & \\ & 0 & 1 & & & \\ & & & \cdots & & \\ & & & & 0 & 1 \\ a_1(t) & a_2(t) & a_3(t) & \ldots & a_{q-1}(t) & a_q(t) \end{bmatrix} \begin{bmatrix} u_1(t) \\ u_2(t) \\ \cdots \\ u_{q-1}(t) \\ u_q(t) \end{bmatrix} + \begin{bmatrix} 0 \\ 0 \\ \cdots \\ 0 \\ b(t) \end{bmatrix}, \quad t>t_0,$$

with initial vector $[u_1(t_0),\ldots,u_q(t_0)]^T=[z_0,z_1,\ldots,z_{q-1}]^T$ by the transformation $u_1(t)=y(t),\ u_2(t)=y'(t),\ldots,u_q(t)=y^{(q-1)}(t)$. In the case of a linear differential system of order N, $y(t)$ and its first $q-1$ derivatives, $z_r(r=0,1,\ldots,q-1)$ and the $u_i(i=2,\ldots,q)$ are all square matrices of order N.

6.2 SPECIAL INITIAL VALUE PROBLEMS

6.2.1 Introduction

Special initial value problems have the form

$$y''(t) = f(t,y(t)), \quad t > t_0; \quad y(t_0) = y_0, \quad y'(t_0) = z_0 . \tag{6.7}$$

They arise frequently in mechanics, particularly orbital mechanics, though a few examples have been found in the biomedical sciences. One such case is in the modelling of endocrine and metabolic systems, a particular example being given in a study by Stear (1975) of a hormonal system. The model equation is given in the following example.

Example 6.3 Assuming that the rate of change of plasma thyroxine level is equal to the sum of the rate of disappearance of thyroxine from plasma and the rate of secretion of thyroxine into the plasma by the thyroid gland, the free plasma thyroxine level, $x(t)$, at time t may be shown to satisfy the special initial value problem

$$x''(t)-(a^2+b^2q/r)x(t) = 0, \quad t > t_0, \quad x(t_0) = x_0, \quad x'(t_0) = z_0$$

where $a>0$, $b>0$, q,r are physiological constants with $a^2+b^2q/r<0$.

A second example has been found in the growth dynamics of oncological disease where simple models assume that growth is represented by a single factor such as the volume of the tumour or the number of cancer cells that comprise it, as in the following example.

Example 6.4 The number of viable tumour cells, $L(t)$, at time t may be represented by the special nonlinear initial value problem

$$L''(t) = (\lambda'+\lambda^2)L-(3\mu\lambda+\mu')L^2+2\mu^2L^3, \quad t>t_0; \quad L(t_0) = x_0, \quad L'(t_0) = z_0$$

where $\lambda=\lambda(t)$ and $\mu=\mu(t)$ may be estimated from clinical data.

Treating λ and μ as constants and thus making their estimation much easier, the initial value problem takes the simplified form

$$L''(t) = \lambda^2 L - 3\mu\lambda L^2 + 2\mu^2 L^3, \quad t > t_0; \quad L(t_0) = x_0, \quad L'(t_0) = z_0.$$

For human multiple myeloma, for instance, a bone marrow cancer, the time taken by the tumour to double its volume is about two months, so that for this particular disease, z_0 is known. (Numerical calculations show that, with $x_0 = 1$ cell, the time taken to produce 1 kg of tumour is then about six and a quarter years.)

Taylor and Runge–Kutta methods exist for the solution of (6.7) but they have not enjoyed the same success as their counterparts for the solution of first-order initial value problems. Linear multistep methods, on the other hand, are more widely used and these will be discussed in the following subsection.

6.2.2 Linear multistep methods and their analyses

The general linear k-step method for the solution of (6.7) has the form

$$\sum_{j=0}^{k} \alpha_j y_{n+j} = l^2 \sum_{j=0}^{k} \beta_j f_{n+j} = l^2 \sum_{j=0}^{k} \beta_j y''_{n+j} , \tag{6.8}$$

where the notation introduced in Chapter 5 has been retained. In equation (6.8) which is analogous to equation (5.29) of Chapter 5, α_k is takent to be $+1$ and α_0 and β_0 do not both vanish. The linear multistep method (LMM) given in (6.8) is said to be *explicit* if $\beta_k = 0$ and to be *implicit* otherwise.

With this LMM can be associated the linear difference operator

$$\mathcal{L}[y(t);l] = \sum_{j=0}^{k} [\alpha_j y(t+jl) - l^2 \beta_j y''(t+jl)] \tag{6.9}$$

where $y(t)$ is continuously differentiable on some closed interval $t_0 \leqslant t \leqslant T$. Then as in Chapter 5, $\mathcal{L}[y(t);l]$ takes the form

$$\mathcal{L}[y(t);l] = C_0 y(t) + C_1 l y'(t) + C_2 l^2 y''(t) + C_3 l^3 y'''(t) + \ldots \tag{6.10}$$

where $C_0, C_1, C_2, \ldots$ are constants.

Definition 6.1 The linear multistep method (6.8) is said to be if *order* p if, in (6.10), $C_0 = C_1 = \ldots = C_p = C_{p+1} = 0$ *and* $C_{p+2} \neq 0$.

Definition 6.2 The LMM (6.8) is said to be *consistent* with (6.7) if it has order $p \geqslant 1$.

The reader will note that, for a method of order p, $C_{p+1} = 0$ also, in contrast to the requirement for first-order differential equations. The difference operator associated with an LMM of order $p \geqslant 1$ is given by

$$\mathscr{L}[y(t);l] = C_{p+2}l^{p+2}y^{(p+2)}(t) + C_{p+3}l^{p+3}y^{(p+3)}(t) + \dots \quad (6.11)$$

The expression on thc right-hand side of (6.11) is the *local truncation error* of (6.8). The *principal local truncation error* at $t=t_n$ is given by $C_{p+2}l^{p+2}y^{(p+2)}(t_n)$ and C_{p+2} is called the *error constant* of the linear multistep method (6.8). Since the local truncation error of (6.8) is $O(l^{p+2})$, the global truncation error is $O(l^p)$.

Example 6.5 Find the error constant of the LMM

$$y_n - 2y_{n+1} + y_{n+2} = \tfrac{1}{9}\, l^2(f_n + 7f_{n+1} + f_{n+2}) \ . \quad (6.12)$$

The linear difference operator associated with (6.12) is

$$\mathscr{L}[y(t);l] = y(t) - 2y(t+l) + y(t+2l) - \tfrac{1}{9}l^2[y''(t) + 7y''(t+l) + y''(t+2l)] \ .$$

Using Taylor's series, the reader can easily show that

$$\mathscr{L}[y(t);l] = -\tfrac{1}{36}\, l^4 y^{(\text{iv})}(t) + \text{higher powers of } l$$

and the error constant of this second-order LMM is $C_4 = -\frac{1}{36}$.

The first and *second characteristic polynomials* of the linear multistep method (6.8) are defined by

$$\rho(r) = \sum_{j=0}^{k} \alpha_j r^j \text{ and } \sigma(r) = \sum_{j=0}^{k} \beta_j r^j \quad (6.13)$$

respectively, an exact comparison with Chapter 5. Theorem 5.3, however, does not apply and is replaced by the following.

Theorem 6.1 The linear multistep method (6.8) is consistent with the initial value problem (6.7) if an only if

$$\rho(1) = \rho'(1) = 0 \quad \text{and} \quad \rho''(1) = 2\sigma(1) \ . \quad (6.14)$$

Corollary The first characteristic polynomial of a consistent LMM always has a double zero at $r = +1$.

This follows directly from the requirement $\rho(1) = \rho'(1) = 0$. This (double) zero is called the *principal zero* of $\rho(r)$ and as in Chapter 5 is usually labelled r_1; clearly, $r_2 = r_1$ and the other zeros $r_3, r_4, \dots, r_k$ are said to be *spurious*.

Theorem 6.2 A consistent LMM cannot have stepnumber $k<2$.

Proof Recalling that $\alpha_k = +1$, the proof is by contradiction. Suppose first of all that $k=0$; then $\rho(r)=\alpha_0$ with $\alpha_0 \neq 0$. Consistency requires $\rho(1)=0$ and so $\alpha_0=0$, a contradiction.

Suppose next that $k=1$; then $\rho(r)=\alpha_0+\alpha_1 r$ with $\alpha_1 \neq 0$ and so $\rho'(r)=\alpha_1$. Consistency requires $\rho'(1)=0$ and so $\alpha_1=0$, a contradiction.

In summary, therefore, k cannot be less than two.

Definition 6.3 The linear multistep method (6.8) is said to be *zero-stable* if no zero of the first characteristic polynomial $\rho(r)$ has modulus greater than unity and if every zero of modulus unity has multiplicity not greater than two.

The definition of convergence is that given in Definition 5.4 of Chapter 5 with the addition of the hypothesis

$$\lim_{l\to 0} [y_{0,r}(l) - y_{0,0}]/l = z_0 \qquad (r=0,1,\ldots,k-1) \ .$$

The Fundamental Theorem of Dahlquist (Theorem 5.4 of Chapter 5) applies to (6.8) and Theorem 5.3 relating to the attainable order of (6.8) is also valid. The optimal two-step method of the form $y_n - 2y_{n+1} + y_{n+2} = l^2[af_n + (1-2a)_{n+1} + f_{n+2}]$ is

$$y_n - 2y_{n+1} + y_{n+2} = \frac{l^2}{12}(f_n + 10f_{n+1} + f_{n+2}) \tag{6.15}$$

which, because it is optimal, has order four; its error constant is $C_6 = -\frac{1}{240}$. This method is known as *Numerov's method* or the *Royal Road Formula*.

In the theory of weak stability the *stability polynomial* $\Pi(r,\bar{l})$ is given by

$$\Pi(r,\bar{l}) = \rho(r) - \bar{l}\sigma(r) \tag{6.16}$$

where, now, $\bar{l} = l^2(\partial f/\partial y)$. Here, two zeros of $\Pi(r,\bar{l})$, to be named r_1 and r_2, tend to the *double* principal zero at $+1$ of $\rho(r)$ as $\bar{l}\to 0$. In Chapter 5 it was noted that $r_1 = \exp(l\partial f/\partial y) + O(l^{p+1})$ where p is the order of the numerical method. With respect to second-order differential equations, it is noted that

$$r_1 = \exp[l(\partial f/\partial y)^{\frac{1}{2}}] + O(l^{p+2}) \quad \text{and} \quad r_2 = \exp[-l(\partial f/\partial y)^{\frac{1}{2}}] + O(l^{p+2})$$

and, since for many methods r_1 and r_2 lie *on* the unit circle, stability intervals are closed intervals and the definitions of absolute and relative stability given in Chapter 5 are modified as follows.

Definition 6.4 The linear multistep method (6.8) is said to be *absolutely stable* for a given $\bar{l}$ if, for that $\bar{l}$, all the zeros r_i $(i=1,2,\ldots,k)$ of (6.16) satisfy $|r_i| \leqslant 1$.

Definition 6.5 An interval $[a,b]$ of the real line is said to be an *interval of absolute stability* if the method is absolutely stable for all $\bar{l} \in [a,b]$.

Definition 6.6 The LMM (6.8) is said to be *relatively stable* for a given $\bar{l}$ if, for that $\bar{l}$, $|r_i| \leqslant \min(|r_1|,|r_2|)$, $i=3,4,\ldots,k$.

Definition 6.7 An interval $[a,b]$ of the real line is said to be an *interval of relative stability* if the method is relatively stable for all $\bar{l} \in [a,b]$.

The consequences of using a value of l which is too large are the same as described in example 5.6 of Chapter 5.

In the case of a system of special equations of the form $\mathbf{y}''(t) = \mathbf{f}(t,\mathbf{y}(t))$, $\bar{l} = l^2\lambda$, where λ is interpreted as an eigenvalue of the Jacobian $\partial\mathbf{f}/\partial\mathbf{y}$. The term $\bar{l}$ may therefore be complex and the intervals of stability defined in Definitions 6.5 and 6.7 become stability regions as in Chapter 5.

It should be noted that, as for first-order differential equations, every consistent zero-stable linear multistep method is absolutely unstable for small positive $\bar{l}$.

Example 6.6 Find the interval of absolute stability of the explicit method

$$y_n - 2y_{n+1} + y_{n+2} = l^2 f_{n+1} \; . \qquad (6.17)$$

Here, $\rho(r) = r^2 - 2r + 1$ and $\sigma(r) = r$ so that $\Pi(r,\bar{l}) = \rho(r) - \bar{l}\sigma(r) = r^2 - 2(1+\frac{1}{2}\bar{l})r + 1$. The *stability equation* $\Pi(r,\bar{l}) = 0$ is of the form $r^2 - 2Br + 1 = 0$ which has roots given by $r = [B \pm (B^2-1)^{\frac{1}{2}}]$ from which it follows that $|r_i| \leqslant 1$ $(i=1,2)$ provided $|B| \leqslant 1$. Clearly, $B = 1 + \frac{1}{2}\bar{l}$ in this example and for stability it is necessary that $-1 \leqslant 1 + \frac{1}{2}\bar{l} \leqslant 1$ giving the interval of absolute stability as $\bar{l} \in [-4,0]$.

In Chapter 5, predictor–corrector combinations in various modes were discussed in detail. The application of linear multistep methods for second-order differential equations is the same as for first-order equations: the theory requires a change from l to l^2. A suitable pair of methods for use as predictor and corrector are given by (6.17) and (6.12) respectively. Milne's device is applicable; the principal local truncation error of a pair of methods with the same order p, such as (6.17) and (6.12), is given at the point $t = t_n$ by

$$C_{p+2}l^{p+2}y^{(p+2)}(t_n) = \frac{C_{p+2}}{C^*_{p+2} - C_{p+2}}\,(y^{[m]}_{n+k} - y^{[0]}_{n+k}) \qquad (6.18)$$

where the notation is that used in Chapter 5.

6.2.3 Periodicity

Periodic initial value problems of the form (6.7) arise following the modelling of systems which have regularly occurring features such as the beating of the heart. The theory associated with the differential equations of such a model is most easily described and understood by resorting to the test equation

$$y''(t) = -\lambda^2 y(t)\ , \quad t > t_0\ ; \quad y(t_0) = y_0\ , \quad y'(t_0) = z_0 \tag{6.19}$$

in which λ is real. The solution of the test problem is

$$y(t) = c_1\cos\lambda t + c_2\sin\lambda t \tag{6.20}$$

which is periodic with period $2\pi/\lambda$ except when the arbitrary constants c_1 and c_2 are both zero, that is, except when the initial conditions y_0 and z_0 are trivial.

Any linear multistep method used to solve (6.19) must generate a solution which follows uniformly and closely the trajectory defined by (6.20). Much of the terminology relating to periodic initial value problems evolved for use in orbital mechanics where, in computing satellite trajectories, for instance, it is necessary that the solution generated stays on orbit. Numerical methods which achieve this are said to be *orbitally stable*, while numerical methods for which the solution spirals inwards or outwards are *orbitally unstable*, and orbital stability is the periodicity property which a linear multistep method must have if it is to be used to solve periodic problems. In examining the periodicity properties of an LMM, the product of the steplength and the angular frequency λ is shown to lie within a certain interval, called the interval of periodicity of the LMM, when certain symmetry conditions are satisfied. Clearly, an LMM must have a nonvanishing interval of periodicity if it is to be used to solve (6.19) or (6.7).

Using the linear multistep method (6.8), which has characteristic polynomials defined by (6.13), to solve the test problem (6.19) gives

$$\sum_{j=0}^{k} (\alpha_j + H^2\beta_j)y_{n+j} = 0\ , \tag{6.21}$$

where $H = l\lambda$. The solution of (6.21) is known to be

$$y_n = \sum_{s=1}^{k} A_s r_s^n \tag{6.22}$$

where the A_s are constant and the r_s $(s = 1,2,\ldots,k)$ are the zeros, assumed distinct, of the *periodicity polynomial*.

$$\Omega(r,H^2) = \rho(r) + H^2\sigma(r)\ . \tag{6.23}$$

Definition 6.8 The LMM (6.8) is said to have *interval of periodicity* $(0,H_0^2)$ if, for all $H^2 \in (0,H_0^2)$, the roots r_i of the *periodicity equation*

$$\Omega(r,H^2) = 0 \tag{6.24}$$

satisfy $r_1 = e^{i\theta(H)}$, $r_2 = e^{-i\theta(H)}$, $|r_s| \leqslant 1$ for $s = 3,4,\ldots,k$, where $\theta(H)$ is real.

Definition 6.9 The LMM is said to be *symmetric* if $\alpha_j = \alpha_{k-j}$ and $\beta_j = \beta_{k-j}$ for $j = 0,1,\ldots,k$.

Some known properties of linear multistep methods which are relevant to the theory on periodicity intervals are contained in the following theorems which first appeared in a paper by Lambert and Watson (1976).

Theorem 6.3 A symmetric LMM cannot have an odd stepnumber.

Theorem 6.4 A symmetric LMM has even order p.

Theorem 6.5 Let the LMM (6.8) have a nonvanishing interval of periodicity. Then the method is symmetric.

Theorem 6.6 Let the LMM (6.8) be symmetric and let $\rho(r)$ have no double zeros on the unit circle other than the principal zero. Then the LMM has a nonvanishing interval of periodicity.

A popular family of linear multistep methods is the family of *Störmer–Cowell methods*, all of which have $\rho(r) = r^k - 2r^{k-1} + r^{k-2}$. Störmer–Cowell methods with stepnumber $k > 2$ have nonvanishing intervals of absolute stability but, since they are not symmetric, they have no intervals of periodicity. In general, therefore, intervals of periodicity are not identical in length with intervals of absolute stability.

Example 6.7 Find the interval of periodicity of the LMM in Example 6.6.

As before, $\rho(r) = r^2 - 2r + 1$ and $\sigma(r) = r$ so that $\Omega(r,H^2) = r^2 - 2(1 - \frac{1}{2}H^2)r + 1 = 0$ when $r = 1 - (B^2 - 1)^{\frac{1}{2}}$ where, now, $B = 1 - \frac{1}{2}H^2$. It follows that $|r_i| < 1$ $(i = 1,2)$ if $-1 < 1 - \frac{1}{2}H^2 < 1$ and the interval of periodicity is consequently $H^2 \in (0,4)$. This linear multistep method is of the symmetric Störmer–Cowell type with $k = 2$.

Definition 6.10 The linear multistep method (6.8) is said to be *P-stable* if its interval of periodicity is $H^2 \in (0, \infty)$.

Theorem 6.7 Suppose the LMM (6.8) is *P*-stable. Then

(1) it is implicit;
(2) all the zeros of its second characteristic polynomial lie on the unit circle;
(3) its order is at most two.

6.2.4 Phase-lag

The phrase 'out of phase' is familiar to all and the phase-lag of a linear multistep method is a measure of how much the solution generated by the linear multistep method lags behind the theoretical solution of the differential equation, or vice versa.

It is not difficult to see that the periodicity polynomial of a symmetric linear multistep method takes the form

$$\Omega(r,H^2) = A(H)r^2 - 2B(H)r + A(H) \tag{6.25}$$

and the interval of periodicity is determined by computing the range of values of H^2 for which r_1 and r_2, the roots of the corresponding periodicity equation (6.24), satisfy

$$r_1 = e^{i\theta(H)} \quad \textit{and} \quad r_2 = e^{-i\theta(H)} \tag{6.26}$$

here $\theta(H)$ is real and is an approximation to H. The *phase-lag* $\Phi(H)$ of the linear multistep method is the leading term in the expansion of

$$|\{\theta(H) - H\}/H| \tag{6.27}$$

(see Brusa and Nigro (1980)). It is easy to deduce from (6.25), (6.24) and (6.26) that

$$\tan\theta(H) = [\{A(H)\}^2 - \{B(H)\}^2]^{\frac{1}{2}}/B(H) \ . \tag{6.28}$$

Then, substituting for $A(H)$ and $B(H)$ gives an expansion for $\tan\theta(H)$ which has the form $\tan\theta(H) = \gamma$; $\theta(H)$ is then found from the expansion

$$\theta(H) = \tan^{-1}\gamma = \gamma - \tfrac{1}{3}\gamma^3 + \tfrac{1}{5}\gamma^5 - \tfrac{2}{35}\gamma^7 + \ldots \tag{6.29}$$

following which the phase-lag $\Phi(H)$ may be found from (6.29).

Example 6.8 Find the phase-lag of the explicit method in Example 6.6.

It was seen in Example 6.7 that $A(H) = 1$ and $B(H) = 1 - \frac{1}{2}H^2$ from which it is easy to show that

$$\tan\theta(H) = H(1 - \tfrac{1}{4}H^2)^{\frac{1}{2}}(1 - \tfrac{1}{2}H^2)^{-1} = H(1 + \tfrac{3}{8}H^2 + \ldots)$$

(using the binomial expansion). Then, using (6.29) it follows that

$$\begin{aligned}\theta(H) &= H(1+\tfrac{3}{8}H^2+\ldots)-\tfrac{1}{3}H^3(1+\tfrac{3}{8}H^3+\ldots)^3+\ldots\\ &= H+\tfrac{3}{8}H^3+\ldots-\tfrac{1}{3}H^3\ldots\\ &= H+\tfrac{1}{24}H^3+\ldots\end{aligned}$$

and, further, that

$$\{\theta(H)-H\}/H = \tfrac{1}{24}H^2+\ldots$$

The phase-lag $\Phi(H)$, as defined by (6.27), is therefore $\frac{1}{24}H^2$.

Using the procedure followed in Example 6.5 it may be shown that the principal local truncation error of this linear multistep method at $t=t_n$ is $\frac{1}{12}l^4y^{(iv)}(t_n)$. The order of this LMM is $p=2$ and the error constant is $C_{p+2}=C_4=\frac{1}{12}$. It is noted that the phase-lag of a linear multistep method of order p with error constant C_{p+2} is given by $\Phi(H)=\frac{1}{2}|C_{p+2}|H^p$.

6.3 GENERAL BOUNDARY VALUE PROBLEMS

6.3.1 Introduction

Boundary value problems arise in the modelling of steady-state or time-dependent systems and consequently x is usually used to represent the independent variable instead of t, which is reserved for time. An increment in x will be denoted by h.

Two-point boundary value problem consist of a differential equation which holds in some interval $a<x<b$, together with conditions imposed on the dependent variable and/or its first derivative at the two points $x=a$ and $x=b$. Suppose $y=y(x)$ is the dependent variable; then there are three types of boundary condition. These are

Type 1: functional boundary conditions $y(a)=A$ and $y(b)=B$ are given,
Type 2: derivative boundary conditions $y'(a)=\alpha$ and $y'(b)=\beta$ are given,
Type 3: mixed boundary conditions are given in the form

$$\begin{aligned}p_0y(a)+q_0y'(a) &= r_0\,,\\ p_1y(b)+q_1y'(b) &= r_1\,,\end{aligned}$$

There are two classes of nonlinear boundary value problem. In the first class the differential equation has the form $y''(x)=f(x,y,y')$ while in the second the differential equation has the form $y''(x)=f(x,y)$. These differential equations give rise, respectively, to *general* and *special boundary value problems*. The following theorem gives conditions that ensure that the general boundary value problem with Type 1 boundary conditions has a unique solution.

Theorem 6.8 Suppose the function f in the *nonlinear boundary value problem*

$$y''(x) = f(x,y,y'), \quad a<x<b, \quad y(a) = A, \quad y(b) = B \tag{6.30}$$

is continuous on the set

$$D = \{(x,y,y') : a \leqslant x \leqslant b, \quad -\infty < y < \infty, \quad -\infty < y' < \infty\}$$

and that $\partial f/\partial y$ and $\partial f/\partial y'$ are also continuous on D. If

(1) $\partial f(x,y,y')/\partial y > 0$ for all $(x,y,y') \in D$, and
(2) a constant M exists with $|\partial f(x,y,y')/\partial y'| \leqslant M$ for all $(x,y,y') \in D$,

then the boundary value problem (6.30) has a unique solution.

When $f(x,y,y') = p(x)y'(x) + q(x)y(x) + r(x)$ the differential equation is said to be linear and Theorem 6.8 can be replaced by the following theorem.

Theorem 6.9 If the *linear boundary value problem*

$$y'(x) = p(x)y'(x) + q(x)y(x) + r(x), \quad a<x<b, \quad y(a) = A, \quad y(b) = B \tag{6.31}$$

satisfies

(1) $p(x)$, $q(x)$ and $r(x)$ are continuous on $[a,b]$,
(2) $q(x) > 0$ *on* $[a,b]$,

then (6.31) has a unique solution.

6.3.2 Shooting methods

Shooting methods have enjoyed considerable success for general boundary value problems with Type 1 boundary conditions and use the solutions to a *sequence* of initial value problems of the form

$$y'(x) = f(x,y,y'), \quad a<x<b, \quad y(a) = A, \quad y'(a) = z_i \tag{6.32}$$

with $i=0,1,2,\ldots$ to find the solution to (6.30). An initial guess z_0 is chosen to approximate $y'(a)$ in a manner which will ensure that

$$\lim_{i\to\infty} y(b,z_i) = y(b) = B \;,$$

where $y(x,z_i)$ denotes the solution to the initial value problem (6.32) and $y(x)$ is the solution to (6.30).

It is probably clear to the reader that 'shooting methods' are so-called because of the analogy of firing missiles at a stationary target. Starting with the parameter z_0, which determines the initial elevation at which the missile is discharged from the point $(x,y)=(a,A)$, the trajectory of the missile is computed by solving the initial value problem given by (6.32) with $i>0$. If the point of landing, $(x,y)=(b,y(b,z_0))$ is not sufficiently close to (b,B), the approximation is corrected by choosing another elevation z_1, and so on, until $y(b,z_k)$ is acceptably close to the 'target' $y(b)=B$.

Methods for solving (6.32) are described in section 6.1 and in Chapter 5. The reader is referred to the specialist text by Keller (1968) for further details of shooting methods.

6.3.3 Finite difference methods

Considering first of all the nonlinear problem (6.30), the derivatives $y''(x)$ and $y'(x)$ will be replaced by their second-order central difference approximants

$$y''(x) = h^{-2}[y(x-h)-2y(x)+y(x+h)]+O(h^2) \; , \qquad (6.33)$$

$$y'(x) = (2h)^{-1}[y(x+h)-y(x-h)]+O(h^2) \; . \qquad (6.34)$$

Discretizing the interval $a \leqslant x \leqslant b$ into $N+1$ subintervals each of width h so that $(N+1)h=b-a$, a numerical method determines a vector $\mathbf{Y}=[y_1,y_2,\ldots,y_N]^{\mathrm{T}}$, where y_n is an approximation to $y(x_n)$ with $n=1,2,\ldots N$. Clearly $y_0=A$ and $y_{n+1}=B$.

Applying (6.30) with (6.33) and (6.34) to each interior point x_n $(n=1,2,\ldots N)$ of the discretization gives

$$-\frac{y_{n-1}-2y_n+y_{n+1}}{h^2}+f\left(x_n,y_n,\frac{y_{n+1}-y_{n-1}}{2h}\right) = 0$$

for $n=1,2,\ldots,N$ with $y_0=A$ and $y_{n+1}=B$. The solution is thus found by solving the nonlinear system

$$\begin{aligned} 2y_1-y_2+h^2f(x_1,y_1,(y_2-A)/2h) &= A \; , \\ -y_1+2y_2-y_3+h^2f(x_2,y_2,(y_3-y_1)/2h) &= 0 \; , \\ \ldots \qquad \ldots & \\ -y_{N-2}+2y_{N-1}-y_N+h^2f(x_{N-1},y_{N-1},(y_N-y_{N-2})/2h) &= 0 \; , \\ -y_{N-1}+2y_N+h^2f(x_N,y_N,(B-y_{N-1})/2h) &= B \; . \end{aligned} \qquad (6.35)$$

The Newton–Raphson method for a nonlinear system may be used to obtain $\mathbf{Y}$.

The same approximations to $y'(x)$ and $y''(x)$ may be used to obtain a numerical solution to the general linear boundary value problem with Type 1 boundary conditions given by (6.31). Applying (6.31) with (6.33) and (6.34) to the points x_n $(n=1,2,\ldots,N)$ of the discretization gives

$$\frac{-y_{n-1}+2y_n-y_{n+1}}{h^2}+p(x_n)\left(\frac{y_{n+1}-y_{n-1}}{2h}\right)+q(x_n)y_n = -r(x_n)$$

with $y_0=A$ and $y_{N+1}=B$. The solution $\mathbf{Y}$ is thus found by solving the linear system

$$\begin{aligned}
&[2+h^2q(x_1)]y_1-[1-\tfrac{1}{2}hp(x_1)]y_2 = -h^2r(x_1)+[1+\tfrac{1}{2}hp(x_1)]A \ , \\
&-[1+\tfrac{1}{2}hp(x_2)]y_1+[2+h^2q(x_2)]y_2-[1-\tfrac{1}{2}hp(x_2)]y_3 = -h^2r(x_2) \ , \\
&\qquad\ldots \qquad\qquad \ldots \\
&-[1+\tfrac{1}{2}hp(x_{N-2})]y_{N-1}+[2+h^2q(x_{N-1})]y_{N-1}-[1-\tfrac{1}{2}hp(x_{N-1})]y_N \\
&\qquad = -h^2r(x_{N-1}) \ , \\
&-[1+\tfrac{1}{2}hp(x_N)]y_{N-1}+[2+h^2q(x_N)]y_N = -h^2r(x_N)+[1-\tfrac{1}{2}hp(x_N)]B \ .
\end{aligned}$$

It is obvious that this linear system has a tridiagonal coefficient matrix and $\mathbf{Y}$ may be obtained using the special algorithm outlined in Chapter 1.

6.3.4 Variational methods

An important linear two-point boundary value problem which arises in the design of a number of structures in bioengineering is given by

$$-\frac{\mathrm{d}}{\mathrm{d}x}\left(p(x)\,\frac{\mathrm{d}y}{\mathrm{d}x}\right)+q(x)y(x) = g(x), \quad 0<x<1, \quad y(0) = y(1) = 0 \ . \tag{6.36}$$

Equation (6.36) describes the deflection $y(x)$ of points on a beam of unit length, fixed at both ends, with variable cross-section $q(x)$. The deflection is due to the applied stresses $p(x)$ and $g(x)$.

Under the assumptions

(1) $p(x)\in C^1[0,1]$,
(2) $q(x),g(x)\in C[0,1]$,
(3) $q(x)\geqslant 0$ for $0\leqslant x\leqslant 1$,
(4) there exists a constant $\delta>0$ such that $p(x)\geqslant\delta>0$ for all $0\leqslant x\leqslant 1$,

it may be shown that (6.36) has a unique solution: these assumptions are, in fact, sufficient conditions for the existence of a unique solution (see Bailey *et al.* (1968)).

The solution of (6.36) satisfies a *variational principle* from which has evolved the *Ritz–Galerkin method* for solving (6.36). This variational property characterizes the solution to (6.36) as the function which minimizes a certain integral over all functions in $C_0^2[0,1]$, the set of these functions in $C^2[0,1]$ which satisfy $y(0) = y(1) = 0$. This characterization is formalized in the following theorem.

Theorem 6.10 Let $p(x) \in C^1[0,1]$ and $q(x), g(x) \in C[0,1]$; suppose that there exists a constant $\delta > 0$ such that $p(x) \geqslant \delta > 0$ and that $q(x) \geqslant 0$ for $0 \leqslant x \leqslant 1$. Then the function $y(x) \in C_0^2[0,1]$ is the unique solution to the differential equation in (6.36) if an only if $y(x)$ is the unique function in $C_0^2[0,1]$ which minimizes the integral

$$I[u(x)] = \int_0^1 \{p(x)[u'(x)]^2 + q(x)[u(x)]^2 - 2g(x)u(x)\}\mathrm{d}x \ . \qquad (6.37)$$

In essence, this theorem says that $y(x) = u(x)$ satisfies (6.36) and minimizes (6.37) simultaneously. In the Ritz–Galerkin method, the integral $I[u]$ is not minimized over all the functions in $C_0^2[0,1]$. It is minimized over a linear combination of a smaller set of function, say $\phi_1(x), \phi_2(x), \ldots, \phi_N(x)$, which are user-dependent and which are called *basis functions*.

The usual choice of low-order basis function is the set of *hat functions* or *chapeau functions* which require forming a partition $0 = x_0 < x < \ldots < x_N < x_{N+1} = 1$, letting $h_i = x_{i+1} - x_i$ for each $i = 0, 1 \ldots, N$, and then defining the ϕ_i $(i = 1, 2, \ldots N)$, which are illustrated in Fig. 6.1, by

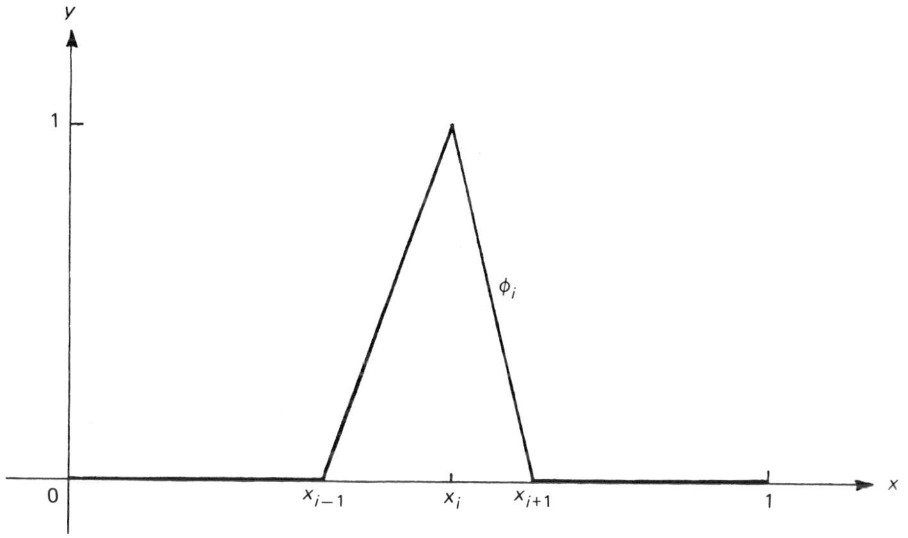

Fig. 6.1 — Hat function.

$$\phi_i(x) = \begin{cases} 0 , & 0 \leqslant x \leqslant x_{i-1} , \\ (x - x_{i-1})/h_{i-1} , & x_{i-1} \leqslant x \leqslant x_i , \\ (x_{i+1} - x)/h_i , & x_i \leqslant x \leqslant x_{i+1} , \\ 0 , & x_{i+1} \leqslant x \leqslant 1 . \end{cases} \tag{6.38}$$

The derivatives of the ϕ_i are given by

$$\phi_i'(x) = \begin{cases} 0 , & 0 < x < x_{i-1} , \\ 1/h_{i-1} , & x_{i-1} < x < x_i , \\ -1/h_i , & x_i < x < x_{i+1} , \\ 0 , & x_{i+1} < x < 1 \end{cases} \tag{6.39}$$

for each $i = 1,2,\ldots,N$. It is easy to see from (6.38), (6.39) and Fig. 6.1 that, since the functions are piecewise linear, the derivatives $\phi_i'(x)$ $(i = 1,2,\ldots,N)$ are constant in the open subinterval (x_m,x_{m+1}) for each $m = 0,1,2,\ldots,N$ of the partition.

Suppose that the linear combination of basis functions is

$$\phi(x) = c_1\phi_1(x) + c_2\phi_2(x) + \ldots + c_N\phi_N(x) \tag{6.40}$$

where the c_i $(i = 1,2,\ldots N)$ are constants: these constants are known as *expansion coefficients*. This function $\phi(x)$ is an approximation to the solution of (6.36) and is found by determining the set of expansion coefficients which minimizes the integral

$$I[\phi(x)] = I[c_1\phi_1(x) + c_2\phi_2(x) + \ldots + c_N\phi_N(x)] . \tag{6.41}$$

Now,

$$\begin{aligned} I[\phi(x)] &= \int_0^1 \{p(x)[\phi'(x)]^2 + q(x)[\phi(x)]^2 - 2g(x)\phi(x)\}\mathrm{d}x \\ &= \int_0^1 \left\{ p(x)\left[\sum_{i=1}^N c_i\phi_i'(x)\right]^2 + q(x)\left[\sum_{i=1}^N c_i\phi_i(x)\right]^2 - 2g(x)\sum_{i=1}^N c_i\phi_i(x)\right\}\mathrm{d}x \end{aligned} \tag{6.42}$$

and, considering I to be a function of the $c_i (i = 1,2,\ldots,N)$ (these, after all, are what are being sought: the ϕ_i have been chosen by the user), the necessary conditions for I to have a minimum value are

$$\frac{\partial I}{\partial c_1} = 0 , \quad \frac{\partial I}{\partial c_2} = 0 , \ \ldots , \ \frac{\partial I}{\partial c_N} = 0 \tag{6.43}$$

simultaneously. Differentiating (6.42) gives

$$\frac{\partial I}{\partial c_n} = \int_0^1 \left\{ 2p(x)\phi_n'(x) \sum_{i=1}^{N} c_i\phi_i'(x) + 2q(x)\phi_n(x) \sum_{i=1}^{N} c_i\phi_i(x) - 2g(x)\phi_n(x) \right\} \mathrm{d}x \tag{6.44}$$

for each $n = 1,2,\ldots,N$. Substituting (6.44) into (6.43) yields

$$\sum_{i=1}^{N} \left[\int_0^1 \{p(x)\phi_i'(x)\phi_n'(x) + q(x)\phi_n(x)\phi_i(x)\}\mathrm{d}x \right] c_i - \int_0^1 g(x)\phi_n(x)\mathrm{d}x = 0 \tag{6.45}$$

for each $n = 1,2,\ldots,N$.

Clearly, (6.45) is a linear system in the expansion constants c_i ($i = 1,2,\ldots,N$) which, because $\phi_i(x)$ and $\phi_i'(x)$ are nonzero only on (x_{i-1},x_{i+1}), is tridiagonal and of order N. It has the form

$$A\mathbf{c} = \mathbf{b} \ , \tag{6.46}$$

where the nonzero elements of A are

$$a_{ii} = \int_0^1 \{p(x)[\phi_i'(x)]^2 + q(x)[\phi_i(x)]^2\}\mathrm{d}x \ , \qquad i = 1,2,\ldots,N$$

$$a_{i,i+1} = \int_0^1 \{p(x)\phi_i'(x)\phi_{i+1}(x)\phi_i(x) + q(x)\phi_i(x)\phi_{i+1}'(x)\}\mathrm{d}x \ , \quad i = 1,2,\ldots,N-1 \ ,$$

$$a_{i,i-1} = \int_0^1 \{p(x)\phi_{i-1}'(x)\phi_i(x) + q(x)\phi_{i-1}(x)\phi_i'(x)\}\mathrm{d} \ , \qquad i = 2,\ldots,N \ ,$$

and the elements of $\mathbf{b}$ are

$$b_i = \int_0^1 g(x)\phi_i(x)\mathrm{d}x \ , \qquad i = 1,2,\ldots,N \ .$$

The vector $\mathbf{c}$ is easily found using the special tridiagonal solver discussed in Chapter 1. This vector contains the expansion coefficients for use in (6.40): the resulting function $\phi(x)$ minimizes I and is thus the solution of (6.36).

Variational techniques are used again in Chapter 9 where finite element methods are used to solve elliptic partial differential equations.

6.4 SPECIAL BOUNDARY VALUE PROBLEMS

6.4.1 Introduction

In special two-point boundary value problems the differential equation has the same form as in special initial value problems, namely $y''(x)=f(x,y(x))$, the function f being independent of $y'(x)$. The boundary conditions are of the Types 1, 2 or 3 as described in subsection 6.3.1; Type 2 does not yield a unique solution. The type which occurs most frequently is probably Type 1, though Type 3 does occur in system form as the following example illustrates.

Example 6.9 Let the elements c_1,c_2,c_3,c_4,c_5 of a vector $\mathbf{c}(x)$, $0 \leqslant x \leqslant a$, represent, respectively, the concentrations of O_2, HbO_2, CO_2, $HbCO_2$ and Hb in blood flowing through a capillary of length L and radius a. Then $\mathbf{c}(x)$ satisfies a differential system of the form $\mathbf{c}''(x)=\mathbf{f}(x,\mathbf{c})$ detailed as follows:

$$D_o c_1'' = k_1 c_1^m c_5 - k_2 c_2 , \qquad D_H c_2'' = -k_1 c_1^m c_5 - k_2 c_2 ,$$

$$D_c c_3'' = m_1 c_3 c_5 - m_2 c_4 , \qquad D_H c_4'' = -m_1 c_3 c_5 + m_2 c_4 ,$$

$$D_H c_5'' = k_1 c_1^m c_5 - k_2 c_2 + m_1 c_3 c_5 - m_2 c_4 ,$$

where $m=2.6478$, D_o is the diffusion coefficient of O_2, D_c is the diffusion coefficient of CO_2 and D_H is the diffusion coefficient of H_b, HbO_2 and $HbCO_2$ (Ulanowicz and Frazier, 1970). Sharan and Singh (1984) give sets of values for k_1,k_2,m_1 and m_2. The boundary conditions relating to this model are derived from physiological conditions and are described as follows:

(1) there are no changes in the concentrations across the central axis of the capillary, giving rise to the flux equalities

$$c_i' = 0 \text{ at } x = 0 \qquad (i = 1,2,3,4,5);$$

(2) the thickness of the capillary wall is small and it may be assumed that the concentration of the two gases O_2 and CO_2 at the wall of the capillary is the same as in the alveolar region. Hence

$$c_1(a) = c_1^* \quad \text{and} \quad c_3(a) = c_3^*$$

where c_1^* and c_3^* are given constants;

(3) Hb molecules are much larger than O_2 and CO_2 molecules. Hence Hb, HbO_2 and $HbCO_2$ cannot penetrate the capillary wall and they have zero flux across it so that

$$c_2'(a) = c_4'(a) = c_5'(a) = 0 .$$

The mathematical model is thus complete and the reader should note that there are ten boundary conditions altogether in (1), (2) and (3), matching the total of ten derivatives in the differential equations (two each of c_1,c_2,c_3,c_4 and c_5).

In the remainder of this chapter, attention will be focused on single two-point boundary value problems with Type 1 boundary conditions, one linear problem and one nonlinear. The reader will see how to use the Numerov method, a typical linear multistep method, to obtain a solution and how to analyse for convergence. The section ends with a subsection on global extrapolation.

6.4.2 The linear problem

Consider the boundary value problem

$$y''(x) = q(x)y(x) + r(x), \quad a<x<b, \quad y(a) = A, \quad y(b) = B \tag{6.47}$$

and assume that all conditions are met which guarantee the existence of a unique solution. Suppose that the interval $a \leqslant x \leqslant b$ is discretized into $N+1$ subintervals each of width h so that $(N+1)h = (b-a)$ as before. A numerical method will determine the vector $\mathbf{Y} = [y_1, y_2, \ldots, y_N]^{\mathrm{T}}$, where y_n is an approximation to $y(x_n)$, at the grid point x_n $(n = 1,2,\ldots,N)$. Clearly, $y_0 = A$ and $y_{N+1} = B$.

Consider, now, Numerov's method, which has order $p = 4$, written in the form

$$-y_{n-1} + 2y_n - y_{n+1} + \frac{h^2}{12}(y''_{n-1} + 10y''_n + y''_{n+1}) = 0 \; ; \tag{6.48}$$

this method has local truncation error

$$t_n \equiv \frac{1}{240}h^6 y^{(\mathrm{vi})}(x_n) + \frac{11}{60480}h^8 y^{(\mathrm{viii})}(x_n) + \ldots \tag{6.49}$$

at the point $x = x_n$. Denoting $q(x_n)$ and $r(x_n)$ by q_n and r_n respectively, substitution for y''_n from (6.47) into (6.48) gives

$$-y_{n-1} + 2y_n - y_{n+1} + \frac{h^2}{12}[q_{n-1}y_{n-1} + r_{n-1} + 10(q_n y_n + r_n) + q_{n+1}y_{n+1} + r_{n+1}] = 0$$

for $n = 1,2,\ldots,N$. The numerical solution is then obtained by solving the linear system

$$\left(2 + \frac{5}{6}h^2 q_1\right)y_1 - \left(1 - \frac{h^2}{12}q_2\right)y_2 = \left(1 - \frac{h^2}{12}q_0\right)A - \frac{h^2}{12}(r_0 + 10r_1 + r_2) \; ,$$

$$-\left(1-\frac{h^2}{12}q_{n-1}\right)y_{n-1}+\left(2+\frac{5}{6}h^2q_n\right)y_n-\left(1-\frac{h^2}{12}q_{n+1}\right)y_{n+1}$$
$$= -\frac{h^2}{12}(r_{n-1}+10r_n+r_{n+1}) \;, \qquad n = 2,\ldots,N-1 \;,$$

$$-\left(1-\frac{h^2}{12}q_{N-1}\right)y_{N-1}+\left(2+\frac{5}{6}h^2q_N\right)y_N=\left(1-\frac{h^2}{12}q_{N+1}\right)B-\frac{h^2}{12}(r_{n-1}+10r_N+r_{N+1}), \tag{6.50}$$

using the linear solver for a tridiagonal system which was described in Chapter 1.

To analyse the convergence of the solution obtained from (6.50) it is necessary to write that system in the vector–matrix form

$$(J+\tfrac{1}{12}h^2MQ)\mathbf{Y} \;=\; \mathbf{c}-\tfrac{1}{12}h^2M\mathbf{r} \;, \tag{6.51}$$

where

$$J=\begin{bmatrix} 2 & -1 & & & 0 \\ -1 & 2 & -1 & & \\ \cdots & \cdots & \cdots & & \\ & & -1 & 2 & -1 \\ 0 & & & -1 & 2 \end{bmatrix}, \qquad \mathrm{M}=\begin{bmatrix} 10 & 1 & & & \\ 1 & 10 & 1 & & 0 \\ & \cdots & \cdots & \cdots & \\ & & 1 & 10 & 1 \\ 0 & & & 1 & 10 \end{bmatrix} \tag{6.52}$$

$Q=\operatorname{diag}\{q_n\}$, $\mathbf{c}=[A-\frac{1}{12}h^2q_0A-\frac{1}{12}h^2r_0,\,0,\ldots,0,\,B-\frac{1}{12}h^2q_{N+1}B-\frac{1}{12}h^2r_{N+1}]^{\mathrm{T}}$ and $\mathbf{r}=[r_1,r_2,\ldots r_N]^{\mathrm{T}}$. It may be shown that $\|J^{-1}\|=(N+1)^2/8$ and that $\|M\|=12$.

Replacing $\mathbf{Y}$ by $\mathbf{y}=\mathbf{y}(x)$ in (6.51) necessitates the introduction of the local truncation error vector $\mathbf{t}=[t_1,t_2,\ldots,t_N]^{\mathrm{T}}$, where t_n $(n=1,2,\ldots,N)$ is given by (6.49), so that

$$(J+\tfrac{1}{12}h^2MQ)\mathbf{y} \;=\; \mathbf{c}-\tfrac{1}{12}h^2M\mathbf{r}+\mathbf{t} \;. \tag{6.53}$$

Defining $\mathbf{z}$ by $\mathbf{z}=\mathbf{y}-\mathbf{Y}$, it follows that $\|\mathbf{z}\|$ must be bounded if convergence is to be attained. Subtracting (6.51) and (6.53) gives

$$(J+\tfrac{1}{12}h^2MQ)\mathbf{z}=\mathbf{t}$$

so that

$$\mathbf{z} = (J + \tfrac{1}{12}h^2MQ)^{-1}\mathbf{t} = (I + \tfrac{1}{12}h^2J^{-1}MQ)^{-1}J^{-1}\mathbf{t}$$

and hence

$$\|\mathbf{z}\| \leqslant \|(I + \tfrac{1}{12}h^2J^{-1}MQ)^{-1}\|\cdot\|J^{-1}\|\cdot\|\mathbf{t}\| \tag{6.54}$$

from which a bound on $\|\mathbf{z}\|$ must be obtained. Lemma 1.4 and the following worked example play an important rôle in establishing the bound on $\|\mathbf{z}\|$.

Example 6.10 Show that

$$[1 - \tfrac{1}{12}h^2\|J^{-1}MQ\|]^{-1} \leqslant [1 - \tfrac{1}{12}h^2\|J^{-1}\|\cdot\|M\|\cdot\|Q\|]^{-1} .$$

Elementary results give

$$\begin{aligned} & \tfrac{1}{12}h^2\|J^{-1}MQ\| \leqslant \tfrac{1}{12}h^2\|J^{-1}\|\cdot\|M\|\cdot\|Q\| \\ \Rightarrow\ & -\tfrac{1}{12}h^2\|J^{-1}MQ\| \geqslant -\tfrac{1}{12}h^2\|J^{-1}\|\cdot\|M\|\cdot\|Q\| \\ \Rightarrow\ & 1 - \tfrac{1}{12}h^2\|J^{-1}MQ\| \geqslant 1 - \tfrac{1}{12}h^2\|J^{-1}\|\cdot\|M\|\cdot\|Q\| \\ \Rightarrow\ & \frac{1}{1 - \tfrac{1}{12}h^2\|J^{-1}MQ\|} \leqslant \frac{1}{1 - \tfrac{1}{12}h^2\|J^{-1}\|\cdot\|M\|\cdot\|Q\|} . \end{aligned} \tag{6.55}$$

Returning to (6.54) and using Lemma 1.4 shows that

$$\|\mathbf{z}\| \leqslant \frac{\|J^{-1}\|\cdot\|\mathbf{t}\|}{1 - \tfrac{1}{12}h^2\|J^{-1}MQ\|}$$

which, using Example 6.10, leads to

$$\|\mathbf{z}\| \leqslant \frac{\|J^{-1}\|\cdot\|\mathbf{t}\|}{1 - \tfrac{1}{12}h^2\|J^{-1}\|\cdot\|M\|\cdot\|Q\|} . \tag{6.56}$$

Values for $\|J^{-1}\|$ and $\|M\|$ have been noted already and (6.49) yields

$$\|\mathbf{t}\| = \max_n |t_n| \leqslant \tfrac{1}{240} h^6 V_6 + \tfrac{11}{60480} h^8 V_8 + \ldots$$

where $V_i = \max_{a \leqslant x \leqslant b} |y^{(i)}(x)|$. It is then easy to show that (6.56) gives

$$\|\mathbf{z}\| \leqslant \frac{(N+1)^2}{8-(N+1)^2 h^2 \|Q\|} \cdot \left(\tfrac{1}{240} h^6 V_6 + \tfrac{11}{60480} h^8 V_8 + \ldots\right) ,$$

and, since $(N+1)h = b - a$, that

$$\|\mathbf{z}\| \leqslant \frac{(b-a)^2}{8-(b-a)^2 \|Q\|} \left(\tfrac{1}{240} h^4 V_6 + \tfrac{11}{60480} h^6 V_8 + \ldots\right) . \tag{6.57}$$

Recalling that norms are non-negative, the denominator in (6.57) must also be non-negative. Furthermore, for convergence, $\|\mathbf{z}\|$ must be finite and the conclusion of the analysis is that $\|\mathbf{z}\| = O(h^4)$, and hence that the Numerov method is fourth-order convergent, provided $\|Q\| = \max_n |q_n| < 8/(b-a)^2$.

6.4.3 The nonlinear problem

Suppose, now, that the Numerov method is applied to the nonlinear two-point boundary value problem

$$y''(x) = f(x, y(x)), \quad a < x < b; \quad y(a) = A, \quad y(b) = B . \tag{6.58}$$

This gives

$$-y_{n-1} + 2y_n - y_{n+1} + \frac{h^2}{12}(f_{n-1} + 10f_n + f_{n+1}) = 0 \tag{6.59}$$

for $n = 1,,\ldots,N$ which, using the boundary conditions, gives

$$2y_1 - y_2 + \frac{h^2}{12}(10f_1 + f_2) = A - \frac{h^2}{12} f_0 \tag{6.60}$$

when $n = 1$, and

$$-y_{N-1} + 2y_N + \frac{h^2}{12}(f_{N-1} + 10f_N) = B - \frac{h^2}{12} f_{N+1} \tag{6.61}$$

when $n = N$. It is clear from (6.60), (6.59) and (6.61) that the solution $\mathbf{Y}$ is obtained by solving the nonlinear algebraic system

$$JY + \tfrac{1}{12}h^2 M\mathbf{f}(x,\mathbf{Y}) \;=\; \mathbf{c} \tag{6.62}$$

where J and M are given in (6.52), $\mathbf{f} = [f_1, f_2, \ldots, f_N]^T$ with $f_n = f(x_n, y_n)$ for $n = 1, 2, \ldots, N$ and $\mathbf{c} = \left[A - \dfrac{h^2}{12}f_0, 0, \ldots, 0, B - \dfrac{h^2}{12}f_{N+1}\right]^T$. The Newton–Raphson method for a nonlinear algebraic system may be used to solve (6.62) (see Chapter 3).

The vector $\mathbf{y} = \mathbf{y}(x) = [y(x_1), y(x_2), \ldots, y(x_N)]^T$ clearly satisfies

$$J\mathbf{y} + \tfrac{1}{12}\, h^2 M\mathbf{f}(x,\mathbf{y}) \;=\; \mathbf{c} + \mathbf{t} \tag{6.63}$$

where $\mathbf{t}$ is the local truncation error vector. Using $\mathbf{z} = \mathbf{y} - \mathbf{Y}$ once again and subtracting (6.62) from (6.63) gives

$$J\mathbf{z} + \tfrac{1}{12}\, h^2 M[\mathbf{f}(x,\mathbf{y}) - \mathbf{f}(x,\mathbf{Y})] \;=\; \mathbf{t}\ . \tag{6.64}$$

Introduce, now, the first-order linearization

$$\mathbf{f}(x,\mathbf{y}) - \mathbf{f}(x,\mathbf{Y}) \;=\; F(\mathbf{y})\mathbf{z} \tag{6.65}$$

where F is the diagonal matrix $F = \text{diag}\{\partial f_n/\partial y(x_n)\}$. Then (6.64) may be written as

$$(J + \tfrac{1}{12}\, h^2 MF)\mathbf{z} \;= \mathbf{t} \tag{6.66}$$

so that

$$\mathbf{z} \;=\; (J + \tfrac{1}{12}\, h^2 MF)^{-1}\mathbf{t}\ . \tag{6.67}$$

An analysis similar to that carried out for the linear problem in subsection 6.4.2 then shows that $\|\mathbf{z}\| = O(h^4)$, and that the Numerov method is fourth-order convergent for the nonlinear problem (6.58), provided $F^* = \max\limits_n\ |\partial f_n/\partial y(x_n)| < 8/(b-a)^2$.

6.4.4 Global extrapolation of the Numerov method

Suppose that the discretization of the interval $a \leqslant x \leqslant b$ in subsection 6.4.3 is called grid G_1. It will be convenient to rename the N points of G_1 $x_1^{(1)}, x_2^{(1)}, \ldots x_N^{(1)}$, to rewrite the vector $\mathbf{Y}$ as $\mathbf{Y}^{(1)} = [y_1^{(1)}, y_2^{(1)}, \ldots, y_N^{(1)}]^T$, the vector $\mathbf{y}(x)$ as $\mathbf{y}^{(1)}(x)$, the local truncation error $\mathbf{t}$ as $\mathbf{t}^{(1)}$, and to rename the matrices M and F as M_h and F_h. It then follows that $\|\mathbf{z}\| = \|\mathbf{z}^{(1)}\|$ satisfies

$$\|\mathbf{z}^{(1)}\| \;\leqslant\; \frac{(b-a)^2}{8 - (b-a)^2 F^*}\,\left(\tfrac{1}{240}\, h^4 V_6 + \tfrac{11}{60480}\, h^6 V_8 + \ldots\right) \tag{6.68}$$

from (6.57).

Suppose, now, that the interval $a \leqslant x \leqslant b$ is subdivided into $2N+2$ subintervals each of width $\frac{1}{2}h$ giving a finer grid G_2 of interior points named $x_1^{(2)}, x_2^{(2)}, \ldots, x_{2N+1}^{(2)}$. Clearly, the points $x_2^{(2)}, x_4^{(2)}, \ldots, x_{2N}^{(2)}$ of G_2 coincide with the points $x_1^{(1)}, x_2^{(1)}, \ldots, x_N^{(1)}$ of the coarse grid G_1.

The nonlinear equations (6.60), (6.59), (6.61) are modified to

$$\begin{aligned} 2y_1^{(2)} - y_2^{(2)} + \tfrac{1}{48}\, h^2(10f_1^{(2)} + f_2^{(2)}) &= A - \tfrac{1}{48}\, h^2 f_0^{(2)} \ , \\ -y_{m-1}^{(2)} + 2y_m^{(2)} - y_{m+1}^{(2)} + \tfrac{1}{48}\, h^2\,(f_n^{(2)} + 10f_n^{(2)} + f_{n+1}^{(2)}) &= 0 \ , \quad m = 2,\ldots 2N \ , \\ +y_{2N}^{(2)} + 2y_{2N+1}^{(2)} + \tfrac{1}{48}\, h^2\,(f_{2N}^{(2)} + 10f_{2N+1}^{(2)}) &= -\tfrac{1}{48}\, h^2 f_{2N+2}^{(2)} \end{aligned}$$

which may be written in matrix–vector form as

$$J_{\frac{1}{2}h}\mathbf{Y}^{(2)} + \tfrac{1}{48}\, h^2 M_{\frac{1}{2}h}\mathbf{f}^{(2)}(x, \mathbf{Y}^{(2)}) = \mathbf{c}^{(2)} \ . \tag{6.69}$$

In (6.69) the matrices $J_{\frac{1}{2}h}$ and $M_{\frac{1}{2}h}$ are of order $2N+1$ and may be written down immediately from (6.52); the vectors $\mathbf{Y}^{(2)} = [y_1^{(2)}, y_2^{(2)}, \ldots y_{2N+1}^{(2)}]^{\mathrm{T}}$, $\mathbf{f}^{(2)} = [f_1^{(2)}, f_2^{(2)}, \ldots f_{2N+1}^{(2)}]^{\mathrm{T}}$ with $f_m^{(2)} = f(x_m^{(2)}, y_m^{(2)})$ for $m = 1, 2, \ldots, 2N+1$, and $\mathbf{c}^{(2)} = [A - \frac{1}{48}\, h^2 f_0^{(2)}, 0, \ldots, 0, B - \frac{1}{48}\, h^2 f_{2N+1}^{(2)}]^{\mathrm{T}}$ are also of order $2N+1$.

The local truncation error is now $\mathbf{t}^{(2)} = [t_1^{(2)}, t_2^{(2)}, \ldots, t_{2N+1}^{(2)}]^{\mathrm{T}}$ with

$$t_m^{(2)} = \tfrac{1}{15360}\, h^6 y^{(\mathrm{vi})}(x_m^{(2)}) + \tfrac{11}{15482880}\, h^8 y^{(\mathrm{viii})}(x_m^{(2)}) + \ldots \ ,$$

so that $\mathbf{z}^{(2)} = \mathbf{y}^{(2)} - \mathbf{Y}^{(2)}$, where $\mathbf{y}^{(2)} = [y(x_1^{(2)}), \ldots, y(x_{2N+1}^{(2)})]^{\mathrm{T}}$, satisfies

$$\|\mathbf{z}^{(2)}\| \leqslant \frac{(b-a)^2}{8 - (b-a)^2 F^*}\left(\tfrac{1}{3840}\, h^4 V_6 + \tfrac{11}{3870720}\, h^6 V_8 + \ldots\right) \tag{6.70}$$

Introduce, now, a vector $\mathbf{z}^{(\mathrm{E})}$ of order N defined by

$$\mathbf{z}^{(\mathrm{E})} = \alpha I_{\frac{1}{2}h}^{h}\mathbf{z}^{(2)} + (1-\alpha)\mathbf{z}^{(1)} \ , \tag{6.71}$$

where $I_{\frac{1}{2}h}^{h}$ is a *fine-to-coarse grid restriction operator* defined by

$$I_{\frac{1}{2}h}^{h}\mathbf{z}^{(2)} = [z_2^{(2)},\ z_4^{(2)}, \ldots, z_{2N}^{(2)}]^{\mathrm{T}} \tag{6.72}$$

(quite simply, $I_{\frac{1}{2}h}^{h}$ picks out the elements of $\mathbf{z}^{(2)}$ which correspond those of $\mathbf{z}^{(1)}$). Defining $\|I_{\frac{1}{2}h}^{h}\|$ to be unity, it follows that

$$\|\mathbf{z}^{(\mathrm{E})}\| \leqslant \alpha\|\mathbf{z}^{(2)}\| + (1-\alpha)\|\mathbf{z}^{(1)}\|$$

$$= \frac{(b-a)^2}{8-(b-a)^2F^*}\left[\left(\frac{\alpha}{3840}+\frac{1-\alpha}{240}\right)h^4V_6+O(h^6)\right]$$
$$= O(h^6) \text{ when } \alpha = \tfrac{16}{15} \,.$$

Now,

$$\begin{aligned}\mathbf{z}^{(E)} &= \alpha I^h_{\frac{1}{2}h}\mathbf{z}^{(2)}+(1-\alpha)\mathbf{z}^{(1)}\\ &= \alpha I^h_{\frac{1}{2}h}\,[\mathbf{y}^{(2)}-\mathbf{Y}^{(2)}]+(1-\alpha)[\mathbf{y}^{(1)}-\mathbf{Y}^{(1)}]\end{aligned}$$

so that

$$\mathbf{z}^{(E)} = \mathbf{y}^{(1)}-[\alpha I^h_{\frac{1}{2}h}\mathbf{Y}^{(2)}+(1-\alpha)\mathbf{Y}^{(1)}] \;,$$

and so the vector

$$\mathbf{Y}^{(E)} = \tfrac{16}{15}\, I^h_{\frac{1}{2}h}\mathbf{Y}^{(2)}-\tfrac{1}{15}\mathbf{Y}^{(1)} \tag{6.73}$$

is a sixth-order convergent numerical method for solving (6.58) or (6.47).

The reader will note that the value $\alpha = \frac{16}{15}$ may be obtained by putting $p = 4$ in equation (5.106) of Chapter 5. Introducing a third grid $\mathbf{G}_3$ with step-size $\frac{1}{3}h$ and introducing a second restriction operator $I^h_{\frac{1}{3}h}$ defined by

$$I^h_{\frac{1}{3}h}\mathbf{Y}^{(3)} = [y_3^{(3)}, y_6^{(3)}, \ldots, y_{3N}^{(3)}]^{\mathrm{T}} \;, \tag{6.74}$$

where $\mathbf{Y}^{(3)} = [y_1^{(3)}, y_2^{(3)}, \ldots, y_{3N+2}^{(3)}]^{\mathrm{T}}$ is the solution obtained using the Numerov method on G_3, it may be shown using a similar analysis to that above that

$$\mathbf{Y}^{(E)} = \tfrac{27}{20}\, I^h_{\frac{1}{3}h}\mathbf{Y}^{(3)}-\tfrac{16}{45}\, I^h_{\frac{1}{2}h}\mathbf{Y}^{(2)}+\tfrac{1}{180}\,\mathbf{Y}^{(1)} \tag{6.75}$$

is an eighth-order method for solving (6.47) or (6.58). The coefficients in (6.75) are also obtained by writing $p = 4$, this time in the equations of (5.111) in Chapter 5.

PROBLEMS

1. The fourth order numerical method

$$y_{n+1} = y_n + l\phi_n, \quad y'_{n+1} = y'_n + l\psi_n \;,$$

where

$$\phi_n = y'_n + \tfrac{1}{6}l(k_1+2k_2) \text{ and } \psi_n = \tfrac{1}{6}\,(k_1+4k_2+k_3)$$

with

$$k_1 = f(y_n), \quad k_2 = f(y_n + \tfrac{1}{2}ly_n' + \tfrac{1}{8}l^2k_1), \quad k_3 = f(y_n + ly_n' + \tfrac{1}{2}l^2k_2) \ ,$$

for solving the second order initial value problem

$$y''(t) \equiv f(y) = -4y, \quad y(0) = 0, \quad y'(0) = 2 \ ,$$

is known as a *Nyström method*. Use this method with a step size $l = 0.1$ to find approximations to $y(0.1)$ and $y'(0.1)$.

2. Show that the second order, singular, initial value problem

$$y''(t) + \frac{\lambda y'(t)}{t} = f(y); \quad t,\lambda > 0, \quad y(0) = y_0, \quad y'(0) = z_0$$

may be written as a system of first order equations of the form

$$\text{(A)} \qquad D\mathbf{w}(t) \equiv \mathbf{w}'(t) = A(t)\mathbf{w}(t) + \boldsymbol{\phi}(\mathbf{w}(t)), \quad \mathbf{w}(0) = \mathbf{w}_0,$$

where $D = \text{diag}\{\text{d/dt}\}$ is a diagonal matrix. Give the precise forms of the vectors $\mathbf{w}(t)$, $\boldsymbol{\phi}(\mathbf{w}(t))$ and of the matrix $A(t)$. Write down the precise form of the initial vector $\mathbf{w}_0$.

System (A) is to be solved by replacing the exponential term in

$$\text{(B)} \qquad \mathbf{w}(t+l) = \exp(lD)\mathbf{w}(t),$$

where l is an increment in t, by its (1,0) Padé approximant. Show that this leads to the numerical method

$$\text{(C)} \qquad [I - lA(t+l)]\mathbf{w}(t+l) - l\boldsymbol{\phi}(\mathbf{w}(t+l)) = \mathbf{w}(t); \quad t = 0, l, 2l, \ldots \ ,$$

where I is the identity matrix of the same order as $A(t)$.

Of the family of (m,k) Padé approximants, what difficulty is experienced in integrating (A) from $t = 0$ to $t = l$ if $k > 0$?

Suppose now that in (A) and (C), the matrix $A = A(t)$ is constant and $\boldsymbol{\phi} \equiv \mathbf{0}$ (the zero vector). Verify that the numerical method is first order accurate and show that it may be extrapolated locally to give second order accuracy.

3. Given the second order initial value problem

$$\text{(B)} \qquad y''(t) = f(t,y); \quad t > t_0, \quad y(t_0) = y_0, \quad y'(t_0) = z_0 \ .$$

The linear multistep method

$$\text{(C)} \qquad y_n - 2y_{n+1} + y_{n+2} = l^2\,(af_n + bf_{n+1} + af_{n+2})$$

is proposed for solving (B). Find the values of the parameters a and b in (C) if the method is to be of order two with error constant equal to $-\frac{1}{36}$. Establish the convergence of the method with these values of a and b, and find the phase-lag.

Determine the interval of periodicity of (C) (if it exists), using the values already determined for a and b, if the method is to be used to solve an equation of type (B) which has a periodic solution with known period.

Formula (C), with the values a and b determined above, is to be used as a corrector formula with the linear multistep method

$$\text{(D)} \qquad y_n - 2y_{n+1} + y_{n+2} = l^2 f_{n+1}$$

as predictor. Determine the interval of absolute stability when this combination is used in *PECE* mode.

4. Given the linear multistep method

$$\text{(A)} \qquad y_n - 2y_{n+1} + y_{n+2} = l^2(ay_n'' + by_{n+1}'' + cy_{n+2}'')$$

for solving the second order initial value problem

$$\text{(B)} \qquad y''(t) = f(t,y); \qquad t > t_0, y(t_0) = y_0, \qquad y'(t_0) = y_0' \;.$$

Find the values of the parameters a,b,c if the method is to be of order two with error constant equal to $+1/12$.

Write down the first and second characteristic polynomials of (A) with these values of a,b,c and discuss consistency, zero-stability and convergence.

Determine the interval of periodicity (if it exists) of (A) with these values of a,b,c if the method is to be used to solve an equation of type (B) which has a periodic solution with known period. Determine also the phase-lag of (A) with these same values of a, b and c.

5. Given the second order initial value problem

$$\text{(A)} \qquad y'' = f(t,y); \qquad t > t_0,\; y(t_0) = y_0, \qquad y'(t_0) = z_0 \;.$$

The linear multistep method

$$\text{(B)} \qquad y_n - 2y_{n+1} + y_{n+2} = l^2(ay_n'' + by_{n+1}'' + ay_{n+2}'')$$

is proposed for solving (A). Find the values of the parameters a,b in (B) if the method is to be of order two with error constant equal to $+1/24$.

Establish the convergence of this method with these values of a,b and find the phase-lag.

Determine the interval of periodicity (if it exists) of (B), using the values already determined for a,b, if the method is to be used to solve an equation of type (A) which has a periodic solution with known period.

6. Given the ordinary differential equation

$$x\frac{d^2y}{dx^2}+\frac{dy}{dx}+\lambda^2xy = 0 , \qquad \lambda \text{ constant} ,$$

with boundary values $y(0)=y(1)=0$, replace the derivatives by the usual central differences, using a step length h, to produce the scheme:

$$(2x_i-h)y_{i-1}+\alpha x_iy_i+(2x_i+h)y_{i+1} = 0 ,$$

where $\alpha=(2h^2\lambda^2-4)$.

Taking $h=1/4$ write down, in matrix form, the set of equations which approximates the differential equation and hence estimate the smallest value for λ, to decimal places.

7. For the ordinary differential equation

$$\frac{d^2y}{dx^2}-2x\frac{dy}{dx}+\lambda^2y = 0 \ (\lambda \text{ constant})$$

with boundary values $y(0)=y(1)=0$, replace the derivatives by the usual central differences, using a step length h, to produce the scheme

$$(1+ih^2)y_{i-1}+(h^2\lambda^2-2)y_i+(1-ih^2)y_{i+1} = 0.$$

Taking $h=0.25$ write down in matrix form the set of equations which approximates the differential equation and hence estimate the smallest value for λ, to decimal places.

8. Given the second order differential equation

$$y'' = -\pi^2y$$

with boundary conditions

$$y(0) = 1, \qquad y'(1) = 1 .$$

Approximating the derivative boundary condition at $x=1$ by a second order replacement, and taking $h=1/4$, set up, in matrix form, the linear system which leads to the numerical solution $y(mh)$ where $m=1,2,3,4$. Do not attempt to solve this system.

9. Given the second order differential equation

$$y' = y$$

with exact boundary values

$$y(0.0) = 2.000\ 00, \qquad y(0.4) = 2.162\ 14\ ,$$

show that the Numerov formula

$$y_{n+2} - 2y_{n+1} + y_n = \tfrac{1}{12} h^2 (y''_{n+2} + 10y''_{n+1} + y''_n) + 0(h^6)$$

with $h = 0.1$ becomes

$$0.999\ 17y_{n+2} - 2.008\ 33y_{n+1} + 0.999\ 17y_n = 0,$$

where the coefficients have been rounded to five decimal places. Hence find approximate values for $y(0,1)$, $y(0.2)$ and $y(0.3)$. Work to five decimal places, giving the answers to four decimal places.

10. The linear multistep method

$$\text{(A)} \qquad -y_{n-1} + 2y_n - y_{n+1} + \frac{h^2}{4}(y''_{n-1} + 2y''_n + y''_{n+1}) = 0$$

is to be used to solve the linear, two-point boundary value problem

$$y''(x) = g(x)y(x) + q(x), \qquad y(a) = A, \qquad y(b) = B,$$

where $g(x)$ and $q(x)$ are given functions.

The interval $a \leqslant x \leqslant b$ is discretized in such a way that $x_m = a + mh$ for $m = 0,1,\ldots,N+1$ with $(N+1)h = b - a$.

Show that the numerical solution vector $\mathbf{Y}$ may be obtained by solving a linear algebraic system of the form

$$(J + \tfrac{1}{4}h^2 EG)\mathbf{Y} = \mathbf{c} - \tfrac{1}{4}h^2 E\mathbf{q}$$

and give the precise forms of J, E, G, $\mathbf{c}$ and $\mathbf{q}$.

Given that, for any square matrix M for which $\|M\|_\infty < 1$, $\|(I+M)^{-1}\|_\infty \leqslant (1 - \|M\|_\infty)^{-1}$, that the modulus of the principal part of the local truncation error of (A) is $\frac{1}{6} h^4 |y^{(iv)}(x)|$, and that $\|J^{-1}\|_\infty = (N+1)^2/8$, find the criterion on G under which (A) is second order convergent.

11. The linear multistep method

$$\text{(A)} \qquad y_{n-1} - 2y_n + y_{n+1} = \frac{h^2}{9}\,(y''_{n+1} + 7y''_n + y''_{n+1})$$

is to be used for solving the two point boundary problem

$$\text{(B)} \qquad y''(x) = g(x)y(x) + k(x); \qquad y(a) = A, \qquad y(b) = B\ ,$$

where $g(x)$, $k(x)$ are given functions.

The interval $x \in [a,b]$ is discretized in such a way that $x_i = a + ih$ for $i = 0,1,\ldots,N+1$ with $h = (b-a)/(N+1)$.

Denoting the solution vector of (A) by $\mathbf{Y}$, show that $\mathbf{Y}$ may be obtained by solving the linear system

$$\text{(C)} \qquad (J + h^2EG)\mathbf{Y} = \mathbf{c} - h^2E\mathbf{k}\ ,$$

where J is the tridiagonal matrix of order N given by

$$J = \begin{bmatrix} 2 & -1 & & & & 0 \\ -1 & 2 & -1 & & & \\ \cdots & \cdots & & \cdots & & \\ & & & -1 & 2 & -1 \\ 0 & & & & -1 & 2 \end{bmatrix}.$$

Give the precise forms of E,G,$\mathbf{c}$ and $\mathbf{k}$.

Given that the principal part of the local truncation error of (A) is $-\frac{1}{36}h^4\,y^{(\mathrm{vi})}(x)$ and that $\|J^{-1}\| = (N+1)^2/8$, find the condition on G under which (A) is second order convergent.

12. Using the basis functions

$$\phi_i(x) = \begin{cases} 0 & ; \quad 0 \leqslant x \leqslant x_{i-1}\ , \\ \dfrac{x - x_i}{h_{i-1}} & ; \quad x_{i-1} \leqslant x \leqslant x_i\ , \\ \dfrac{x_{i+1} - x}{h_i} & ; \quad x_i \leqslant x \leqslant x_{i+1}\ , \\ 0 & ; \quad x_{i+1} \leqslant x \leqslant 1\ , \end{cases}$$

where $h_i = x_{i+1} - x_i$, the Ritz–Galerkin method is to be used to solve the two-point boundary value problem

$$-4y''+3y = x; \qquad y(0) = y(1) = 0$$

at the points $x_i(i=1,2,\ldots,N)$ with $0<x_1<x_2<\ldots<x_N<1$. Show that this method leads to a system of equations of the form $A\mathbf{c}=\mathbf{b}$, that the non zero entries of A are

$$a_{ii} = \int_0^1 \{4[\phi_i'(x)]^2+3[\phi_i(x)]^2\}\,\mathrm{d}x; \qquad i = 1,2,\ldots,N\ ,$$

$$a_{i,i+1} = \int_0^1 \{4\phi_i'(x)\phi_{i+1}'(x)+3\phi_i(x)\phi_{i+1}(x)\}\,\mathrm{d}x; \qquad i = 1,2,\ldots,N-1\ ,$$

$$a_{i,i+1} = \int_0^1 \{4\phi_i'(x)\phi_{i-1}'(x)+3\phi_i(x)\phi_{i-1}(x)\}\,\mathrm{d}x; \qquad i = 2,3,\ldots,N\ ,$$

and that

$$b_i = \int_0^1 x\phi_i(x)\,\mathrm{d}x; \qquad i = 1,2,\ldots,N\ .$$

Find expressions for the $a_{i,j}$, b_i in terms of the h_i,x_i. Deduce, in matrix form, the linear system which gives c_1,c_2 relating to A with $N=2$, $x_1=\frac{1}{4}$, $x_2=\frac{1}{2}$.

13. The Ritz–Galerkin method with basis functions

$$\phi_i(x) = \begin{cases} 0 & , \quad 0 \leqslant x \leqslant x_{i-1} \\ \dfrac{x-x_{i-1}}{h_{i-1}}, & \quad x_{i-1} \leqslant x \leqslant x_i \\ \dfrac{x_{i+1}-x}{h_i} & , \quad x_i \leqslant x \leqslant x_{i+1} \\ 0 & , \quad x_{i+1} \leqslant x \leqslant 1 \end{cases}$$

where $h_i=x_{i+1}-x_i$, is to be used to solve the boundary value problem

$$\text{(A)} \qquad -y''+y = x, \qquad 0<x<1; \qquad y(0) = 0, \qquad y(1) = 0$$

at the points $x_i(i=1,2,\ldots,N)$ with $0<x_1<x_2<\ldots<x_N<1$. Show that, in $A\mathbf{c}=\mathbf{b}$, the non-zero entries of A are

$$a_{ii} = \int_0^1 \{[\phi_i'(x)]^2 + [\phi_i(x)]^2\}\,\mathrm{d}x \ , \qquad i = 1,2,\ldots,N$$

$$a_{i,i+1} = \int_0^1 \{\phi_i'(x)\phi_{i+1}'(x) + \phi_i(x)\phi_{i+1}(x)\}\,\mathrm{d}x \ , \qquad i = 1,2,\ldots,N-1 \ ,$$

$$a_{i,i-1} = \int_0^1 \{\phi_i'(x)\phi_{i-1}'(x) + \phi_i(x)\phi_{i-1}(x)\}\,\mathrm{d}x \ , \qquad i = 2,\ldots,N \ ,$$

and that

$$b_i = \int_0^1 x\phi_i(x)\,\mathrm{d}x \ , \qquad i = 1,2,\ldots,N \ ;$$

the vector **c** is the vector of expansion coefficients.

Find expressions for the a_{ij}, b_i in terms of the h_i, x_i and hence find, in matrix form, the linear system which gives c_1, c_2 relating to (A) with $n=2$, $x_1=0.25$, $x_2=0.5$.

7

Hyperbolic equations

7.1 INTRODUCTION TO FIRST-ORDER EQUATIONS

Given an $M \times M$ real matrix A with real eigenvalues and M linearly independent eigenvectors, and a vector $\mathbf{u}(x,t)$ with M components, the general first-order hyperbolic system is given by

$$\frac{\partial \mathbf{u}}{\partial t} + A\frac{\partial \mathbf{u}}{\partial x} = \mathbf{0} \tag{7.1}$$

where $\mathbf{0}$ is the zero-vector of order M. The matrix A is not necessarily symmetric. Given, additionally, the initial conditions

$$\mathbf{u}(x,0) = \mathbf{g}(x) \ ; \qquad -\infty < x < \infty \ , \tag{7.2}$$

(7.1) together with (7.2) form the *Cauchy initial value problem* for a first-order hyperbolic system.

Explicit finite difference methods of solution, in which the solution at a single point at some time $t + l$ (l being an appropriate increment in t) is expressed in terms of the solution at points at time t, may be used to solve this problem. The solution is computed in a triangular domain and may be obtained when A has positive, negative, or a mixture of positive and negative eigenvalues.

If, on the other hand, the problem is an *initial-boundary value problem*, a solution can only be obtained when the problem is *well posed* or *properly posed*. In general, a problem is well posed if

(1) the solution is unique when it exists,
(2) the solution depends continuously on the initial data,
(3) a solution always exists for initial data that are arbitrarily close to initial data for which no solution exists.

A problem which is not well posed is said to be *ill posed*.

There are three types of well posed *initial-boundary value* problem. They are related to the eigenvalues of A as follows:

Type 1 A positive definite (the outflow problem).

Here the problem is given in the form

$$\begin{aligned} &\frac{\partial \mathbf{u}}{\partial t} + A\frac{\partial \mathbf{u}}{\partial x} = \mathbf{0} \; ; \quad x>0, \; t>0 \\ &\mathbf{u}(0,t) = \mathbf{v}(t) \quad ; \quad t>0 \\ &\mathbf{u}(x,0) = \mathbf{g}(x) \quad ; \quad x \geqslant 0 \end{aligned} \tag{7.3}$$

and the solution is obtained in the first quadrant. Explicit or implicit methods (where the solution at more than one point at the advanced time $t+l$ is involved) may be used to obtain the solution.

Type 2 A negative definite (the inflow problem).

Here the problem is given in the form

$$\begin{aligned} &\frac{\partial \mathbf{u}}{\partial t} + A\frac{\partial \mathbf{u}}{\partial x} = \mathbf{0} \; ; \quad x<0, \; t>0 \\ &\mathbf{u}(0,t) = \mathbf{v}(t) \quad ; \quad t>0 \\ &\mathbf{u}(x,0) = \mathbf{g}(x) \quad ; \quad x \leqslant 0 \end{aligned} \tag{7.4}$$

and the solution is obtained in the second quadrant. This problem may also be solved using explicit or implicit finite difference methods.

Type 3 A neither positive nor negative definite.

Here the initial-boundary value problem is much more difficult to solve. The solution is obtained in an open rectangle and implicit methods are used in the computation. The problem must be given in the form

$$\begin{aligned} &\frac{\partial \mathbf{u}}{\partial t} + A\frac{\partial \mathbf{u}}{\partial x} = \mathbf{0} \; ; \quad L_0<x<L_1, \; t>0 \\ &\mathbf{u}(x,0) = \mathbf{g}(x) \quad ; \quad L_0 \leqslant x \leqslant L_1, \end{aligned} \tag{7.5}$$

any k components of $\mathbf{u}$ given on the line $x = L_0, t>0$, the other $M-k$ components of $\mathbf{u}$ given on the line $x = L_1$ $(t>0)$. The numerical solution is obtained in the open rectangle $L_0<x<L_1, t>0$.

In each of these three types of problem, a total of $2M$ boundary and/or initial conditions are specified, thus matching the total number of derivatives (M with respect to each of x and t) in the hyperbolic system (7.1). There exist numerical methods which require all components of the vector **u** to be given on both boundaries $x = L_0$ and $x = L_1$ as well as the initial vector **g**, thus raising the number of specified conditions to $3M$. As the number of derivatives in the hyperbolic system is still only $2M$, the problem becomes *overposed*.

First-order hyperbolic problems may be solved using one-step methods of characteristic, explicit (as described above), or implicit type. In cases where the matrix A in (7.1) is constant, the finite difference methods to be developed in the following sections of the chapter may be applied to a system of the form (7.1) with only minor alterations.

To a certain extent this is also true when the matrix A is a function of x or a function of x and t. In these cases the matrix A is evaluated at the mesh points involved in the finite difference scheme.

Systems in which the matrix A is constant, or a function of x, or a function of t, or a function of x and t, are examples of *linear* first-order hyperbolic systems. *Nonlinear* hyperbolic systems arise in conservation problems where the system is in the *conservation form*

$$\frac{\partial \mathbf{u}}{\partial t} + \frac{\partial \mathbf{f}(\mathbf{u})}{\partial x} = \mathbf{0} \tag{7.6}$$

and **u** and **f** have M components.

The nonlinear system (7.6) can be written in a form analogous to (7.1) by writing

$$A(\mathbf{u}) \equiv \frac{\partial \mathbf{f}}{\partial \mathbf{u}} = \frac{\partial(f_1, f_2, \ldots, f_M)}{\partial(u_1, u_2, \ldots, u_M)} \; . \tag{7.7}$$

The matrix $A(\mathbf{u})$ in (7.7) is the Jacobian of **f** with respect to **u**. System (7.6) can now be written in the form

$$\frac{\partial \mathbf{u}}{\partial t} + A(\mathbf{u})\frac{\partial \mathbf{u}}{\partial x} = \mathbf{0} \; . \tag{7.8}$$

and finite difference methods may be applied as in the linear case. A significant difference is that the final solution is obtained using a method for solving a nonlinear algebraic system (such as the Newton–Raphson method for a system which was described in Chapter 3), whereas for linear hyperbolic systems, implicit finite difference methods require the use of methods for solving a linear algebraic system (as in Chapter 1).

Finite difference methods for a single hyperbolic equation (that is, $M = 1$) will be developed and analysed in section 7.3; such methods are easily adapted to problems in which $M > 1$. Furthermore, only outflow problems will be considered; generalization to other problems is then immediate.

7.2 THE RÔLE OF CHARACTERISTICS

7.2.1 The Method of Characteristics

The *Method of Characteristics* is universally regarded as the most effective method for solving hyperbolic problems. Indeed, a study of other methods of solution should not be undertaken until the reader has become familiar with the rôle played by characteristics in the solution of hyperbolic equations.

However, it is important to study other methods for solving hyperbolic equations in one space variable (and time), for these often generalize in a natural way to equations with two or three space dimensions where characteristic methods are less satisfactory.

Consider, then, the general first-order hyperbolic equation

$$a\frac{\partial u}{\partial x} + b\frac{\partial u}{\partial t} = c \tag{7.9}$$

where a,b,c are functions of x,t,u but not of $\partial u/\partial x$ nor of $\partial u/\partial t$. The notations $p \equiv \partial u/\partial x$ and $q \equiv \partial u/\partial t$ will be used for convenience so that (7.9) may be written in the form

$$ap + bq = c \ . \tag{7.10}$$

The total differential formula is

$$\mathrm{d}u = \frac{\partial u}{\partial x}\mathrm{d}x + \frac{\partial u}{\partial t}\mathrm{d}t = p\mathrm{d}x + q\mathrm{d}t \tag{7.11}$$

and eliminating p from (7.10) and (7.11) gives

$$\mathrm{d}u = \frac{(c - bq)}{a}\mathrm{d}x + q\mathrm{d}t$$

which can be written in the form

$$q(a\mathrm{d}t - b\mathrm{d}x) + (c\mathrm{d}x - a\mathrm{d}u) = 0 \ . \tag{7.12}$$

Note, it is possible to derive an equation independent of p because a,b,c are independent of p. Equation (7.12) can be made independent of q by stipulating that it hold only along some curve C in the (x,t)-plane which satisfies the equation

$$a\mathrm{d}t - b\mathrm{d}x = 0 \ . \tag{7.13}$$

It follows from (7.12) and (7.13) that, along C, the relation

$$c\mathrm{d}x - a\mathrm{d}u = 0 \tag{7.14}$$

holds also and, since a and b are uniquely determined at any point (x,t), $\mathrm{d}t/\mathrm{d}x$ has only one value at (x,t). This means that C is the only curve which passes through (x,t). From (7.13) and (7.14) it follows that

$$\frac{\mathrm{d}x}{a} = \frac{\mathrm{d}t}{b} = \frac{\mathrm{d}u}{c} \tag{7.15}$$

along C.

7.2.2 Implementation of the method

Suppose, now, that the value of u is known at N discrete points $T_m^{(1)}$ $(m = 1,2,\ldots,N)$ on some initial curve L in the (x,t)-plane; the curve L must not be a characteristic. Then through the point $T_m^{(1)}$ the characteristic C_m may be drawn. The method of characteristics for first-order hyperbolic equations, which provides a procedure for integrating along the characteristics, approximates C_m by the piecewise straight lines $T_m^{(k)}T_m^{(k+1)}$ $(k = 1,2,\ldots)$ where the points $T_m^{(k)}$ $(k = 1,2,\ldots)$ lie on C_m; these points need not be equally spaced along C_m. The notation $x_m^{(k)}$ will be used to denote the value of x at the point $T_m^{(k)}$ $(k = 1,2,\ldots)$ on $C_m (m = 1,2,\ldots,N)$, with similar notations for t,u,a,b,c and, when required, for $p,q,f \equiv \mathrm{d}t/\mathrm{d}x = b/a$ at that point.

The algorithm advances the integration from $T_m^{(k)}$ to $T_m^{(k+1)}$ as follows:

(1) a,b,c at the base point $T_m^{(k)}$ are evaluating using the forms in which they appear in the differential equation (7.9);

(2) assume that $x_m^{(k+1)}$ is known; then a predicted estimate of $t_m^{(k+1)}$ may be obtained by adapting (7.13) to give

$$a_m^{(k)}(t_m^{(k+1)} - t_m^{(k)}) - b_m^{(k)}(x_m^{(k+1)} - x_m^{(k)}) = 0, \tag{7.16}$$

in which $t_m^{(k+1)}$ is the only unknown;

(3) a predicted value of $u_m^{(k+1)}$ is obtained by adapting (7.14) to give

$$c_m^{(k)}(x_m^{(k+1)} - x_m^{(k)}) - a_m^{(k)}(u_m^{(k+1)} - u_m^{(k)}) = 0, \tag{7.17}$$

in which $u_m^{(k+1)}$ is the only unknown;

(4) a,b,c at $T_m^{(k+1)}$ are evaluated using $x_m^{(k+1)}$, $t_m^{(k+1)}$, $u_m^{(k+1)}$ in the forms in which a,b,c appear in the differential equation (7.9);

(5) a corrected value of $t_m^{(k+1)}$ is obtained by writing (7.16) in the form

$$\tfrac{1}{2}(a_m^{(k+1)}+a_m^{(k)})(t_m^{(k+1)}-t_m^{(k)})-\tfrac{1}{2}(b_m^{(k+1)}+b_m^{(k)})(x_m^{(k+1)}-x_m^{(k)})=0$$

in which $t_m^{(k+1)}$ is the only unknown;

(6) a corrected value of $u_m^{(k+1)}$ is obtained by writing (7.17) in the form

$$\tfrac{1}{2}(c_m^{(k+1)}+c_m^{(k)})(x_m^{(k+1)}-x_m^{(k)})-\tfrac{1}{2}(a_m^{(k+1)}+a_m^{(k)})(u_m^{(k+1)}-u_m^{(k)})=0$$

in which $u_m^{(k+1)}$ is the only unknown.

An iterative procedure is set up using stages (4), (5), (6) of the algorithm and is carried out until each of $t_m^{(k+1)}$ and $u_m^{(k+1)}$ converge to a predetermined accuracy.

In the event that q,p,f are required at $T_m^{(k+1)}$, they may now be calculated from (7.12), (7.10) and the equation $f=b/a$, respectively.

7.3.3 Discontinuities

Certain discontinuities can arise across the characteristic through a point on the initial curve. The first such difficulty arises with discontinuities in initial data and the second arises with discontinuities in one, or both, of the first derivatives at an initial point.

In both cases the discontinuities are propagated along the characteristic curve through the point at which the discontinuities occur. It is reasonable to conclude therefore that, for such problems, the method of characteristics should be used to carry out the integration of the hyperbolic equation. For systems of hyperbolic equations, or for a single equation in more than one space variable, it is more difficult to write computer codes for characteristic methods and the user may well be advised to use finite difference methods to obtain a numerical solution. A description of how the two types of discontinuity are propagated is given in Chapter 4 of Twizell (1984).

7.3 BACKWARD DIFFERENCE METHODS

7.3.1 Development

Backward difference methods may be used to compute numerical solutions of the properly posed initial value problem

$$\frac{\partial u}{\partial t}+a\frac{\partial u}{\partial x}=0\ ;\quad a>0,\ x>0,\ t>0 \tag{7.18}$$

(the outflow problem) with initial conditions given by

$$u(x,0)=g(x)\ ;\quad x\geqslant 0 \tag{7.19}$$

and boundary conditions by

$$u(0,t)=\upsilon\ ;\quad t>0 \tag{7.20}$$

in which υ is a constant. If $g(0) \neq \upsilon$ there is a discontinuity between initial conditions and boundary conditions at the origin.

Suppose that the solution $u(x,t)$ of (7.18) is sought in some region $R = [0 < x \leqslant X, t > 0]$ of the first quadrant. The interval $0 \leqslant x \leqslant X$ is divided into N equal parts each of width h so that $Nh = X$. The x-coordinates of the N points of this discretization are $x_m = mh$ $(m = 1,2,\ldots,N)$; clearly, $x_0 = 0$ is the x-coordinate of every point on the t-axis.

The vector $\mathbf{u}(t) = [u_1(t), u_2(t), \ldots, u_N(t)]^{\mathrm{T}}$, T denoting transpose, will represent the solution vector at time $t > 0$ and $\mathbf{U}(t)$ will denote the theoretical solution of an approximating numerical method. The time variable t will be discretized by a constant time step l such that $t = nl$ $(n = 0,1,2,\ldots)$.

Replacing the space derivative in (7.18) by the backward difference formula

$$\frac{\partial u}{\partial x} = h^{-1}\{u(x,t) - u(x-h,t)\} + O(h) \tag{7.21}$$

and applying (7.18) with (7.21) to every one of the N discrete points at time level t leads to the system of first-order ordinary differential equations given by

$$\frac{\mathrm{d}\mathbf{U}(t)}{\mathrm{d}t} = -aC\mathbf{U}(t) + a\mathbf{c} \ . \tag{7.22}$$

This approach of transforming a partial differential equation in this way into a system of ordinary differential equations is called the *Method of Lines*; it will be met again in section 7.7 for second-order hyperbolic equations and in Chapter 8 for parabolic equations.

In (7.22), C is a lower bi-diagonal square matrix of order N given by

$$C = h^{-1}\begin{bmatrix} 1 & & 0 & & \\ -1 & 1 & & & \\ & -1 & 1 & & \\ & & & \cdots & \\ 0 & & & -1 & 1 \end{bmatrix} \tag{7.23}$$

and $\mathbf{c}$ is a vector with N components given by

$$\mathbf{c} = h^{-1}[\upsilon, 0, \ldots, 0]^{\mathrm{T}} \ . \tag{7.24}$$

The N eigenvalues of the matrix C each have the value $1/h$.

The solution of (7.22) with (7.19) is

$$\mathbf{U}(t) = C^{-1}\mathbf{c} + \exp(-atC)\{\mathbf{g} - C^{-1}\mathbf{c}\} \tag{7.25}$$

where $\mathbf{g} = [g(h), g(2h), \ldots, g(Nh)]^T$ is the vector of initial conditions. It is easy to show that this vector $\mathbf{U}(t)$ satisfies

$$\mathbf{U}(t+l) = C^{-1}\mathbf{c} + \exp(-alC)\{\mathbf{U}(t) - C^{-1}\mathbf{c}\} \ . \tag{7.26}$$

To obtain numerical solutions $\mathbf{U}(t+l)$, the exponential matrix function $\exp(-alC)$ must be replaced by a valid approximant.

One approximant is given by

$$\exp(-alC) = (I + \theta alC)^{-1}[I - (1-\theta)alC] \tag{7.27}$$

in which $0 \leqslant \theta \leqslant 1$ is a parameter and I is the identity matrix of order N; clearly (7.27) is rational if $\theta \neq 0$. This leads, in (7.26), to the family of methods

$$(I + \theta alC)\mathbf{U}(t+l) = [I - (1-\theta)alC]\mathbf{U}(t) + al\mathbf{c} \ . \tag{7.28}$$

Setting $\theta = 0, 1, \frac{1}{2}$ in (7.28) gives the (0,1), (1,0) and (1,1) Padé approximants to $\exp(-alC)$ and it is these three values which will be discussed further.

Taking $\theta = 0$ first of all yields

$$\mathbf{U}(t+l) = (I - alC)\mathbf{U}(t) + al\mathbf{c} \tag{7.29}$$

from (7.26). Applying (7.29) to the mesh point $(x_m, t_n) = (mh, nl)$ gives the three-point explicit scheme

$$U_m^{n+1} = (1 - ar)U_m^n + arU_{m-1}^n \tag{7.30}$$

at the mesh point (mh, nl). The computed solution of the difference scheme (the solution actually obtained, which may be subject to errors to be discussed in subsection 7.3.3) will be denoted by $\tilde{U}_m^n$ at this point. For $m = 1$, $n = 0$, equation (7.30) uses the origin as one of its mesh points so that a discontinuity between initial conditions and boundary conditions will be propagated as the computation proceeds.

It is constructive at this point to consider the *Courant–Friedrichs–Lewy* (*CFL*) *condition* for first-order hyperbolic equations. This condition states that *the characteristic through the advanced point of an explicit difference scheme must intersect the line through the points used by the scheme at the base time level within the range of these points*.

From section 7.1 the slope of the characteristic through the advanced point $(mh, (n+1)l)$ is $1/a$ and it is easy to verify geometrically that the CFL condition for (7.30) is $0 \leqslant ar \leqslant 1$.

Writing, next, $\theta = 1$ in (7.27) and using the resulting approximant in the recurrence relation (7.28) gives

$$(I + alC)\mathbf{U}(t + l) - al\mathbf{c} = \mathbf{U}(t) \quad . \tag{7.31}$$

Applying (7.31) to the mesh point (mh,nl) gives the three-point formula

$$(1 + ar)U_m^{n+1} - arU_{m-1}^{n+1} = U_m^n \quad . \tag{7.32}$$

Equation (7.32) uses more than one point at the advanced time level $t = (n + 1)l$ and would, therefore, be normally classed as implicit (as it uses only one point at the base time level, it is further classed as *fully implicit*). The reader will remember, however, that the boundary condition associated with the hyperbolic equation is given along the t-axis, so that, for any n, applying (7.32) to the mesh point (h,nl) means that the mesh points (h,nl), $(0,(n + 1)l)$ and $(h,(n + 1)l)$ are used. The only unknown in the application of (7.32) to (h,nl) is, therefore, U_1^{n+1}. This means that (7.32) is being used explicitly after all. Obviously, for all other $m = 2,3,\ldots,N$ the scheme is also used explicitly. The CFL condition may not, however, be used in analysing the scheme. An attractive feature of the method is that any discontinuity between initial and boundary conditions is not used by the method.

The third method of the family to be discussed is obtained by writing $\theta = \frac{1}{2}$ in (7.27). This gives

$$(I + \tfrac{1}{2}alC)\mathbf{U}(t + l) - \tfrac{1}{2}al\mathbf{c} = (I - \tfrac{1}{2}alC)\mathbf{U}(t) + \tfrac{1}{2}al\mathbf{c} \tag{7.33}$$

which, when applied to the general mesh point (mh,nl), gives the four-point implicit scheme

$$(1 + \tfrac{1}{2}ar)U_m^{n+1} - \tfrac{1}{2}arU_{m-1}^{n+1} = (1 - \tfrac{1}{2}ar)U_m^n + \tfrac{1}{2}arU_{m-1}^n \quad . \tag{7.34}$$

Like (7.32), this scheme may also be used explicitly in the first quadrant of the (x,t)-plane.

7.3.2 Local truncation errors and consistency

The *local truncation error* of a finite difference scheme at a mesh point (mh,nl) is the difference between the finite difference scheme and the differential equation it is replacing. The *local discretization error* is the difference between the theoretical solutions of the differential and difference equations and is represented at the mesh point (mh,nl) by $z_m^n = u_m^n - U_m^n$ where $u_m^n = u(mh,nl)$. If the local truncation error tends to zero as the mesh is refined, that is as $h,l \to 0$ simultaneously, the difference equation is said to be *consistent* with the differential equation.

Consider the general finite difference method arising from (7.28). It is easy to see that this method takes the form

$$\frac{U_m^{n+1} - U_m^n}{l} + a\frac{\{\theta(U_m^{n+1} - U_{m-1}^{n+1}) + (1 - \theta)(U_m^n - U_{m-1}^n)\}}{h} = 0 \tag{7.35}$$

from which it follows that the first term is an approximation to $\partial u/\partial t$ while the second term is a weighted approximation to $a\partial u/\partial x$. The local truncation error of (7.35) is therefore given by

$$L(x,t) = \frac{u(x,t+l)-u(x,t)}{l}$$

$$+ a\frac{\theta\{u(x,t+l)-u(x-h,t+l)\}+(1-\theta)\{u(x,t)-u(x-h,t)\}}{h}$$

$$-\left(\frac{\partial u}{\partial t}+a\frac{\partial u}{\partial x}\right) . \tag{7.36}$$

Expanding about $u(x,t)$ using Taylor's series, it may be shown that

$$L(x,t) = \left(-\tfrac{1}{2}ah\frac{\partial^2 u}{\partial x^2}+C_2 l\frac{\partial^2 u}{\partial t^2}\right)+C_3 l^2\frac{\partial^3 u}{\partial t^3} + \ldots \tag{7.37}$$

with $C_2=\frac{1}{2}-\theta$ and $C_3=\frac{1}{6}-\frac{\theta}{2}$. (Note: for (7.18), $a\partial^2u/\partial x\partial t = -\partial^2u/\partial t^2$ and $a\partial^3u/\partial x\partial t^2 = -\partial^3u/\partial t^3$.) In (7.37), the term in parentheses on the right-hand side is the *principal part* of the local truncation error. The component $-\frac{1}{2}ah\partial^2u/\partial x^2$ is due to the space discretization and to the use of (7.21) in (7.18); this term will appear in the principal part of the local truncation error of every finite difference method arising from (7.28) in which $0\leqslant\theta\leqslant 1$.

The constant C_2 in (7.37) is known as the *time error constant* of the finite difference method. Setting $\theta=0$ gives $C_2=\frac{1}{2}$ which is the time error constant of method (7.30) while setting $\theta=1$ gives $C_2=-\frac{1}{2}$ which is the time error constant of method (7.32). These two methods are said to be first-order methods, meaning first-order accurate in time. Setting $\theta=\frac{1}{2}$ causes C_2 to vanish in (7.37) and the time error constant of method (7.34) is $C_3=-\frac{1}{12}$: this method is said to be a second-order method.

It is easy to see from (7.37) that $L(x,t)\rightarrow 0$ as $h,l\rightarrow 0$ simultaneously so that all members of the family of methods (7.28) are consistent with the hyperbolic equation (7.18).

7.3.3 Stability

The concept of *stability* is concerned with the boundedness of the solution of the finite difference equations and this is examined by finding conditions under which $Z_m^n = U_m^n - \widetilde{U}_m^n$ remains bounded as n increases, for fixed h and l. A stability analysis considers the growth of perturbations in initial data or the growth of errors introduced at mesh points at a given time level. It will be convenient to define the vector $\mathbf{U}^n=[U_1^n,U_2^n,\ldots,U_N^n]^{\mathrm{T}}$, with similar definitions of $\mathbf{Z}^n$ and $\widetilde{\mathbf{U}}^n$.

There are three common methods of investigating stability: the *energy method*, the *matrix method*, and the *von Neumann* or *Fourier method*. The energy method is powerful in dealing with boundary conditions, variable coefficient problems and nonlinear problems. Its application can, unfortunately, become extremely complicated and its successful use will be due in no small part to the ingenuity of the user. Besides proving the stability of a finite difference scheme, the energy method can indicate the correct choice of a method. However, the method provides only sufficient conditions for stability which may be far removed from what is necessary in certain initial-boundary value problems.

The method calculates the sum of the squares of the errors Z_m^n at a time level $t = nl$. This sum of squared errors is called the *energy*, from which the method gets its name, and the method determines criteria under which this energy remains bounded as the computation proceeds. The reader should note that this conserved quantity (the energy) is not the physical energy of the system.

The matrix method of analysis is applicable to initial-boundary value problems and is very popular among users. Unfortunately it can give erroneous results. By way of example, consider the explicit scheme given in matrix form by (7.29). It is clear that a perturbation $\mathbf{Z}^0 = \mathbf{U}^0 - \widetilde{\mathbf{U}}^0$ of the initial conditions satisfies

$$\mathbf{Z}^{n+1} = (I - alC)\mathbf{Z}^n \tag{7.38}$$

and since every eigenvalue of C has the value $1/h$ each eigenvalue of the *amplification matrix* $W = I - alC$ has the value $1 - ar$. Clearly, $\mathbf{Z}^{n+1}$ will not grow in magnitude as the computation proceeds provided

$$\|W\| \leqslant 1 \ , \tag{7.39}$$

The success of the matrix method depends on being able to obtain a suitable estimate for $\|W\|$. The spectral norm, which is equal to the spectral radius when W is a symmetric matrix, is the usual choice of norm. The spectral radius is usually used to approximate the spectral norm when W is nonsymmetric, leading occasionally to an erroneous stability interval, as will now be seen.

The stability criterion (7.39) when applied to (7.29) leads to

$$-1 \leqslant 1 - ar \leqslant 1$$

from which it follows that (7.29) is stable provided $0 \leqslant ar \leqslant 2$. This restriction on ar is, however, too lenient (Morton, 1980).

The correct stability interval for (7.29) may be found using the von Neumann or Fourier method developed by J. von Neumann in the early 1940s. A Fourier stability analysis determines the criterion governing the growth of a function which reduces to this Fourier series for $t = 0$.

The Fourier series can be written in terms of sines and cosines in the usual way but it soon becomes apparent that the algebraic manipulation involved is reduced if the exponential form is used. The errors at the $N + 2$ mesh points along the initial line $t = 0$ are written in terms of the error function

$$Z_m^0 = \sum_{r=0}^{N+1} A_r \exp(\mathrm{i}\beta_r mh) \tag{7.40}$$

where $m = 0, 1, \ldots, N+1$, $\mathrm{i} = +\sqrt{-1}$, and the $A_r (r = 0, 1, \ldots, N+1)$ are the Fourier coefficients. The $N+2$ equations in (7.40) are sufficient to determine the $N+2$ unknowns $A_r (r = 0, 1, \ldots, N+1)$ uniquely.

Assuming that the finite difference method to be analysed is linear, so that separate solutions will be additive, it is only necessary to consider the propagation of the error due to a single typical term. Consider the term associated with the typical frequency $|\beta_s| = |\beta|$. The coefficient A_s is constant and can be neglected.

To examine the propagation of this single typical error as $t \rightarrow \infty$, the solution of the finite difference scheme must be found which reduces to $\exp(\mathrm{i}\beta mh)$ when $t = 0$. Let such a solution be

$$\mathrm{e}^{\alpha t}\mathrm{e}^{\mathrm{i}\beta x} = \mathrm{e}^{\alpha nl}\mathrm{e}^{\mathrm{i}\beta mh}$$

where α is complex. The original error component will not grow with time (as n increases, that is) if

$$|\mathrm{e}^{\alpha l}| \leqslant 1 \tag{7.41}$$

for all α. This is *von Neumann's criterion for stability*, sometimes called the *von Neumann necessary condition for stability*.

In deriving this condition, a number of assumptions were made and so a number of important points must be noted. The first is that the Fourier method applies only if the coefficients of the linear difference scheme are constant. If the difference scheme has variable coefficients the method can still be applied at every mesh point, individually, of the mesh (this is known as examining the local stability) and it is reasonable to assume that the difference scheme is stable globally if it is stable at *every* mesh point. The von Neumann condition is necessary only for three-level schemes (see subsection 7.7.4); for two-level schemes the condition is sufficient as well as necessary. This also applies to problems with more than one space variable. Strictly speaking, the von Neumann or Fourier method applies only to pure initial value problems with periodic initial data. In practice, however, it is used to analyse finite difference schemes applied to initial-boundary value problems also.

Returning to (7.30), the error Z_m^n satisfies

$$Z_m^{n+1} = (1 - ar)Z_m^n + arZ_{m-1}^n \ . \tag{7.42}$$

Substituting $\mathrm{e}^{\alpha nl}\mathrm{e}^{\mathrm{i}\beta mh}$ for Z_m^n, with similar substitutions for Z_m^{n+1} and Z_{m-1}^n and dividing by $\mathrm{e}^{\alpha nl}\mathrm{e}^{\mathrm{i}\beta mh}$ leaves

$$\xi = 1 - ar + are^{-\mathrm{i}\beta h}$$

where $\xi = e^{\alpha l}$ is the *amplification factor*. This leads to

$$\xi = 1 - 2ar\sin^2\tfrac{1}{2}\beta h + iar\sin\beta h$$

so that $|\xi|^2 = 1 - 4ar(1-ar)\sin^2\tfrac{1}{2}\beta h$. It follows that $|\xi| \leqslant 1$ provided $0 < ar \leqslant 1$ (the possibility $ar = 0$ can be ignored).

This restriction on ar is in agreement with the CFL condition for the method. This example is probably the most widely used to illustrate the failure of the matrix method for analysing stability.

The matrix method is successful in analysing the method given by (7.32). A perturbation $\mathbf{Z}^0$ of the initial vector $\mathbf{g}$ satisfies

$$\mathbf{Z}^{n+1} = (I + alC)^{-1}\mathbf{Z}^n \tag{7.43}$$

(from (7.31)) and since the eigenvalues of C all have the value $1/h$, the eigenvalues of the amplification matrix $(I + alC)^{-1}$ are all $1/(1+ar)$ and thus never exceed unity for non-negative values of r. The initial perturbation is therefore damped as t increases, from which it is concluded that the scheme is *unconditionally stable*. This gives the method an obvious advantage over (7.29), as larger time steps may be used.

Turning, finally, to the method given by (7.34), a perturbation $\mathbf{Z}^0$ of the initial vector $\mathbf{g}$ satisfies

$$\mathbf{Z}^{n+1} = (I + \tfrac{1}{2}alC)^{-1}(I + \tfrac{1}{2}alC)\mathbf{Z}^n \tag{7.44}$$

(from (7.33)). The eigenvalues of the amplification matrix $(\mathrm{I} + \tfrac{1}{2}\mathrm{alC})^{-1}(\mathrm{I} + \tfrac{1}{2}\mathrm{alC})$ all have the value $(1 - \tfrac{1}{2}ar)/(1 + \tfrac{1}{2}ar)$: this value is less than $+1$ for small positive values of ar and greater than -1 for large positive values of ar. An initial perturbation $\mathbf{Z}^0$ will not, therefore, grow as time increases, but for $ar > 2$ such a perturbation will oscillate about $\mathbf{0}$.

7.3.4 Convergence

The problem of convergence of a finite difference scheme for solving a partial differential equation (hyperbolic, parabolic or elliptic) consists of finding the criteria under which the local discretization error $z_m^n = u_m^n - U_m^n$ at a *fixed mesh point* tends to zero uniformly as the mesh is refined. For a first-order hyperbolic equation such as (7.18) and a finite difference scheme such as (7.30), this refinement means that $h,l \to 0$ as $m,n \to \infty$. In carrying out the convergence analysis, it is convenient to assume that h,l do not tend to zero independently but according to a relationship of the form $l = rh^\alpha$, where r is a constant and $\alpha \geqslant 1$ is some parameter.

It is easy to see from (7.30) that (7.37) multiplied by l, ((7.30) was divided by l to calculate (7.37)), (7.18), and the definition of z_m^n, satisfies

$$z_m^{n+1} = (1 - ar)z_m^n + arz_{m-1}^n + 0(lh + l^2)\ . \tag{7.45}$$

It was seen in subsection 7.3.3 that the stability criterion for (7.30) is $0 < ar \leqslant 1$. Assuming, therefore, that the stability criterion is satisfied, it follows that

$$|z_m^{n+1}| \leqslant (1 - ar)|z_m^n| + ar|z_{m-1}^n| + O(lh + l^2)$$

$$\leqslant \hat{z}^{(n)} + k^{(n)}(lh + l^2)$$

where $\hat{z}^{(n)} = \max|z_m^n|$ for $m = 1,2,\ldots,N$ and $k^{(n)} = \max(|-\tfrac{1}{2}| + |C_2|)\{|\partial^2 u/\partial x^2|,|\partial^2 u/\partial t^2|\}$, that is, $k^{(n)} = M^{(n)}$ where $M^{(n)} = \max\{|\partial^2 u/\partial x^2|,|\partial^2 u/\partial t^2|\}$ at time $t = nl$. Hence

$$\hat{z}^{(n+1)} \leqslant z^{(n)} + k^{(n)}l(h + l) \leqslant z^{(n-1)} + (k^{(n)} + k^{(n-1)})l(h + l)$$

$$\leqslant \ldots \leqslant z^{(0)} + (k^{(n)} + k^{(n-1)} + \ldots + k^{(0)})l(h + l)$$

and, assuming that the same initial conditions are used with both differential equation and difference equation so that $z^{(0)} = 0$, it follows that

$$\hat{z}^{(n+1)} \leqslant (k^{(n)} + k^{(n-1)} + \ldots + k^{(0)})l(h + l)$$

$$\rightarrow 0 \text{ as } h,l \rightarrow 0$$

and convergence is thus ensured provided $0 < ar \leqslant 1$.

An alternative approach for establishing the convergence of a finite difference scheme will be discussed in subsection 7.7.5.

7.4 CENTRAL DIFFERENCE METHODS

7.4.1 The Lax–Wendroff method

Consider again the first-order hyperbolic equation

$$\frac{\partial u}{\partial t} + a\frac{\partial u}{\partial x} = 0 \; ; \quad a > 0 \tag{7.46}$$

defined this time as

(1) the overposed initial-boundary value problem in the open rectangular region $R = [0 < x < X] \times [t > 0]$ with initial conditions

$$u(x,0) = g(x) \; ; \quad 0 \leqslant x \leqslant X \qquad (7.47$$

and boundary conditions

$$u(0,t) = v \ ; \quad t > 0 \ , \tag{7.48}$$

$$u(X,t) = w \ ; \quad t > 0 \ , \tag{7.49}$$

or as

(2) the pure initial value problem in the upper half plane $R = [-\infty < x < \infty] \times [t > 0]$ with initial distribution

$$u(x,0) = g(x) \ ; \quad -\infty < x < \infty \ . \tag{7.50}$$

The Lax–Wendroff finite difference method is given by

$$U_m^{n+1} = \tfrac{1}{2}ar(1 + ar)U_{m-1}^n + (1 - a^2r^2)U_m^n - \tfrac{1}{2}ar(1 - ar)U_{m+1}^n \tag{7.51}$$

and has local truncation error

$$L(x,t) = \tfrac{1}{6}ah^2\frac{\partial^3 u}{\partial x^3} + \tfrac{1}{6}l^2\frac{\partial^3 u}{\partial t^3} + \text{higher-order terms} \ . \tag{7.52}$$

It may be shown using the Fourier method of analysis that the method is stable for $0 < ar \leqslant 1$; further analysis shows that the method is convergent for the same range of ar. Overall, the method is to be preferred to the lower-order methods of section 7.3, but for the properly posed initial-boundary value problem (7.18), (7.19), (7.20) the Lax–Wendroff method gives a solution only in the right-angled triangle bounded by the lines $x = 0$, $t = 0$ *and* $t = -ar(x - X)$.

The notation and discretizations used above were those introduced in section 7.3.

7.4.2 A family of central difference methods

Considering the overposed initial-boundary value problem (7.46)–(7.49), inclusive, with the interval $0 \leqslant x \leqslant X$ divided into $N + 1$ subintervals each of width h, so that $(N + 1)h = X$, and the time discretization $t = nl$ $(n = 0,1,\ldots)$ as in section 7.3, the hyperbolic problem is easily transformed into the first-order initial value problem

$$\frac{\mathrm{d}\mathbf{U}(t)}{\mathrm{d}t} = -\tfrac{1}{2}aB\mathbf{U}(t) + \tfrac{1}{2}a\mathbf{b} \ , \quad t > 0, \quad \mathbf{U}(0) = \mathbf{g} \tag{7.53}$$

by replacing the space derivative in (7.46) by

$$\frac{\partial u}{\partial x} = (2h)^{-1}\{u(x + h,t) - u(x - h,t)\} + O(h^2) \ . \tag{7.54}$$

In (7.53), the matrix B is of order N and is given by

$$B = h^{-1}\begin{bmatrix} 0 & 1 & & & & 0 \\ -1 & 0 & 1 & & & \\ & & \cdots & & & \\ & & & -1 & 0 & 1 \\ 0 & & & & -1 & 0 \end{bmatrix} \tag{7.55}$$

while the vector $\mathbf{b}$ also of order N is given by

$$\mathbf{b} = h^{-1}[v,0,\ldots,0,-w]^{\mathrm{T}} . \tag{7.56}$$

The solution of (7.54) is

$$\mathbf{U}(t) = B^{-1}\mathbf{b} + \exp(-\tfrac{1}{2}atB)\{\mathbf{g} - B^{-1}\mathbf{b}\} \tag{7.57}$$

which satisfies the recurrence relation

$$\mathbf{U}(t+l) = B^{-1}\mathbf{b} + \exp(-\tfrac{1}{2}alB)\{\mathbf{U}(t) - B^{-1}\mathbf{b}\} . \tag{7.58}$$

Replacing the exponential term in (7.58) by

$$\exp(-alB) = (I + \tfrac{1}{2}\theta alB)^{-1}[I - \tfrac{1}{2}(1-\theta)alB] , \tag{7.59}$$

where $0 \leqslant \theta \leqslant 1$, as in section 7.3, and I is the identity matrix of order N, leads to the family of methods

$$(I + \tfrac{1}{2}\theta alB)\mathbf{U}^{n+1} = [I - \tfrac{1}{2}(1-\theta)alB]\mathbf{U}^{n} + \tfrac{1}{2}al\mathbf{b} . \tag{7.60}$$

Taking $\theta = 0$ gives the explicit method

$$U_m^{n+1} = \tfrac{1}{2}arU_{m-1}^{n} + U_m^{n} - \tfrac{1}{2}arU_{m+1}^{n} ; \tag{7.61}$$

taking $\theta = 1$ gives the fully implicit method

$$-\tfrac{1}{2}arU_{m-1}^{n+1} + U_m^{n+1} + \tfrac{1}{2}arU_{m+1}^{n+1} = U_m^{n} ; \tag{7.62}$$

and taking $\theta = \tfrac{1}{2}$ gives the implicit method

$$-\tfrac{1}{4}arU_{m-1}^{n+1} + U_m^{n+1} + \tfrac{1}{4}arU_{m+1}^{n+1} = \tfrac{1}{4}arU_{m-1}^{n} + U_m^{n} - \tfrac{1}{4}arU_{m+1}^{n} . \tag{7.63}$$

The local truncation errors of these three schemes are given in the general form

$$L(x,t) = \left(\tfrac{1}{6}ah^2\frac{\partial^3 u}{\partial x^3} + C_2 l\frac{\partial^2 u}{\partial t^2}\right) + C_3 l^2\frac{\partial^3 u}{\partial t^3} + \ldots \tag{7.64}$$

in which $C_2 = \frac{1}{2} - \theta$ and $C_3 = \frac{1}{6} - \frac{\theta}{2}$, so that (7.61) and (7.62) are first order accurate (in time) while (7.63) is second-order accurate. All members of the family are second-order accurate in space.

Stability analyses show that (7.61) is unconditionally unstable and that (7.62) and (7.63) are both unconditionally stable. Convergence analyses show that (7.62) and (7.63) are convergent methods.

Numerical results produced by (7.63) and (7.64) are highly oscillatory in nature and, in spite of their stability properties, these two methods produce poor results for problems with smooth solutions whereas the lower-order backward difference methods of section 7.3 model smooth solutions very well. Further discussion of the oscillatory solutions just mentioned is given in the Ph.D. thesis of Khaliq (1983) and in the text by Smith (1985).

7.5 INTRODUCTION TO SECOND-ORDER EQUATIONS

Hyperbolic equations, first and second order, usually arise when shocks, waves or vibrations occur in a physical system, and mathematical modelling of such a system invariably involves the solution of a hyperbolic equation or system of equations.

It is seen in Appendix A that the general partial differential equation of the second order may be written in the form

$$a\frac{\partial^2 u}{\partial x^2} + b\frac{\partial^2 u}{\partial x \partial t} + c\frac{\partial^2 u}{\partial t^2} = e \tag{7.65}$$

in which a,b,c,e are functions of $u,x,t,\partial u/\partial x,\partial u/\partial t$, but not of the second-order derivatives. The abbreviations p,q,r,s,y will be used for $\partial u/\partial x, \partial u/\partial t, \partial^2 u/\partial x^2, \partial^2 u/\partial x\partial t, \partial^2 u/\partial t^2$, respectively. Equation (7.65) is hyperbolic whenever $b^2 - 4ac > 0$.

The simplest second-order hyperbolic equation is the wave equation

$$\frac{\partial^2 u}{\partial x^2} - \frac{\partial^2 u}{\partial t^2} = 0$$

which has a general solution

$$u(x,t) = F(x+t) + G(x-t) \ ,$$

where F and G are arbitrary differentiable functions. If the initial conditions $u(x,0) = f(x)$ and $\partial u(x,0)/\partial t = g(x)$ for $-\infty < x < \infty$ are specified, the wave equation together with these initial conditions form the *Cauchy initial value problem* which has solution

$$u(x,t)=\tfrac{1}{2}\left[f(x+t)+f(x-t)+\int_{x-t}^{x+t}g(\theta)\mathrm{d}\theta\right].$$

The simple wave equation will receive further treatment in section 7.7.

7.6 THE RÔLE OF CHARACTERISTICS

7.6.1 The Method of Characteristics

It is seen in Appendix A that u,p,q satisfy the total differentials

$$\mathrm{d}u=p\mathrm{d}x+q\mathrm{d}t, \tag{7.66}$$

$$dp=r\mathrm{d}x+s\mathrm{d}t, \tag{7.67}$$

$$dq=s\mathrm{d}x+y\mathrm{d}t, \tag{7.68}$$

respectively, and that a,b,c satisfy

$$a\left(\frac{\mathrm{d}t}{\mathrm{d}x}\right)^2-b\left(\frac{\mathrm{d}t}{\mathrm{d}x}\right)+c=0 \tag{7.69}$$

Whenever $b^2-4ac>0$ in (7.69) at a point (x,t) in the x,t-plane (that is, whenever (7.65) is hyperbolic), $\mathrm{d}t/\mathrm{d}x$ has two distinct values at (x,t). This means that two curves, called *characteristic curves* or *characteristics* (as in section 7.2) may be drawn through (x,t); these curves have slopes

$$f=\{b+(b^2-4ac)^{\frac{1}{2}}\}/(2a) \tag{7.70}$$

and

$$g=\{b-(b^2-4ac)^{\frac{1}{2}}\}/(2a) \tag{7.71}$$

and consequently the two characteristics will be referred to as *f-type* and *g-type characteristics*.

For the simple wave equation, $a=1, b=0, c=1$ and $e=0$ so that $f=-g=1$ and the characteristic curves through a specific point $T(x_1,t_1)$ are the straight lines $t-t_1=x-x_1$ and $t-t_1=-x+x_1$. These characteristics cut the x-axis (where $u(x,0)=f(x)$ and $\partial u(x,0)/\partial t=g(x)$ are given in the Cauchy initial value problem) at the points where $x=x_1-t_1$ and $x=x_1+t_1$; let these points be named A and B. The solution of the Cauchy initial value problem at the point T depends on the initial data between A and B, inclusive, and takes the value

$$u(x_1,t_1) = \tfrac{1}{2}\left[f(x_1+t_1) + f(x_1-t_1) + \int_{x_1-t_1}^{x_1+t_1} g(\theta)\mathrm{d}\theta \right] .$$

The triangle *TAB* is called the *domain of dependence* of the point *T*, and characteristics drawn within TAB form a *characteristic mesh* or *characteristic grid*.

It is easy to see that *f* and *g* do not have constant values for the general hyperbolic equation (7.65) and so the characteristic mesh may not be uniform.

Multiplying equation (7.67) by $a\mathrm{d}t$, equation (7.68) by $c\mathrm{d}x$ and adding gives

$$a\mathrm{d}t\mathrm{d}p + c\mathrm{d}x\mathrm{d}q = (ar + cy)\mathrm{d}x\mathrm{d}t + as(\mathrm{d}t)^2 + cs(\mathrm{d}x)^2 \ ,$$

$$= (-bs + e)\mathrm{d}x\mathrm{d}t + as(\mathrm{d}t)^2 + cs(\mathrm{d}x)^2 \ , \qquad \text{from (7.65))}$$

$$= e\mathrm{d}x\mathrm{d}t + s\{a(\mathrm{d}t)^2 - b\mathrm{d}x\mathrm{d}t + c(\mathrm{d}x)^2\} \ , \tag{7.72}$$

so that on curves for which (7.69) holds

$$a\mathrm{d}t\mathrm{d}p + c\mathrm{d}x\mathrm{d}q = e\mathrm{d}x\mathrm{d}t \ . \tag{7.73}$$

Curves in the (x,t)-plane for which (7.69) and (7.73) hold are thus the characteristic curves.

Suppose, then, that (7.65) is hyperbolic and that values of u,p,q are given at M points $T_m^{(1)}$ $(m = 1,2,\ldots,M)$, which are not necessarily equally spaced, along some initial curve L in the (x,t)-plane which is *not* a characteristic. The exclusion of characteristic curves as initial curves is necessary because the solution would then be unique along the initial curve but nowhere else. Furthermore, it is impossible in such cases to use the method of characteristics to advance the solution from the M points on the initial curve to points off it. As was seen above, two distinct characteristics with slopes given by (7.70) and (7.71) pass through each of these initial points.

The domain of dependence is a curvilinear triangle. The f-type characteristic through the initial point $T_m^{(1)}$ meets the g-type characteristic through $T_{m+1}^{(1)}$ at the point $T_m^{(2)}$ $(m = 1,2,\ldots,M-1)$; the f-type characteristic through the mesh point $T_m^{(2)}$ meets the g-type characteristic through $T_{m+1}^{(2)}$ at the point $T_m^{(3)}(m = 1,2,\ldots,M-2)$; in general, the f-type characteristic through the mesh point $T_m^{(n)}$ meets the g-type characteristic through the mesh point $T_{m+1}^{(n)}$ at the mesh point $T_m^{(n+1)}$ $(n = 1,\ldots, M-1; m = 1,\ldots,M-n)$ giving rise to the curvilinear characteristic mesh of Fig. 7.1.

The values of u,p,q at the mesh points of this grid are dependent only on those values of u,p,q at the initial points $T_m^{(1)}(m = 1,2,\ldots,M)$ on the curve L. Should u,p,q be required at points not on the mesh, inside or outside the domain of dependence, the values of u,p,q at further points on L will be required. At the general mesh point $T_m^{(n)}$ the value of u will be denoted by $u_m^{(n)}$, with similar notations for x,t,p,q,a,b,c,e,f,g at that point.

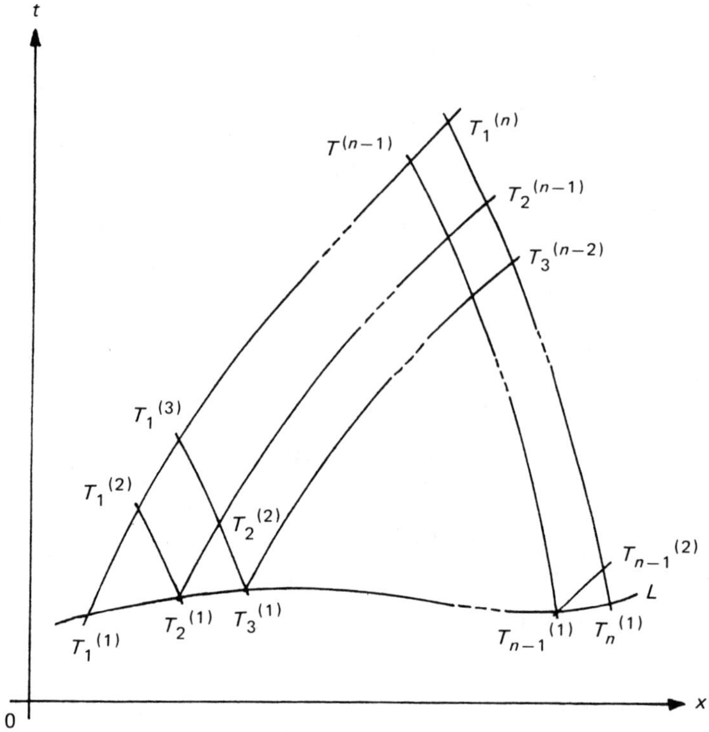

Fig. 7.1 — General curvilinear mesh.

7.6.2 Implementation of the method

The method approximates the curves $T_m^{(n)}T_{m+1}^{(n)}T_m^{(n+1)}$ by the straight lines

$$t - t_m^{(n)} = f_m^{(n)}(x - x_m^{(n)}) \tag{7.74}$$

and

$$t - t_{m+1}^{(n)} = g_{m+1}^{(n)}(x - x_{m+1}^{(n)}) \ . \tag{7.75}$$

The point of intersection of these lines gives a first approximation to the grid point $T_m^{(n+1)}$ which has coordinates $(x_m^{(n+1)}, t_m^{(n+1)})$. Now equation (7.73) is divided by dx and applied along $T_m^{(n)}T_m^{(n+1)}$ to give

$$a_m^{(n)}(p_m^{(n+1)} - p_m^{(n)})f_m^{(n)} + c_m^{(n)}(q_m^{(n+1)} - q_m^{(n)}) = e_m^{(n)}(t_m^{(n+1)} - t_m^{(n)}) \tag{7.76}$$

and along $T_{m+1}^{(n)}T_m^{(n+1)}$ to give

$$a_{m+1}^{(n)}(p_m^{(n+1)} - p_{m+1}^{(n)})g_{m+1}^{(n)} + c_{m+1}^{(n)}(q_m^{(n+1)} - q_{m+1}^{(n)}) = e_{m+1}^{(n)}(t_m^{(n+1)} - t_{m+1}^{(n)}) \ . \tag{7.77}$$

Equations (7.76) and (7.77) are solved simultaneously to give first approximations to their unknowns $p_m^{(n+1)}$ and $q_m^{(n+1)}$.

A first approximation to $u_m^{(n+1)}$ is now determined from the total differential formula (7.66) giving

$$\begin{aligned} U_m^{(n+1)} = U_m^{(n)} &+ \tfrac{1}{2}(x_m^{(n+1)} - x_m^{(n)})(p_m^{(n+1)} - p_m^{(n)}) \\ &+ \tfrac{1}{2}(t_m^{(n+1)} - t_m^{(n)})(q_m^{(n+1)} - q_m^{(n)}) \end{aligned} \tag{7.78}$$

where, as in previous sections, upper case U is used to distinguish the theoretical solution of an approximating method from that of the differential equation.

The coefficients a,b,c,e of the hyperbolic equation (7.65) can now be evaluated at the mesh point $T_m^{(n+1)}$ and substitution of these values of a,b,c into (7.70) and (7.71) gives first approximations to the slopes of the two characteristics advancing from this point.

The reader who has studied chapter 5 will recognize equations (7.74), (7.75) as applications of Euler's predictor formula: in a similar fashion (7.76), (7.77) are predictor equations and the values of u,a,b,c,e,f and g just calculated are predicted values. In the special case of constant coefficients a,b,c,e in (7.65), the values of x,t,p,q,u,a,b,c,e,f,g which have so far been determined at the grid point $T_m^{(n+1)}$ cannot be improved. In the case when a,b,c,e are functions of x,t,p,q, only $U_m^{(n+1)}$ need not be calculated just yet. In cases where a,b,c,e are functions of x,t,p,q,u, the full predictive procedure just described must be carried out. Except in constant coefficient problems, these predicted values can be corrected as follows.

Equations (7.74) and (7.75) become

$$t - t_m^{(n)} = \tfrac{1}{2}(f_m^{(n+1)} + f_m^{(n)})(x - x_m^{(n)}) \tag{7.79}$$

and

$$t - t_{m+1}^{(n)} = \tfrac{1}{2}(g_m^{(n+1)} + g_{m+1}^{(n)})(x - x_{m+1}^{(n)}) \tag{7.80}$$

which are solved simultaneously to give corrected values of $x_m^{(n+1)}$ and $t_m^{(n+1)}$. Equations (7.79) and (7.80) are applications of Euler's corrector formula, sometimes called the modified Euler formula or trapezoidal rule.

Using the corrected coordinates of the mesh point $T_m^{(n+1)}$, equations (7.76) and (7.77) are written in the forms

$$\begin{aligned} &\tfrac{1}{2}(a_m^{(n+1)} + a_m^{(n)})(p_m^{(n+1)} - p_m^{(n)}).\tfrac{1}{2}(f_m^{(n+1)} + f_m^{(n)}) \\ &\qquad + \tfrac{1}{2}(c_m^{(n+1)} + c_m^{(n)})(q_m^{(n+1)} - q_m^{(n)}) \\ &\qquad = \tfrac{1}{2}(e_m^{(n+1)} + e_m^{(n)})(t_m^{(n+1)} - t_m^{(n)}) \end{aligned} \tag{7.81}$$

and

$$\tfrac{1}{2}(a_m^{(n+1)}+a_m^{(n)})(p_m^{(n+1)}-p_{m+1}^{(n)})\cdot\tfrac{1}{2}(g_m^{(n+1)}+g_{m+1}^{(n)})$$

$$+\tfrac{1}{2}(c_m^{(n+1)}+c_{m+1}^{(n)})(q_m^{(n+1)}-q_{m+1}^{(n)})$$

$$=\tfrac{1}{2}(e_m^{(n+1)}+e_{m+1}^{(n)})(t_m^{(n+1)}-t_{m+1}^{(n)}) \qquad (7.82)$$

which are solved simultaneously to give corrected values of $p_m^{(n+1)}$ and $q_m^{(n+1)}$. In the case when a,b,c,e are functions of all of x,t,p,q,u, the corrected values of x,t,p,q,u are then used to find a corrected value of $u_m^{(n+1)}$ from (7.78).

Corrected values of a,b,c,e are now computed using their forms given in (7.65) and used in (7.70) and (7.71) to calculate corrected values of the slopes of the two characteristics through the mesh point $T_m^{(n+1)}$. An iterative procedure using equations (7.79), (7.80), (7.81), (7.82), (7.78), (7.65), (7.70) and (7.71) is thus established and is carried out until all of x,t,p,q,u at $T_m^{(n+1)}$ have converged. In variable coefficient problems when a,b,c,e are functions of x,t,p,q only, equation (7.78) may be omitted from the predictor–corrector combination and need not be used to compute $U_m^{(n+1)}$ until x,t,p,q have converged.

7.6.3 An application in arterial blood pressure

The pressure u at a point x in an artery of length X (assumed thin) at time t following a single heart beat is given by

$$\frac{\partial^2 u}{\partial t^2}=c^2(u)\frac{\partial^2 u}{\partial x^2}\ ; \quad 0<x<X\ , \quad t>0\ , \qquad (7.83)$$

where c is the wave speed and it is assumed that a heart beat takes place at time $t=0$. In (7.83), $c^2(u)$ is given by

$$c^2(u)=\frac{EH}{2R\rho(1-\sigma^2)}-\frac{u}{\rho} \qquad (7.84)$$

where E is Young's modulus for the arterial wall, σ is Poisson's ratio for the arterial wall, ρ is the density of blood, and H,R represent the thickness and inner radius of the artery, which is assumed to be a hollow cylindrical tube (see McDonald (1974)).

The method of characteristics is applicable to (7.83) for which initial conditions and boundary conditions must be specified. The irregular mesh generated by the method of characteristics for nonlinear problems such as (7.83) with (7.84) makes it difficult to monitor exactly any fixed point along the length of the artery. To do so, it would be advisable to turn to one of the methods of the following sections.

7.7 FINITE DIFFERENCE METHODS FOR LINEAR PROBLEMS

7.7.1 Derivation of the methods: method of lines approach

Consider the linear initial-boundary value problem consisting of the second-order hyperbolic equation

$$\frac{\partial^2 u}{\partial t^2} = c^2 \frac{\partial^2 u}{\partial x^2} \; ; \qquad 0 < x < X \; , \qquad t > 0 \; , \tag{7.85}$$

in which the wave speed c is constant, together with the initial displacement

$$u(x,0) = f(x) \; ; \qquad 0 \leqslant x \leqslant X \; , \tag{7.86}$$

the initial velocity

$$\frac{\partial u}{\partial t}(x,0) = g(x) \; ; \qquad 0 \leqslant x \leqslant X \tag{7.87}$$

and the boundary conditions

$$u(0,t) = u(X,t) = 0 \; ; \qquad t > 0 \; . \tag{7.88}$$

The region in which (7.85) is defined is $R = [0 < x < X] \times [t > 0]$; the boundary of R is defined by the lines $x = 0, x = X, t = 0$ and is denoted by ∂R. This problem models the vibrations of a string fixed at both ends which has initial displacement and velocity given by (7.86) and (7.87).

The interval $0 \leqslant x \leqslant X$ will be divided into $N+1$ subintervals each of width h so that $(N+1)h = X$, and the space derivative in (7.85) will be replaced by

$$\frac{\partial^2 u}{\partial x^2} = h^{-2}\{u(x-h,t) - 2u(x,t) + u(x+h,t)\} + O(h^2) \; . \tag{7.89}$$

The time variable t will be discretized in steps of length l so that t may be written in the form $t = nl$ where $n = 0,1,2,\ldots$. The region R and its boundary ∂R have thus been discretized at the points (mh,nl) where $m = 0,1,\ldots,N+1$ and $n = 0,1,2,\ldots$. The solution of (7.85) at (mh,nl) is $u(mh,nl)$, which will be denoted by u_m^n while the theoretical solution of a finite difference scheme will be denoted by U_m^n or by $U_m(t)$, where $t = nl$, as convenient.

Consider, now, the general time level $t = nl$ at which there are N interior mesh points. At every one of these points, equation (7.85) is applied with the space derivative replaced by (7.89). At point m on this level $(m = 1,2,\ldots,N)$ equation (7.85) becomes

$$\frac{d^2U_m(t)}{dt^2}=c^2h^{-2}\{U_{m-1}(t)-2U_m(t)+U_{m+1}(t)\} \ . \tag{7.90}$$

Recalling that $U_0(t)=U_{N+1}(t)=0$ from (7.88), the N equations of the form (7.90) may be written in system form as

$$\frac{d^2\mathbf{U}(t)}{dt^2}=c^2A\mathbf{U}(t) \ , \tag{7.91}$$

where $\mathbf{U}(t)=[U_1(t),U_2(t),\ldots,U_N(t)]^{\mathrm{T}}$ and the square matrix A of order N is given by

$$A=h^{-2}\begin{bmatrix} -2 & 1 & & 0 & \\ 1 & -2 & 1 & & \\ & & \cdots & & \\ & 1 & -2 & 1 \\ 0 & & 1 & -2 \end{bmatrix} \ . \tag{7.92}$$

The eigenvalues of A are given by

$$\lambda_s=-4h^{-2}\sin^2[s\pi/\{2(N+1)\}] \ ; \quad s=1,2,\ldots,N \ . \tag{7.93}$$

Solving (7.91) subject to the initial conditions (7.86), (7.87) gives

$$\mathbf{U}(t)=\tfrac{1}{2}\exp(ctB)\ \{\mathbf{f}+c^{-1}B^{-1}\mathbf{g}\}+\tfrac{1}{2}\exp(-ctB)\{\mathbf{f}-c^{-1}B^{-1}\mathbf{g}\} \tag{7.94}$$

in which B is a matrix such that $B^2=A$ and $\mathbf{f},\mathbf{g}$ are the vectors of initial conditons. It will be seen that the elements of B need not be found explicitly.

It is easy to show that, for a time step l, equation (7.94) satisfies the recurrence relation

$$\mathbf{U}(t+l)-\{\exp(clB)+\exp(-clB)\}\mathbf{U}(t)+\mathbf{U}(t-l)=\mathbf{0} \tag{7.95}$$

where $t=l,2l,\ldots$ and $\mathbf{0}$ is the zero-vector of order N.

Replacing the exponential terms in (7.95) by their (0,2) Padé approximants $\exp(\pm clB)=I\pm clB+\tfrac{1}{2}c^2l^2B^2+O(l^3)$ leads to

$$\mathbf{U}(t+l)-(2I+c^2l^2A)\mathbf{U}(t)+\mathbf{U}(t-l)+O(l^4)=\mathbf{0} \ , \tag{7.96}$$

recalling that $B^2=A$, which, when applied to the mesh point (mh,nl) gives the well-known five-point, three-time-level, explicit scheme

$$U_m^{n+1} = 2(1 - c^2r^2)U_m^n + c^2r^2(U_{m-1}^n + U_{m+1}^n) - U_m^{n-1} \tag{7.97}$$

where $r = l/h$.

Replacing the exponential terms in (7.95) by their (1,1) Padé approximants $\exp(\pm clB) = (I \mp \tfrac{1}{2}clB)^{-1}(I \pm \tfrac{1}{2}clB) + O(l^3)$ leads, after premultiplication by $(I - \tfrac{1}{2}clB)(I + \tfrac{1}{2}clB)$, to

$$(I - \tfrac{1}{4}c^2l^2A)\mathbf{U}(t+l) - (2I + \tfrac{1}{2}c^2l^2A)\mathbf{U}(t) + (I - \tfrac{1}{4}c^2l^2A)\mathbf{U}(t-l) + O(l^4) = \mathbf{0}\ . \tag{7.98}$$

Applying (7.98) to the general mesh point (mh,nl) gives the well-known nine-point, three-level, implicit finite difference method

$$\begin{aligned}-\tfrac{1}{4}c^2r^2U_{m-1}^{n+1} + (1 + \tfrac{1}{2}c^2r^2)U_m^{n+1} - \tfrac{1}{4}c^2r^2U_{m+1}^{n+1}\\ = \tfrac{1}{2}c^2r^2U_{m-1}^n + (2 - c^2r^2)U_m^n + \tfrac{1}{2}c^2r^2U_{m+1}^n\\ + \tfrac{1}{4}c^2r^2U_{m-1}^{n-1} - (1 + \tfrac{1}{2}c^2r^2)U_m^{n-1} + \tfrac{1}{4}c^2r^2U_{m+1}^{n-1}\ .\end{aligned} \tag{7.99}$$

Another implicit method may be derived by replacing the exponential terms by their (1,2) Padé approximants $\exp(\pm clB) = (I \mp \tfrac{1}{3}clB)^{-1}\ (I \pm \tfrac{2}{3}clB + \tfrac{1}{6}c^2l^2B^2) + \boldsymbol{O}(l^4)$. After premultiplication by $(I - \tfrac{1}{3}clB)(I + \tfrac{1}{3}clB)$, this leads to

$$(I - \tfrac{1}{9}c^2l^2A)\mathbf{U}(t+l) - (2I + \tfrac{7}{9}c^2l^2A)\mathbf{U}(t) + (I - \tfrac{1}{9}c^2l^2A)\mathbf{U}(t-l) + \boldsymbol{O}(l^4) = \mathbf{0} \tag{7.100}$$

which, following application to (mh,nl), gives

$$\begin{aligned}-\tfrac{1}{9}c^2r^2U_{m-1}^{n+1} + (1 + \tfrac{2}{9}c^2r^2)U_m^{n+1} - \tfrac{1}{9}c^2r^2U_{m+1}^{n+1}\\ = \tfrac{7}{9}c^2r^2U_{m-1}^n + (2 - \tfrac{14}{9}c^2r^2)U_m^n + \tfrac{7}{9}c^2r^2U_{m+1}^n\\ + \tfrac{1}{9}c^2r^2U_{m-1}^{n-1} - (1 + \tfrac{2}{9}c^2r^2)U_m^{n-1} + \tfrac{1}{9}c^2r^2U_{m+1}^{n-1}\ .\end{aligned} \tag{7.101}$$

This nine-point implicit method is not as well known as (7.99) but, as will be seen in subsection 7.7.3, it is more accurate than (7.99) and is neither more difficult nor more expensive to implement, the solution vector $\mathbf{U}(t+l)$ being obtained by solving a tridiagonal linear system in each case.

7.7.2 Solution at the first time step

It has been noted that the recurrence relation (7.95) is applied with $t = l, 2l, \ldots$ and that the accuracy depends on the chosen approximation to the exponential maxtrix functions $\exp(\pm clB)$. The remaining difficulty then is to find a sufficiently accurate approximation to $\mathbf{U}(l)$. This is not contained explicitly in the initial conditions, and an approximation to it must be found which is sufficiently accurate to be used with a scheme such as (7.96), (7.98) or (7.100).

Considering the theoretical solution given by (7.94), replacing the exponential matrix functions with their (0,2) Padé approximants and writing $t=l$ gives

$$\mathbf{U}(l)=\tfrac{1}{2}[\exp(clB)+\exp(-clB)]\mathbf{f}+\tfrac{1}{2}[\exp(clB)-\exp(-clB)]B^{-1}\mathbf{g}$$

$$=\tfrac{1}{2}[I+clB+\tfrac{1}{2}c^2l^2B^2+I-clB+\tfrac{1}{2}c^2l^2B^2+O(l^3)]\mathbf{f}$$

$$+\tfrac{1}{2}[I+clB+\tfrac{1}{2}c^2l^2B^2-I+clB-\tfrac{1}{2}c^2l^2B^2+O(l^3)]B^{-1}\mathbf{g}\ ,$$

that is

$$\mathbf{U}(l)=(I+\tfrac{1}{2}c^2l^2A)\mathbf{f}+l\mathbf{g}+\mathbf{O}(l^3)\ . \tag{7.102}$$

Equation (7.102) has error proportional to l^3 in time (recall that the space discretization has an error proportional to h^2). It is equation (7.102) which is usually used with (7.96), (7.98) and (7.100), all of which have error $=O(l^4)$ in time. Indeed, if many time steps are to be taken with (7.96), (7.98) or (7.100) then (7.102) may be used with confidence. However, an approximation to $\mathbf{U}(l)$ which has error at least $O(l^4)$ should be used if only a few time steps are to be taken with the general scheme. In addition, (7.102) becomes unreliable for large values of l.

Replacing the matrix exponential functions in (7.94) with their (0,4) Padé approximants $\exp(\pm clB)=I\pm clB+\tfrac{1}{2}c^2l^2B^2\pm\tfrac{1}{6}c^3l^3B^3+\tfrac{1}{24}c^4l^4B^4+O(l^5)$ leads to

$$\mathbf{U}(l)=(I+\tfrac{1}{2}l^2A+\tfrac{1}{24}l^4A^2)\mathbf{f}+l(I+\tfrac{1}{6}l^2A)\mathbf{g}+\mathbf{O}(l^5) \tag{7.103}$$

which may be used to give better accuracy at the first time step than the general schemes do at subsequent time steps. The reader is cautioned that (7.103) also becomes unreliable for large values of l.

7.7.3 Local truncation errors and consistency

Writing the explicit scheme (7.97) as

$$\frac{U_m^{n+1}-2U_m^n+U_m^{n-1}}{l^2}-c^2\frac{U_{m-1}^n-2U_m^n+U_{m+1}^n}{h^2}=0$$

it is clear that the first term is an approximation to the time derivative $\partial^2u/\partial t^2$ in (7.85) while the second term is an approximation to $c^2\partial^2u/\partial x^2$. The local truncation error of (7.97) is therefore given by

$$L(x,t)=\frac{u(x,t+l)-2u(x,t)+u(x,t-l)}{l^2}$$

$$-c^2\frac{u(x-h,t)-2u(x,t)+u(x+h,t)}{h^2}-\left(\frac{\partial^2 u}{\partial t^2}-c^2\frac{\partial^2 u}{\partial x^2}\right).$$

Expanding about $u(x,t)$ using Taylor's series, it may be shown that

$$L(x,t)=\left(-\tfrac{1}{12}c^2h^2\frac{\partial^4 u}{\partial x^4}+C_4l^2\frac{\partial^4 u}{\partial t^4}\right)+\text{higher-order terms,} \tag{7.104}$$

with $C_4=\frac{1}{12}$, verifying that (7.97) is second-order accurate in both space and time.

It is easy to show, for the constant coefficient wave equation $\partial^2u/\partial t^2=c^2\partial^2u/\partial x^2$, that $\partial^4u/\partial t^4=c^4\partial^4u/\partial x^4$, $\partial^6u/\partial t^6=c^6\partial^6u/\partial t^6$, etc. Including some of the higher-order terms, $L(x,t)$ may be written in the alternative form

$$h^2c^2\left\{\tfrac{1}{12}(c^2r^2-1)\frac{\partial^4 u}{\partial x^4}+\tfrac{1}{360}h^2(c^4r^4-1)\frac{\partial^6 u}{\partial x^6}+\ldots\right\}$$

from which it is easy to see that the local truncation error vanishes when $cr=1$. The explicit difference scheme (7.97) then takes the form

$$U_m^{n+1}=U_{m-1}^n+U_{m+1}^n-U_m^{n-1} \tag{7.105}$$

which is therefore an *exact* representation of the wave equation (7.85) when used on a grid for which $l/h=1/c$.

The term in parentheses on the right-hand side of (7.104) is the principal part of the local truncation error, the first component of which arises from the space discretization and the use of (7.89) in (7.85); this component will be present in every finite difference method arising from (7.95). The second component in the principal part of $L(x,t)$ is due to the use of the (0,2) Padé approximants. The constant C_4 is the time error constant and it is not a difficult exercise to show that, for the implicit method (7.99), $C_4=-\frac{1}{6}$ while for (7.101), $C_4=-\frac{1}{36}$.

Clearly the local truncation error of all three methods derived from the recurrence relation (7.95) tends to zero as $h,l\to 0$ simultaneously and the three methods are therefore consistent with the hyperbolic equation (7.85).

7.7.4 Stability

The Fourier method of anlysing stability was described in subsection 7.3.3. In the case of the explicit method (7.97), the error $Z_m^n=U_m^n-\widetilde{U}_m^n$, where $\widetilde{U}$ is the computed solution actually obtained, satisfies the equation

$$Z_m^{n+1}=2(1-c^2r^2)\,Z_m^n+c^2r^2(Z_{m-1}^n+Z_{m+1}^n)-Z_m^{n-1} \tag{7.106}$$

and substituting the typical term $e^{\alpha nl}e^{i\beta mh}$ for Z_m^n, where β is real and α is complex, leads after cancellation to

$$\xi^2 - \{2(1-c^2r^2) + c^2r^2(e^{-i\beta h} + e^{i\beta h})\}\xi + 1 = 0 \tag{7.107}$$

in which $\xi = e^{\alpha l}$. Using Euler's relations $e^{\pm i\theta} = \cos\theta \pm i\sin\theta$ and the half-angle formula $\cos\theta = 1 - 2\sin^2\frac{1}{2}\theta$ for any real θ gives

$$\xi^2 - 2(1 - 2c^2r^2\sin^2\tfrac{1}{2}\beta h)\xi + 1 = 0 \ . \tag{7.108}$$

This is the stability equation of the explicit difference scheme (7.97) and von Neumann's stability criterion requires both values of ξ to satisfy $|\xi| \leqslant 1$; the condition on the mesh ratio $r = l/h$ to ensure stability is now determined.

In general, three-level difference schemes, explicit and implicit, have stability equations of the form

$$a\xi^2 - 2b\xi + a = 0 \ ; \tag{7.109}$$

here, a,b are functions of r,c. The roots of (7.109) are easily seen to be $\xi = [b \pm (b^2 - a^2)^{\frac{1}{2}}]/a$, so that $|\xi| \leqslant 1$ when $|b| \leqslant |a|$. In the case of (7.108) this gives the three-part inequality

$$-1 \leqslant 1 - 2c^2r^2\sin^2\tfrac{1}{2}\beta h \leqslant 1 \ . \tag{7.110}$$

The right-hand side of (7.110) is satisfied for all values of cr (recall that c,β,h are all real); the left-hand side is satisfied only for $c^2r^2 \leqslant 1$ since $0 \leqslant \sin^2\frac{1}{2}\beta h \leqslant 1$. The right-hand side of (7.110) suggests that the finite difference scheme is unconditionally stable but the left-hand side imposes the condition $c^2r^2 \leqslant 1$. The stability criterion for the method is the intersection of the two ranges of values, namely $0 \leqslant c^2r^2 \leqslant 1$.

The matrix method of analysing stability is applicable to initial boundary value problems, but it can only be used for second-order hyperbolic equation when the difference scheme has been expressed in two-time level form. Suppose that the difference scheme is in the general form

$$G\mathbf{U}^{n+1} = B\mathbf{U}^n + C\mathbf{U}^{n-1} + \mathbf{b} \tag{7.111}$$

where $\mathbf{b}$ is a vector which depends on the boundary conditions and G,B,C are square matrices of order N (recall, N is the number of mesh points at each time level). In the case of a time-dependent problem, the matrices G,B,C are evaluated at times $(n+1)l, nl, (n-1)l$, respectively; in constant coefficient problems the matrices G,B,C are constant and G and C are usually equal and nonsingular.

Writing (7.111) as

$$\mathbf{U}^{n+1} = G^{-1}B\mathbf{U}^n + G^{-1}C\mathbf{U}^{n-1} + G^{-1}\mathbf{b} \tag{7.112}$$

it follows that a perturbation $\mathbf{Z}^0$ of the initial conditions will satisfy

$$\mathbf{Z}^{n+1} = G^{-1}B\mathbf{Z}^n + G^{-1}C\mathbf{Z}^{n-1} \ . \tag{7.113}$$

This may be written as

$$\begin{bmatrix} \mathbf{Z}^{n+1} \\ \mathbf{Z}^n \end{bmatrix} = \begin{bmatrix} G^{-1}B & G^{-1}C \\ I & 0 \end{bmatrix} \begin{bmatrix} \mathbf{Z}^n \\ \mathbf{Z}^{n-1} \end{bmatrix} \tag{7.114}$$

which is of the form

$$\mathbf{E}^{n+1} = W\mathbf{E}^n \tag{7.115}$$

where $\mathbf{E}^{n+1} = [(\mathbf{Z}^{n+1})^{\mathrm{T}}, (\mathbf{Z}^n)^{\mathrm{T}}]^{\mathrm{T}}$ and W is the amplification matrix.

It follows that

$$\|\mathbf{E}^{n+1}\| \leqslant \|W\| . \|\mathbf{E}^n\| \ , \tag{7.116}$$

where $\|.\|$ denotes a suitable norm. Equation (7.115) is a two-time-level equation and the von Neumann necessary condition for the stability of a difference scheme such as (7.111), using a constant time step and letting $n \to \infty$, is

$$\|W\| \leqslant 1 \ . \tag{7.117}$$

The success of the matrix method for second-order hyperbolic equations thus depends on being able to obtain a suitable estimate for $\|W\|$. The spectral norm, which is equal to the spectral radius when W is a symmetric matrix, is the usual choice of norm. The spectral radius is often used to approximate the spectral norm, leading occasionally to an incorrect stability interval as was seen in subsection 7.3.3. Often, the matrices $G,B,C,G^{-1}B,G^{-1}C$ have the same system of linearly independent eigenvectors and the eigenvalues $\mu_i (i = 1,2,\ldots,2N)$ of W are given by

$$\det \begin{bmatrix} \alpha_k^{-1}\beta_k - \mu & \alpha_k^{-1}\gamma_k \\ 1 & -\mu \end{bmatrix} = 0$$

where $\alpha_k \beta_k, \gamma_k$ $(k = 1,2,\ldots,N)$ are the eigenvalues of G,B,C.

It is easy to verify that the matrix method gives the same stability criterion for the explicit scheme (7.97) as that determined by the Fourier method. Taking the scheme in its matrix form (7.96), it may be written in two-time-level form as

$$\mathbf{U}^{n+1} = (2I + c^2 l^2 A)\mathbf{U}_n - \mathbf{V}^n \ ,$$

$$\mathbf{V}^{n+1}=\mathbf{U}^n$$

so that the error $\mathbf{E}$ satisfies

$$\mathbf{E}^{n+1}=\begin{bmatrix} 2I+c^2l^2A & -I \\ I & 0 \end{bmatrix}\mathbf{E}^n \ . \tag{7.118}$$

The $2N$ eigenvalues of the amplification matrix in (7.118) are determined from the N quadratic equations

$$\det\begin{bmatrix} 2-4c^2r^2\sin^2[s\pi/\{2(N+1)\}]-\mu & -1 \\ 1 & -\mu \end{bmatrix}=0$$

for $s=1,2,\ldots,N$ (the eigenvalues of the matrix A are given in (7.93)). The eigenvalues of the amplification matrix are thus given by

$$\mu_s^2-2(1-2c^2r^2\sin^2\theta_s)\mu_s+1=0 \tag{7.119}$$

for $s=1,2,\ldots,N$, where $\theta_s=s\pi/\{2(N+1)\}$. Clearly, equation (7.119) corresponds to (7.108) and it follows from (7.117) that the stability criterion is $0\leqslant c^2r^2\leqslant 1$ as obtained by the Fourier method.

Turning, now, to the implicit method (7.99) which is based on the use of the (1,1) Padé approximant in (7.95), the matrix method for analysing stability proceeds from (7.98) and leads to the amplification matrix

$$W=\begin{bmatrix} (I-\tfrac{1}{4}c^2l^2A)^{-1}(2I+\tfrac{1}{2}c^2l^2A) & -I \\ I & 0 \end{bmatrix}$$

which has $2N$ eigenvalues found from the N quadratic equations

$$\det\begin{bmatrix} (1-\tfrac{1}{4}c^2l^2\lambda_s)^{-1}(2+\tfrac{1}{2}c^2l^2\lambda_s)-\mu & -1 \\ 1 & -\mu \end{bmatrix}=0 \ .$$

These give the stability equations

$$(1-\tfrac{1}{4}c^2l^2\lambda_s)\mu_s^2-2(1+\tfrac{1}{4}c^2l^2\lambda_s)\mu_s+(1-\tfrac{1}{4}c^2l^2\lambda_s)=0 \tag{7.120}$$

where λ_s $(s=1,2,\ldots,N)$ are the eigenvalues of the matrix A given by (7.93). Equation (7.120) is of the form (7.109) for each $s=1,2,\ldots,N$ and it follows that

$$-1+\tfrac{1}{4}c^2l^2\lambda_s \leqslant 1+\tfrac{1}{4}c^2l^2\lambda_s \leqslant 1-\tfrac{1}{4}c^2l^2\lambda_s \tag{7.121}$$

must be satisfied for stability. It is an easy exercise to show from (7.121) that (7.99) is unconditionally stable. The reader should verify this result, as an exercise, using the Fourier method. It may also be shown that the more accurate implicit method (7.102) is, unfortunately, stable only for $0 \leqslant cr \leqslant 3\sqrt{5}/5$.

Convergence

An alternative approach for establishing convergence to that used in subsection 7.3.4 is to assume that the theoretical solution of the difference scheme at the fixed mesh point (mh,nl) has the form

$$U_m^n = \mathrm{e}^{im\theta}\mathrm{e}^{in\psi} \neq 0 \tag{7.122}$$

where θ is real and ψ is complex, and to derive the condition or conditions under which such a solution remains bounded. The arguments used are very similar to those used in the Fourier method for examining the stability of a linear finite difference scheme with constant coefficients.

Substituting (7.122) into (7.97), for instance, and dividing by $\mathrm{e}^{im\theta}\mathrm{e}^{in\psi}$ gives

$$\mathrm{e}^{\mathrm{i}\psi} = 2(1-c^2r^2) + c^2r^2(\mathrm{e}^{-\mathrm{i}\theta}+\mathrm{e}^{\mathrm{i}\theta}) - \mathrm{e}^{-\mathrm{i}\psi} \tag{7.123}$$

Using the Euler relations $\mathrm{e}^{\pm\mathrm{i}\theta} = \cos\theta \pm \mathrm{i}\sin\theta$ and the half-angle formula $\cos\theta = 1-2\sin^2\tfrac{1}{2}\theta$, equation (7.123) becomes

$$\sin^2\tfrac{1}{2}\psi = c^2r^2\sin^2\tfrac{1}{2}\theta\ .$$

Now, c,r,θ are all real and as $0 \leqslant \sin^2\tfrac{1}{2}\psi \leqslant 1$ whenever $0 \leqslant c^2r^2 \leqslant 1$ since $0 \leqslant \sin^2\tfrac{1}{2}\theta \leqslant 1$. This implies that ψ is real whenever $c^2r^2 \leqslant 1$ which, in turn, implies that U_m^n is bounded whenever $c^2r^2 \leqslant 1$. It is assumed that u_m^n is bounded everywhere in the region R and it follows that the local discretization error is bounded at the fixed grid point (mh,nl) provided $c^2r^2 \leqslant 1$. This condition for convergence coincides with the non Neumann necessary condition for the stability of (7.97).

It is easy to verify that the solution of (7.97) does not remain bounded at the mesh point (mh,nl) if $c^2r^2 > 1$, for then $\sin^2\tfrac{1}{2}\psi > 1$ for some value or values of θ. Hence ψ occurs in complex conjugate pairs $\psi_{1,2} = \mu \pm \mathrm{i}\nu$, say, where μ and ν $(\nu \neq 0)$ are real and:

$$\text{for } \psi_1,\ U_m^n = \mathrm{e}^{im\theta}\mathrm{e}^{in(\mu+\mathrm{i}\nu)} = \mathrm{e}^{im\theta}\mathrm{e}^{in\mu}\mathrm{e}^{-in\nu}\ ;$$

$$\text{for } \psi_2,\ U_m^n = \mathrm{e}^{im\theta}\mathrm{e}^{in(\mu-\mathrm{i}\nu)} = \mathrm{e}^{im\theta}\mathrm{e}^{in\mu}\mathrm{e}^{in\nu}\ .$$

This means that for one of ψ_1 or ψ_2, U_m^n grows exponentially as $n \to \infty$ since either $\mathrm{e}^{-n\nu}$ or $\mathrm{e}^{n\nu}$ grows exponentially as $n \to \infty$ for real values of ν $(\nu \neq 0$, otherwise ψ is real). This exponential growth of U_m^n occurs whenever $c^2r^2 < 1$, irrespective of the initial conditions $U_m^0 = e^{im\theta}$ which may be arbitrarily small.

It must be concluded therefore that U_m^n is bounded only for $0 \leqslant c^2r^2 \leqslant 1$. Since u_m^n *is* bounded, the local discretization error $z_m^n = u_m^n - U_m^n$ is not bounded if $c^2r^2 > 1$; z_m^n does not then tend to zero as the mesh is refined and convergence does not take place.

It is important at this point to consider the *Courant–Friedrichs–Lewy* (*CFL*) *condition* for second-order hyperbolic equations. This condition states that *the domain of dependence of an explicit difference scheme must include the domain of dependence of the differential equation*. This means that the characteristics through the advanced point, when $t = (n+1)l$, must intersect the line through the points used by the scheme at the base time level ($t = nl$) within the range of those points, otherwise the difference scheme is not convergent.

The slopes of the characteristics of (7.85) through $(mh,(n+1)l)$ are $+1/c$ and $-1/c$ and it is thus easy to verify geometrically that the CFL condition for (7.97), for instance, is $0 \leqslant cr \leqslant 1$.

7.8 FINITE DIFFERENCE METHODS FOR NONLINEAR PROBLEMS

Consider again the nonlinear wave equation (7.83) with initial and boundary conditions given by (7.86), (7.87) and (7.88). Assuming that the solution is unique in the region R defined in subsection 7.7.1 and that the discretization of R and its boundary ∂R, as defined in that subsection, are retained, replacement of the space derivative in (7.83) by (7.89) leads to the system of nonlinear second-order ordinary differential equations given by

$$D^2\mathbf{U}(t) \equiv \frac{\mathrm{d}^2\mathbf{U}(t)}{\mathrm{d}t^2} = C(\mathbf{U})A\mathbf{U}(t) \ , \tag{7.124}$$

where D is the diagonal differential matrix $D = \mathrm{diag}\{\mathrm{d}/\mathrm{d}t\}$, $C(\mathbf{U})$ is the diagonal matrix of order N given by $C(\mathbf{U}) = \mathrm{diag}\{c^2U_m(t)\}$, and A is the matrix given by (7.92).

The numerical methods are derived from the relations

$$\mathbf{U}(t+l) - \{\exp(lD) + \exp(-lD)\}\mathbf{U}(t) + \mathbf{U}(t-l) = \mathbf{0} \tag{7.125}$$

which is analogous to (7.95). Replacing the exponential terms in (7.125) by their (0,2) Padé approximants gives

$$\mathbf{U}(t+l) - (2I + l^2D^2)\mathbf{U}(t) + \mathbf{U}(t-l) = \mathbf{0}$$

which, using (7.124), gives

$$\mathbf{U}^{n+1} = \{2I + l^2C(\mathbf{U}^n)A\}\mathbf{U}^n - \mathbf{U}^{n-1} \tag{7.126}$$

which may be used explicitly to determine $\mathbf{U}^{n+1}$.

Replacing the exponential terms by their (1,1) Padé approximants leads to

$$(I-\tfrac{1}{4}l^2D^2)\mathbf{U}(t+l)-(2I+\tfrac{1}{2}l^2D^2)\mathbf{U}(t)+(I-\tfrac{1}{4}l^2D^2)\mathbf{U}(t-l)=\mathbf{0}$$

which, using (7.124), gives

$$\begin{aligned}&\{I-\tfrac{1}{4}l^2C(\mathbf{U}^{n+1})A\}\mathbf{U}^{n+1}-\{2I+\tfrac{1}{2}l^2C(\mathbf{U}^n)A\}\mathbf{U}^n\\&+\{I-\tfrac{1}{4}l^2C(\mathbf{U}^{n-1})A\}\mathbf{U}^{n-1}=\mathbf{0}\ .\end{aligned} \tag{7.127}$$

This is a nonlinear difference scheme in $\mathbf{U}^{n+1}$ which has the form

$$\mathbf{F}(\mathbf{U}^{n+1})=\mathbf{0}\ . \tag{7.128}$$

The solution is obtained from (7.128) by solving this nonlinear algebraic system using, say, the Newton–Raphson method for a system as described in Chapter 3. Each component of $\mathbf{F}(\mathbf{U}^{n+1})$ is given by

$$\begin{aligned}F_m=&-\tfrac{1}{4}r^2(c_m^{n+1})^2U_{m-1}^{n+1}+[1+\tfrac{1}{2}r^2(c_m^{n+1})^2]U_m^{n+1}-\tfrac{1}{4}r^2(c_m^{n+1})^2U_{m+1}^{n+1}\\&-\tfrac{1}{2}r^2(c_m^n)^2U_{m-1}^n-[2-r^2(c_m^n)^2]U_m^n-\tfrac{1}{2}r^2(c_m^n)^2U_{m+1}^n\\&-\tfrac{1}{4}r^2(c_m^{n-1})^2U_{m-1}^{n-1}+[1+\tfrac{1}{2}r^2(c_m^{n-1})^2]U_m^{n-1}-\tfrac{1}{4}r^2(c_m^{n-1})^2U_{m+1}^{n-1},\end{aligned} \tag{7.129}$$

with $m=1,2,\ldots,N$ and $c_m^n=c(U_m^n)$, in which $U_0^n=U_{N+1}^n=0$ $(n=0,1,\ldots)$. The Jacobian $\partial\mathbf{F}/\partial\mathbf{U}^{n+1}$ is tridiagonal with

$$\begin{aligned}&\partial F_m/\partial U_{m-1}^{n+1}=\partial F_m/\partial U_{m+1}^{n+1}=-\tfrac{1}{4}r^2(c_m^{n+1})^2\ ,\\&\frac{\partial F_m}{\partial U_m^{n+1}}=1+\tfrac{1}{2}r^2(c_m^{n+1})^2+\tfrac{1}{2}r^2U_m^{n+1}\frac{\mathrm{d}(c_m^{n+1})^2}{\mathrm{d}U_m^{n+1}}\ .\end{aligned} \tag{7.130}$$

An alternative approach to solving (7.128) is to use a predictor–corrector combination. One such algorithm is given by

$$\text{P:}\ \mathbf{U}_{(0)}^{n+1}=2\mathbf{U}^n+l^2C(\mathbf{U}^n)A\mathbf{U}^n-\mathbf{U}^{n-1}\ ,$$

$$\text{E:}\ D^2\mathbf{U}_{(k-1)}^{n+1}=C(\mathbf{U}_{(k-1)}^{n+1})A\mathbf{U}_{(k-1)}^{n+1}\ ,$$

$$\text{C:}\ \mathbf{U}_{(k)}^{n+1}=\tfrac{1}{4}l^2D^2\mathbf{U}_{(k-1)}^{n+1}+2\mathbf{U}^n+\tfrac{1}{2}l^2C(\mathbf{U}^n)A\mathbf{U}^n \qquad k=1,2,\ldots$$

$$-\mathbf{U}^{n-1}+\tfrac{1}{4}l^2C(\mathbf{U}^{n-1})A\mathbf{U}^{n-1}\ ,$$

$$\text{E: } D^2\mathbf{U}^{n+1}_{(k)}=C(\mathbf{U}^{n+1}_{(k)})A\mathbf{U}^{n+1}_{(k)}\ ,$$

in which $n=0,1,\ldots$ and $k=1,2,\ldots$ is the number of times the corrector is applied. This algorithm is given in $P(EC)^kE$ mode (see Chapter 5). The predictor formula is, of course, (7.126) while the corrector formula is (7.127). Correcting to convergence threfore means that the excellent stability properties of (7.127) are retained.

PROBLEMS

1. The function u satisfies the equation

$$2(x+1)^{\frac{1}{2}}\frac{\partial u}{\partial x}+\frac{u}{2t}\frac{\partial u}{\partial t}=1,\qquad x>-1,\qquad t>0$$

and the given condition $u(x,1)=1$ for $0\leqslant x<\infty$. Calculate the Cartesian equation of the characteristic through the point (0,1) and find an approximation to the value of t at the point P on this characteristic where $x=0.5$. Using the method of characteristics for first order equations determine a first approximation to u at P giving the answer to three decimal places.

Find, to three decimal places, corrected values of t and u at P and use them to determine the slope of the characteristic through P.

2. The function $u=u(x,t)$ satisfies the nonlinear hyperbolic equation

$$\sqrt{x}\frac{\partial u}{\partial x}+u\frac{\partial u}{\partial t}=-u^2\ ;\qquad x>0,\qquad t>0$$

and the given condition $u(x,0)=1$ for $0<x<\infty$. Calculate the Cartesian equation of the characteristic through the point (1,0) and find an approximation to the value of t at the point P on this characteristic where $x=1.5$. Using the method of characteristics for first order equations determine a first approximation to u at P giving the answer to three decimal places.

Find, to three decimal places, corrected values of t and u at P and use them to determine the slope of the characteristic through P.

3. Show that the characteristic of the hyperbolic equation

$$x^2e^t\frac{\partial u}{\partial x}+\frac{\partial u}{\partial t}=-e^tu\ ,$$

with initial condition $u(x,0)=1$, $0<x<\infty$, through the point $(X,0)$ in the (x,t) plane, is given by

$$t = \frac{1}{X} - \frac{1}{x} .$$

4. Show that the equation

$$(x^2+1)\frac{\partial^2 u}{\partial x^2} - \lambda^2(t^2+1)\frac{\partial^2 u}{\partial t^2} = 6(x - x^3 - \lambda^2 t - \lambda^2 t^3)$$

is hyperbolic for all real x,t and all real non-zero values of the parameter λ.

Two points T_1 and T_2 lie on some initial curve L in the (x,t) plane which is not a characteristic. The values of x,t,u,p,q at these points are given in the following table:

	x	t	u	p	q
T_1	0.15	0.30	0.030375	0.0675	0.2700
T_2	0.20	0.28	0.029952	0.1200	0.2352

Determine first approximations to the coordinates of T_2, at which the characteristic of positive slope through T_1 intersects the characteristic of negative slope through T_2, for the case $\lambda = 1$.

Given that $p = 0.0818$ and $q = 0.2988$ are first approximations to the values of p,q at T_3, determine second approximation s to the coordinates of T_3.

Find an approximation to u at T_3.

5. Show that the equation

$$\frac{\partial^2 u}{\partial x^2} - (u^2+1)\frac{\partial^2 u}{\partial t^2} + 16(x^2+t^2)^2 = 0$$

is always hyperbolic for real x,t and $u(x,t)$.

Two points T_1 and T_2 lie on some initial curve L in the (x,t) plane which is not a characteristic; the values of x,t,u,p,q at these points are given in the following table:

	x	t	u	p	q
T_1	0.1	0.2	0.10	0.4	0.8
T_2	0.2	0.2	0.16	0.8	0.8

Calculating answers to four decimal places

(i) determine first approximations to the coordinates of the point T_3, at which the characteristic of positive gradient through T_1 meets the characteristic of negative gradient through T_2

(ii) given that $p = 0.6$ and $q = 1.0$ are first approximations to the values of p and q at T_3, determine a first approximation to u at T_3.

6. Show that the equation

$$\frac{\partial^2 u}{\partial x^2} - (u^2+1)\frac{\partial^2 u}{\partial t^2} + 2x^2 + 2t^2 = 0$$

is always hyperbolic for real x,t and $u(x,t)$.

Values of $u, p \equiv \partial u/\partial x$ and $q \equiv \partial u/\partial t$ are given at all points on part of some initial curve L in the (x,t) plane which is not a characteristic. Write down expressions for the slopes of the characteristics through any point (x,t) in the domain of dependence and describe the geometry of this domain.

Two points T_1 and T_2 lie on L; the values of x,t,u,p,q at these points are given in the following table:

	x	t	u	p	q
T_1	0.1	0.2	0.05	0.2	0.4
T_2	0.2	0.2	0.08	0.4	0.4

Determine first approximations to the coordinates of the point T_3, at which the characteristic of positive slope through T_1 intersects the characteristic of negative slope through T_2.

Given that $p = 0.3$ and $q = 0.5$ are first approximations to the values of p,q at T_3, determine a first approximation to u at T_3.

Determine second approximations to the coordinates of T_3.

7. In (7.85) with $c = 1$, the differential equation is approximated by the finite difference scheme

$$(I - \tfrac{1}{9}l^2A)\mathbf{U}^{n+1} - (2I + \tfrac{7}{9}l^2A)\mathbf{U}^n + (I - \tfrac{1}{9}l^2A)\mathbf{U}^{n-1} = \mathbf{0}$$

where I is the identity matrix of order N, l is an increment in t, and A is the square matrix of order N given by (7.72).

Show that this gives the nine point implicit formula

$$-\tfrac{1}{9}r^2(U_{m-1}^{n+1} + U_{m+1}^{n+1}) + (1 + \tfrac{2}{9}r^2)U_m^{n+1}$$

$$= \tfrac{7}{9}r^2(U_{m-1}^n + U_{m+1}^n) + 2(1 - \tfrac{7}{9}r^2)U_m^n + \tfrac{1}{9}r^2(U_{m-1}^{n-1} + U_{m+1}^{n-1}) - (1 + \tfrac{2}{9}r^2)U_m^{n-1} ,$$

where $r = l/h$.

Find, in its simplest form, the local truncation error of this formula, and use the von Neumann method to find the restriction on r (if any) if the scheme is to be stable.

8. The differential equation

$$\frac{\partial^2 u}{\partial t^2} = \frac{\partial^2 u}{\partial x^2}$$

has initial conditions

$$u(x,0) = x, \frac{\partial u}{\partial t}(x,0) = \tfrac{1}{2}x \ \text{ for } \ 0 \leqslant x \leqslant 1$$

and boundary conditions

$$u(0,t) = 0 \ ; \quad u(1,t) = 1, \quad t > 0 \ .$$

Find the values of U for $x = 0.00, 0.25, 0.50, 0.75, 1.00$ when $t = 0.25$ and $t = 0.50$ using $h = l = 0.25$ and an appropriate finite difference scheme.

9. The function $u(x,y,t)$ satisfies the two dimensional wave equation

$$\frac{\partial^2 u}{\partial t^2} = \frac{\partial^2 u}{\partial x^2} + \frac{\partial^2 u}{\partial y^2} \ ; \qquad 0 < x,y < X, t > 0$$

with appropriate initial and boundary conditions specified. Both the intervals $0 \leqslant x \leqslant X$ and $0 \leqslant y \leqslant X$ are discretized into N subintervals each of width h and the time variable t is discretized in steps of length l so that $t = nl (n = 0,1,2, \ldots)$.

The wave equation, with its derivatives replaced by the second order approximations derived above, is applied to the point (kh,mh,nl) with $k,m = 1,2, \ldots, N$ and $n = 0,1,2, \ldots$. Show that the principal part of the local truncation error of the resulting explicit finite difference scheme is

$$\tfrac{1}{12}l^2 \frac{\partial^4 u}{\partial t^4} - \tfrac{1}{12}h^2 \left(\frac{\partial^4 u}{\partial x^4} + \frac{\partial^4 u}{\partial y^4} \right) \ .$$

Estimate the principal part of the local truncation error of the analogous scheme in three space dimensions.

Assuming a solution of the form $U^n_{k,m} = e^{ik\theta} e^{im\phi} e^{in\lambda}$ at the point (kh,mh,nl), where $i = +\sqrt{-1}$, θ and ϕ are real and λ is complex, show that

$$\sin^2 \tfrac{1}{2}\lambda = r^2 (\sin^2 \tfrac{1}{2}\theta + \sin^2 \tfrac{1}{2}\phi) \ ,$$

where $r = l/h$. Hence discuss the boundedness of U.

8

Parabolic equations

8.1 THE SIMPLE HEAT EQUATION AND ITS NUMERICAL SOLUTION

Second-order parabolic equations arise in many mathematical models of biomedical systems; for example, in the study of heat flow in the human body, in neurophysiological problems, in genetics, and in drug absorption problems. These examples will be discussed in detail later in the chapter.

Parabolic equations in one space variable are one of the three classes of the general partial differential equation

$$a\frac{\partial^2 u}{\partial x^2}+b\frac{\partial^2 u}{\partial x\partial t}+c\frac{\partial^2 u}{\partial t^2} = e \tag{8.1}$$

in which a, b, c, e are functions of x, t, u, $\partial u/\partial x$, $\partial u/\partial t$ but not of the second-order derivatives; for parabolic equations, $b^2-4ac = 0$ (see Appendix A).

The most widely used parabolic equation is obtained from (8.1) by writing $a = 1$, $b = c = 0$ and $e = \partial u/\partial t$ to give

$$\frac{\partial u}{\partial t} = \frac{\partial^2 u}{\partial x^2} \tag{8.2}$$

in some region R. Equation (8.2) is known as the *one (space)-dimensional heat equation* and is the model or test equation generally used to test the reliability and accuracy of a numerical method.

Writing $a = 1/\mathrm{Re}$, $b = c = 0$ and $e = \partial u/\partial t+\mu\partial u/\partial x$ gives

$$\frac{\partial u}{\partial t}+\mu\frac{\partial u}{\partial x} = \frac{1}{\mathrm{Re}}\frac{\partial^2 u}{\partial x^2} \tag{8.3}$$

which is the one-dimensional *diffusion–convection equation*, Re being the Reynolds number and $\mu > 0$ being the convection coefficient. The convection term $\mu \partial u/\partial x$ relates, for example, to the flow of heat by a moving fluid such as blood.

Writing $a = \kappa$, $b = c = 0$ and $e = \partial u/\partial t - f(u)$, where f is some linear or nonlinear function of u, gives the one-dimensional *reaction–diffusion equation*

$$\frac{\partial u}{\partial t} = \kappa \frac{\partial^2 u}{\partial x^2} + f(u) \; ; \tag{8.4}$$

this important equation will be met again later in the chapter.

The heat flow equation (8.2) is also an example of the parabolic equation

$$\beta(x,t)\frac{\partial u}{\partial t} = \frac{\partial}{\partial x}\left(\alpha(x,t)\frac{\partial u}{\partial x}\right) - \gamma(x,t)u \tag{8.5}$$

which is itself an example of (8.1) with $a = \alpha(x,t)$, $b = c = 0$ and $e = \beta(x,t)\partial u/\partial t + \gamma(x,t)u - (\partial\alpha/\partial x)(\partial u/\partial x)$. In (8.5) the conditions $\alpha(x,t) > 0$, $\beta(x,t) > 0$ and $\gamma(x,t) \geqslant 0$ must hold at every point of the region R. The functions α, β and γ are independent of u and thus (8.5) is a *linear* parabolic expression. Similarly, equation (8.1) is said to be linear if a, b, c, e do not depend on u, otherwise (8.1) is *nonlinear*. When a, b, c, e depend on x and/or t only, equation (8.1) is said to be of the *variable coefficient* type (and still linear) and when a, b, c, e have constant values, (8.1) is said to be a *constant coefficient* parabolic equation. Looking again at (8.5) this equation is said to be *self-adjoint* whenever α is a function of x only and β is constant.

The region R is usually one of three types with the associated initial conditions and boundary conditions as specified:

(1) the open rectangle $X_0 \leqslant x \leqslant X_1, t > 0$. Here, a properly posed problem consists of a parabolic equation together with the initial distribution

$$u(x,0) = g(x) \;, \qquad X_0 \leqslant x \leqslant X_1$$

and boundary conditions of the form

$$a_0(X_0,t)u + b_0(X_0,t)\frac{\partial u}{\partial x} = c_0(X_0,t) \;, \qquad t > 0$$

$$a_1(X_1,t)u + b_1(X_1,t)\frac{\partial u}{\partial x} = c_1(X_1,t) \;, \qquad t > 0$$

with $a_0(x,t) \geqslant 0, a_1(x,t) \geqslant 0, b_0(x,t) \leqslant 0, b_1(x,t) \geqslant 0$ and with $a_0(x,t) - b_0(x,t) > 0$, $a_1(x,t) - b_1(x,t) > 0$.

The open rectangle is a frequently-occurring region in the numerical solution of parabolic equations, particularly the simple heat equation (8.2); the associated boundary conditions usually have $a_0 = a_1 = 1$ and $b_0 = b_1 = 0$. This formulation is usually used in the analyse of a numerical method;

(2) the semi-infinite plane $-\infty < x < \infty, t > 0$. Here, a properly posed problem consists of a parabolic equation together with the initial distribution

$$u(x,0) = g(x) , \qquad 0 \leqslant x < \infty$$

and the formulation is known as the *Cauchy initial value problem*;

3. the first quadrant $x > 0, t > 0$ of the (x,t)-plane. In this final case a properly posed problem consists of a parabolic equation together with the initial distribution

$$u(x,0) = g(x) , \qquad 0 \leqslant x < \infty$$

and the boundary condition

$$a_0(0,t)u + b_0(0,t)\frac{\partial u}{\partial x} = c_0(0,t) , \qquad t > 0 ,$$

where, as in (1), $a_0(x,t) \geqslant 0$, $b_0(x,t) \leqslant 0$ and $a_0(x,t) - b_0(x,t) > 0$.

To give the reader some familiarity with numerical methods for parabolic equations, the simple diffusion problem consisting of the differential equation

$$\frac{\partial u}{\partial t} = \frac{\partial^2 u}{\partial x^2} ; \qquad X_0 < x < X_1 , \qquad t > 0 \tag{8.6}$$

together with the boundary conditions

$$u(X_0,t) = u(X_1,t) = 0 , \qquad t > 0 \tag{8.7}$$

and the initial distribution

$$u(x,0) = g(x) ; \qquad X_0 \leqslant x \leqslant X_1 , \tag{8.8}$$

will be discussed first of all. In (8.8), $g(x)$ is a given continuous function of x and it is not specified that $g(X_0) = 0$ nor that $g(X_1) = 0$ so that discontinuities between initial conditions and boundary conditions may exist.

The interval $X_0 \leqslant x \leqslant X_1$ is divided into $N+1$ subintervals each of width h so that $(N+1)h = X_1 - X_0$ and the time variable t is discretized in steps of length l. The open region $R = [X_0 < x < X_1] \times [t > 0]$ and its boundary ∂R consisting of the lines $x = X_0$, $x = X_1$ and the axis $t = 0$ have thus been covered by a rectangular mesh, the mesh points having coordinates $x_m = X_0 + mh$ and $t_n = nl$ where $m = 0, 1, \ldots, N+1$ and $n = 0, 1, 2, \ldots$. The solution $u(x_m, t_n)$ of (8.6) at the mesh point (x_m, t_n) will be denoted by u_m^n, while the theoretical solution of an approximating difference scheme (numerical method) will be denoted by U_m^n. The computed solution of the difference scheme (the solution actually obtained, which may be subject to the errors discussed in Chapter 7) at this mesh point will be denoted by $\widetilde{U}_m^n$.

The approach to be followed is the method of lines introduced in Chapter 7. The space derivative in (8.6) is replaced by

$$\frac{\partial^2 u}{\partial x^2} = h^{-2}\{u(x-h,t) - 2u(x,t) + u(x+h,t)\} + 0(h^2) \tag{8.9}$$

and (8.6) with (8.9) are applied to all N interior mesh points at time level $t = nl (n = 0, 1, \ldots)$ to produce a system of N first-order, linear, ordinary differential equations. Noting from (8.7) that $U_0^n = U_{N+1}^n = 0$, this system is of the form

$$\frac{\mathrm{d}\mathbf{U}(t)}{\mathrm{d}t} = A\mathbf{U}(t) \; ; \qquad \mathbf{U}(0) = \mathbf{g} \tag{8.10}$$

where $\mathbf{U}(t) = [U_1(t), U_2(t), \ldots, U_N(t)]^{\mathrm{T}}$ is the vector of approximations to the solution of (8.6) at time $t = nl$, $\mathbf{g}$ is the vector of initial conditions obtained from (8.8), and A is the tridiagonal matrix given by

$$A = h^{-2}\begin{bmatrix} -2 & 1 & & & 0 \\ 1 & -2 & 1 & & \\ & & \cdots & \cdots & \cdots \\ & & 1 & -2 & 1 \\ 0 & & & 1 & -2 \end{bmatrix} \tag{8.11}$$

of which the eigenvalues are

$$\lambda_s = -4h^{-2}\sin^2[s\pi/\{2(N+1)\}] \; ; \qquad s = 1, 2, \ldots, N \; . \tag{8.12}$$

A practical difficulty with (8.10) is that the system is stiff (see Chapter 5) because a large value of N must be used in some biomedical problems, leading to a large range of magnitude of eigenvalues.

Solving (8.10) gives

$$\mathbf{U}(t) = \exp(tA)\mathbf{g} \tag{8.13}$$

which satisfies the recurrence relation

$$\mathbf{U}(t+l) = \exp(lA)\mathbf{U}(t) \; ; \qquad t = 0, l, 2l, \ldots . \tag{8.14}$$

To obtain numerical solutions $\mathbf{U}(t+1)$, the exponential matrix function $\exp(lA)$ must be replaced by a valid approximants.

One such replacement is the rational approximant

$$\exp(lA) = [I - \theta lA]^{-1}[I + (1-\theta)lA] \tag{8.15}$$

in which $0 \leqslant \theta \leqslant 1$ is a parameter. This leads, in (8.14), to the family of methods

$$[I - \theta lA]\mathbf{U}(t+l) = [I + (1-\theta)lA]\mathbf{U}(t) \; . \tag{8.16}$$

Setting $\theta = 0, 1, \frac{1}{2}$, in (8.15) gives the $(0,1)$, $(1,0)$ and $(1,1)$ Padé approximants to $\exp(lA)$ and using these three approximants in (8.16) leads to three well-known methods for solving (8.6).

Writing $\theta = 0$, first of all, yields

$$\mathbf{U}(t+l) = (I + lA)\mathbf{U}(t) \tag{8.17}$$

from (8.16). Applying (8.17) to the mesh point (x_m, t_n) gives the four-point explicit scheme

$$U_m^{n+1} = pU_{m-1}^n + (1-2p)U_m^n + pU_{m+1}^n \tag{8.18}$$

where $p = l/h^2$, $m = 1, 2, \ldots, N$ and $n = 0, 1, 2, \ldots$.

Writing, next, $\theta = 1$ in (8.16) yields

$$(I - lA)\mathbf{U}(t+l) = \mathbf{U}(t) \tag{8.19}$$

which, when applied to the internal mesh point (x_m, t_n), gives

$$-pU_{m-1}^{n+1} + (1+2p)U_m^{n+1} - pU_{m+1}^{n+1} = U_m^n \; . \tag{8.20}$$

This method is often referred to as the *fully implicit method* and its solution is easily obtained from (8.19) using a numerical method for solving a tridiagonal linear algebraic system (see Chapter 1).

The third well-known method is obtained by writing $\theta = \frac{1}{2}$ in (8.16). This gives

$$(I - \tfrac{1}{2}lA)\mathbf{U}(t+l) = (I + \tfrac{1}{2}lA)\mathbf{U}(t) \tag{8.21}$$

which, when applied to the general mesh point (x_m, t_n) ($m = 1, 2, \ldots, N$; $n = 0, 1, \ldots$), gives the six-point, two-time-level implicit formula

$$\begin{aligned} -\tfrac{1}{2}pU_{m-1}^{n+1} &+ (1+p)U_m^{n+1} - \tfrac{1}{2}pU_{m+1}^{n+1} \\ &= \tfrac{1}{2}pU_{m-1}^n + (1-p)U_m^n + \tfrac{1}{2}pU_{m+1}^n \ . \end{aligned} \tag{8.22}$$

This formula is known as the *Crank–Nicolson method* (1947) and has enjoyed great popularity since its publication. It will be seen in section 8.2 that it is unconditionally stable but, as shown by Lawson and Morris (1978), it is, nevertheless, necesary to impose a restriction on the ratio $r = l/h$.

The computed solution of the Crank–Nicholson scheme may be obtained efficiently from the vector form (8.21) by rewriting it as

$$(I - \tfrac{1}{2}lA)\mathbf{U}^* = 2\mathbf{U}(t) \ ,$$

$$\mathbf{U}(t+l) = \mathbf{U}^* - \mathbf{U}(t) \ .$$

This approach requires the storage of the intermediate vector $\mathbf{U}^*$ but needs $3N-2$ fewer multiplications than using a tridiagonal solver to find $\mathbf{U}(t+1)$ directly from (8.21).

Applying the explicit scheme (8.18) or the Crank–Nicholson method (8.22) to the mesh points $(x_1, 0)$ or $(x_N, 0)$ involves using the points $(X_0, 0)$ or $(X_1, 0)$ where there may be discontinuities between initial conditions and boundary conditions. The effects of such discontinuities will be propagated as the computation proceeds, giving poor numerical results near the boundaries for all values of t. This phenomenon does not arise when the fully implicit scheme (8.20) is used to obtain the numerical solution, for this method does not use the points of discontinuity.

Applications of (8.18), (8.20) and (8.22) to variable coefficient problems is considered in Twizell (1984) to which the reader is referred.

The behaviour of each of these numerical methods is ably illustrated by the model problem

$$\frac{\partial u}{\partial t} = \frac{\partial^2 u}{\partial x^2} \ ; \qquad 0 < x < 2, \ t > 0$$

subject to $u(0,t) = u(2,t) = 0$ for $t > 0$ and $u(x,0) = 1$ for $0 \leqslant x \leqslant 2$. This problem, which has theoretical solution

$$u(x,t) = \sum_{k=1}^{\infty} [1 - (-1)^k] \frac{2}{k\pi} \sin(\tfrac{1}{2}k\pi x) \exp(-\tfrac{1}{4}k^2\pi^2 t) \ ,$$

is due to Lawson and Morris (1978).

In the first of two numerical experiments, the problem is solved for $h = 0.1$ and $l = 0.005$, giving $p = \frac{1}{2}$ and $r = 0.05$. The ratios p and r both satisfy the restrictions placed on them in subsection 8.2.2 and the computed solutions at time $t = 1.0$, which are depicted in Fig. 8.1, are satisfactory. To the scale used, the graph of the

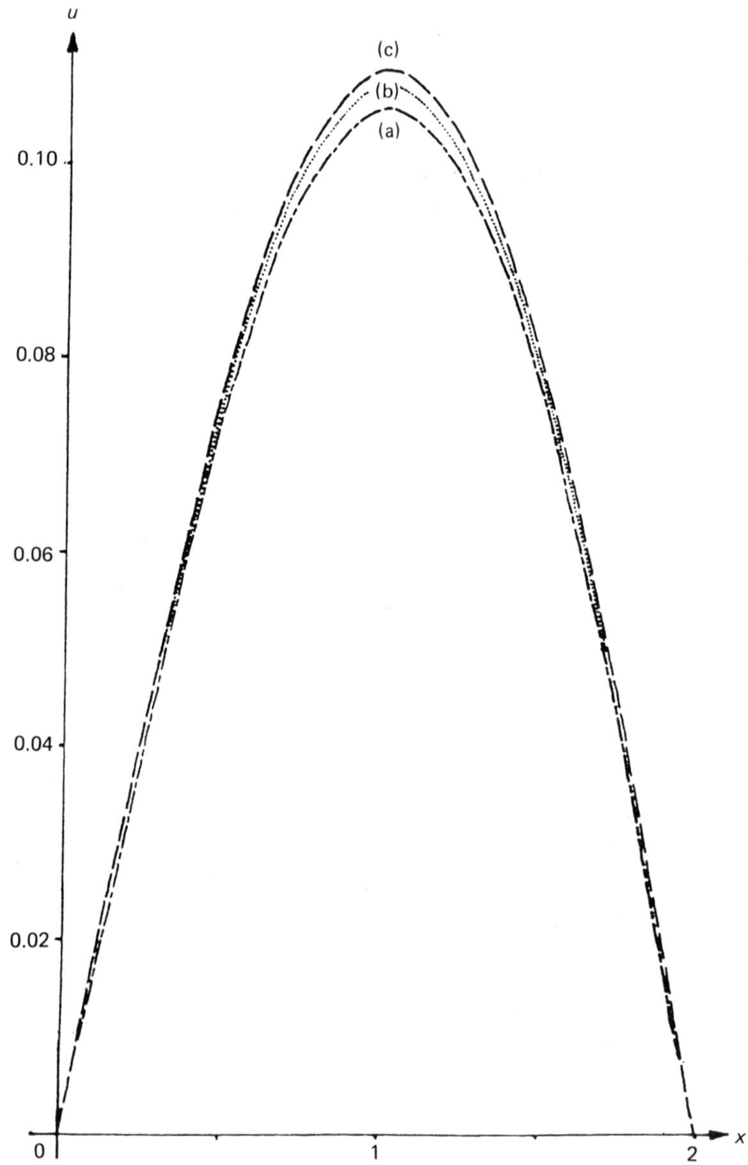

Fig. 8.1 — Solution of model problem at time $t = 1.0$ with $p = \frac{1}{2}$: (*a*))0, 1) Padé approximant, (b) (1, 1) Padé approximant, (c) (1, 0) Padé approximant.

theoretical solution is indistinguishable from that relating to the Crank–Nicholson method and is consequently omitted from Fig. 8.1.

In the second numerical experiment, the problem is solved using $h = 0.05$ and $l = 0.1$, giving $p = 40$ and $r = 2$. Here, the valuc of p is much too large for the explicit method (8.18) to give a meaningful solution; the computed solution oscillates widely near both boundaries while at $x = 1.0$, $t = 1.0$ it has the value 1.0, the theoretical solution being 0.108 approximately. As can be seen from Fig. 8.2, the Crank–

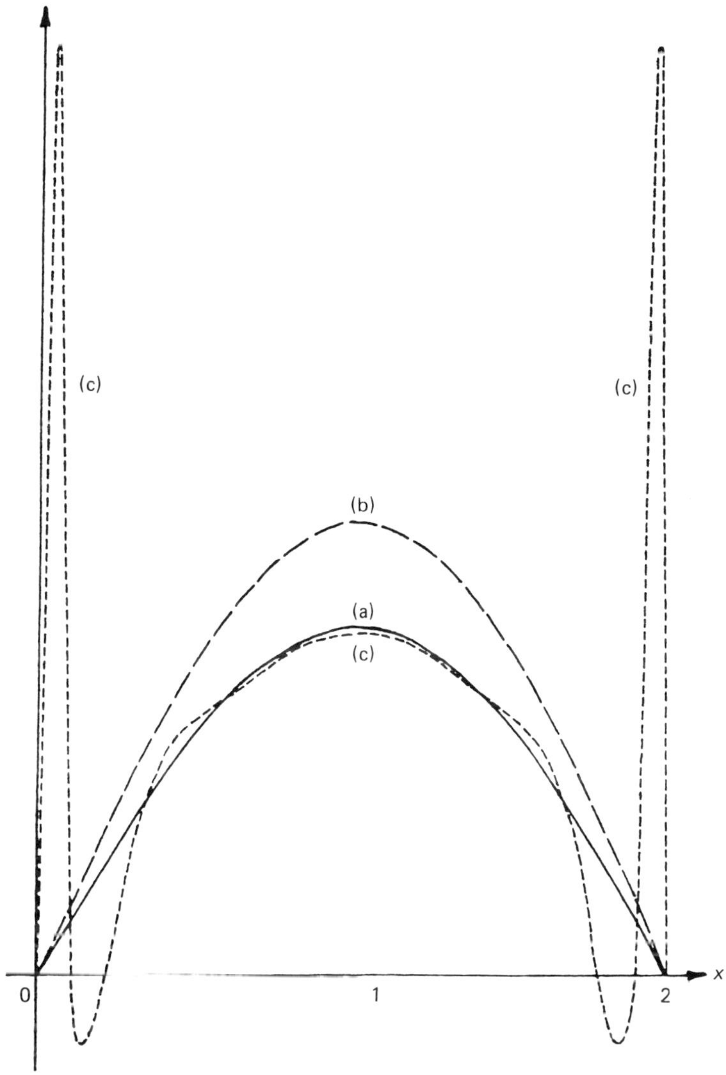

Fig. 8.2 — Solution of model problem at time $t = 1.0$ with $p = 40$: (a) theoretical solution, (b) (1, 0) Padé approximant, (c) (1, 1) Padé approximant.

Nicolson scheme is accurate in the interior of the region but has a large residual error near the boundary. This phenomenon arises because the ratio $r = l/h$ violates a bound on its to be discussed in subsection 8.2.2; the magnitude of this residual error will be increased by refining h. The fully implicit method (8.20) behaves smoothly over the whole interval $0 \leqslant x \leqslant 2$ but is not as accurate near the centre of the region as the Crank–Nicholson method.

8.2 ANALYSES OF THE METHODS

In this section, the tests which finite difference schemes for solving parabolic equations must pass before being used are discussed. As a result of the analyses, the behaviour of some of the methods developed in section 8.1 will be clarified.

8.2.1 Truncation errors and consistency

Consider the general finite difference method arising from (8.16). It is easy to see that this method takes the form

$$\frac{U_m^{n+1}+U_m^n}{l} = \frac{\theta(U_{m-1}^{n+1}-2U_m^{n+1}+U_{m+1}^{n+1})+(1-\theta)(U_{m-1}^n-2U_m^n+U_{m+1}^n)}{h^2} \tag{8.23}$$

from which it follows that the left-hand side approximates $\partial u/\partial t$ while the right-hand side gives a weighted approximation to $\partial^2 u/\partial x^2$. The *local truncation error* $L(x,t)$ of (8.23) is therefore given by

$$\begin{aligned} L(x,t) &= \frac{u(x,t+l)-u(x,t)}{l} \\ &- \frac{\theta\{u(x-h,t+l)-2u(x,t+l)+u(x+h,t+l)\}}{h^2} \\ &- \frac{(1-\theta)\{u(x-h,t)-2u(x,t)+u(x+h,t)\}}{h^2} - \left(\frac{\partial u}{\partial t}-\frac{\partial^2 u}{\partial x^2}\right) . \end{aligned} \tag{8.24}$$

Expanding about $u(x,t)$ using Taylor's series, it is left as an exercise to the reader to show that $L(x,t)$ takes the form

$$L(x,t) = \left(-\frac{1}{12}h^2\frac{\partial^4 u}{\partial x^4}+C_2 l\frac{\partial^2 u}{\partial t^2}\right)+C_3 l^2\frac{\partial^3 u}{\partial t^3}+\ldots \tag{8.25}$$

with $C_2 = \frac{1}{2}-\theta$ and $C_3 = \frac{1}{6}-\frac{\theta}{2}$. (Note: for (8.6), $\partial^2 u/\partial t^2 = \partial^3 u/\partial x^2\partial t$ and $\partial^3 u/\partial t^3 = \partial^4 u/\partial x^2/\partial t^2$.)

In (8.25) the term in parentheses on the right-hand side is the *principle part* of the local truncation error. The component $-\frac{1}{12}h^2\partial^4 u/\partial x^4$ is due to the space discretization and to the use of (8.9) in (8.6); this term will appear in the principal part of the local truncation error of every finite difference method arising from (8.16) in which $0 \leqslant \theta \leqslant 1$.

The constant C_2 in (8.25) is the *time error constant* of the finite difference method. Setting $\theta = 0$ gives $C_2 = \frac{1}{2}$ which is the time error constant of the explicit method (8.18) while setting $\theta = 1$ gives $C_2 = -\frac{1}{2}$ which is the time error constant of the fully implicit method (8.20). As noted in Chapter 7, these two methods are first-order accurate in time. Setting $\theta = \frac{1}{2}$ causes C_2 to vanish in (8.25) and the time error constant of the Crank–Nicholson method (8.22) is consequently $C_3 = -\frac{1}{12}$: this method is second-order accurate in time.

Comparing the error constants of (8.18), (8.20) and (8.22) it is evident that the Crank–Nicholson scheme is more accurate than the explicit and fully implict methods. This explains the closeness of the Crank–Nicholson curve to the graph of the theoretical solution in the first numerical experiment in section 8.1; the behaviour of the Crank–Nicolson method near the boundaries in the second numerical experiment will be described in subsection 8.2.2. The error constants for the explicit and fully implicit methods are opposite in sign. This explains why their curves are on opposite sides of the graph of the theoretical solution in Fig. 8.1.

It is easy to see from (8.25) that $L(x,t) \to 0$ as $h, l \to 0$ simultaneously. Every finite difference method arising from (8.16) with $0 \leqslant \theta \leqslant 1$ is therefore consistent with (8.6).

8.2.2 Conventional stability analyses

Brief decriptions of the energy, Fourier and matrix methods of analysing the numerical stability of finite difference schemes were given in Chapter 7. These methods are, or course, applicable to parabolic partial differential equations and will now be considered.

Analysing numerical stability using the energy method can be a formidable task and for a discussion of the application of the method to the explicit scheme (8.18) for solving (8.6) the reader is referred to Twizell (1984). Jones and Sleeman (1983) apply the energy method to the reaction–diffusion equation (8.4). These authors take $x_0 = 0$ and $X_1 = 1$ and consider initial conditions of the form (8.8) together with boundary conditions in the form (8.7) and also in the zero flux form

$$\frac{\partial}{\partial x}u(0,t) = \frac{\partial}{\partial x}u(1,t) = 0 \ ; \qquad t > 0 \ . \tag{8.26}$$

Jones and Sleeman also apply the energy method to the FitzHugh–Nagumo equations

$$\frac{\partial u}{\partial t} = \frac{\partial^2 u}{\partial x^2} + u(1-u)(u-a) - w \ ; \qquad x > 0, t > 0, 0 < a < 1 \tag{8.27}$$

$$\frac{\partial w}{\partial t} = bu - cw \ ; \qquad x > 0, t > 0, b > 0, c > 0 \tag{8.28}$$

(see FitzHugh (1961) and Nagumo *et al.* (1962)) with initial conditions

$$u(x,0) = w(x,0) = 0 , \qquad x \geqslant 0 \tag{8.29}$$

and boundary conditions

$$u(0,t) = P(t) , \qquad w(0,t) = b\int_0^t P(s)\,\mathrm{e}^{-cs}\mathrm{d}s , \tag{8.30}$$

which are used to model the global behaviour of nerve impulse transmissions (see section 8.6). The function $P(t)$ represents the stimulus emanating at $x = 0$.

The Fourier method is quite adequate for analysing the stability of constant coefficient finite difference schemes (see Chapter 7). Considering the general scheme (8.23), the difference between the theoretical and computed solutions of the difference scheme at the mesh point (x_m, t_n) is considered to have typical term $\mathrm{e}^{\alpha nl}\,\mathrm{e}^{\mathrm{i}\beta mh}$ where α is complex and β is real. This error satifies the difference scheme and so, after cancelling $\mathrm{e}^{\alpha nl}\,\mathrm{e}^{\mathrm{i}\beta mh}$, the error *amplification factor* $\xi = \mathrm{e}^{\alpha l}$ is found to be

$$\xi = \frac{1-4(1-\theta)p\sin^2\frac{1}{2}\beta h}{1+4\theta p\sin^2\frac{1}{2}\beta h} . \tag{8.31}$$

This factor governs the growth of a perturbation of the initial conditions and must therefore satisfy the *von Neumann necessary condition for stability*, namely $|\xi| \leqslant 1$. In the case of the explicit method (8.18), putting $\theta = 0$ in (8.31) leads to

$$-1 \leqslant 1-4p\sin^2\tfrac{1}{2}\beta h \leqslant 1$$

and, recalling that β is real, the left-hand side is satisfied for $p \leqslant \frac{1}{2}$ while the right-hand side is satisfied provided $p \geqslant 0$ (which it is by definition). The stability interval is given by the intersection of these two restrictions on p and the conclusion is that (8.18) is stable in the conventional sense provided $p \leqslant \frac{1}{2}$. The reader will recall that this restriction was satisfied in the first numerical experiment of section 8.1 but was violated in the second numerical experiment.

Turning, now, to the fully implicit method (8.20) and using (8.31) with $\theta = 1$, the amplification factor is found to be

$$\xi = \frac{1}{1+4p\sin^2\frac{1}{2}\beta h} . \tag{8.32}$$

Obviously, ξ, as defined by (8.32), satisfies the von Neumann criterion for all $p \geqslant 0$ and the fully implicit method is unconditionally stable.

In the case of the Crank–Nicolson method (8.22), a conventional stability analysis by the matrix method shows that a perturbation $\mathbf{Z}^0$ of the initial data satisfies the equation

$$\mathbf{Z}^{n+1} = (I-\tfrac{1}{2}lA)^{-1}(I+\tfrac{1}{2}lA)\mathbf{Z}^n \ ; \qquad n = 0,1,2,\ldots \ . \tag{8.33}$$

Since the eigenvalues λ_s $(s = 1,2,\ldots,N)$ of the matrix A, given by (8.12), are all negative, it is easy to show that all the eigenvalues of the amplification matrix $(I-\frac{1}{2}lA)^{-1}(I+\frac{1}{2}lA)$ are less than or equal to unity in modulus, so that the perturbation errors $\mathbf{Z}^n$ $(n = 0,1,\ldots)$ in (8.33) do not grow as $n \to \infty$.

It would seem therefore that the Crank–Nicholson method is unconditionally stable but, as can be seen from Figs. 8.1 and 8.2, some restriction must be placed on l if the method is to be used to solve problems with discontinuities between initial conditions and boundary conditions. This restriction is not made evident by the matrix method of analysis, nor by the Fourier method, and further analysis must be carried out.

The eigenvalues of the amplification matrix in (8.33) are given by $\mu_s = (1+\frac{1}{2}l\lambda_s)/(1-\frac{1}{2}l\lambda_s)$, where λ_s are the eigenvalues of the matrix A. It is then easy to see that μ_N is close to -1 for small h (recall: $h = (X_1 - X_0)/(N+1)$). Suppose, now, that the initial vector $\mathbf{g}$ can be written as a linear combination of the normalized eigenvectors $\mathbf{v}_s$ $(s = 1,2,\ldots,N)$ of the matrix A, that is

$$\mathbf{g} = \sum_{s=1}^{N} c_s\mathbf{v}_s \ ,$$

where the c_s $(s = 1,2,\ldots,N)$ are constants. Then from (8.13) and (8.14)

$$\mathbf{U}(l) = \exp(lA)\mathbf{g} \simeq \sum_{s=1}^{N} c_s(I-\tfrac{1}{2}lA)^{-1}(I+\tfrac{1}{2}lA)\mathbf{v}_s = \sum_{s=1}^{N} c_s\mu_s\mathbf{v}_s \ ,$$

$$\mathbf{U}(2l) = \exp(lA)\mathbf{U}(l) \simeq \sum_{s=1}^{N} c_s\mu_s(I-\tfrac{1}{2}lA)^{-1}(I+\tfrac{1}{2}lA)\mathbf{v}_s = \sum_{s=1}^{N} c_s\mu_s^2\mathbf{v}_s \ ,$$

and so on, so that, for $n = 0, 1, \ldots,$

$$\mathbf{U}^n = \sum_{s=1}^{N} c_s\mu_s^n\mathbf{v}_s \ . \tag{8.34}$$

It follows that the components $c_N\mathbf{v}_N, c_{N-1}\mathbf{v}_{N-1}, \ldots$ of the initial conditions are preserved at subsequent time steps. In view of μ_N being close to -1 for large N, these components will alternate in sign and, for problems with discontinuities between initial conditions and boundary conditions, the effect will be to propagate the discontinuities and distort $\mathbf{U}$, if too large a value of l is used, by producing oscillations.

Not all components are oscillatory; those corresponding to the eigenvalues of A for which the numerator of the amplification factor $(1+\frac{1}{2}l\lambda_s)/(1-\frac{1}{2}l\lambda_s)$ is non-negative, are non-oscillatory. Oscillatory components do not have a grave effect on the solution of the difference scheme provided they decay faster than the others.

Refering to (8.34) it is clear that the component $c_s\mathbf{v}_s(s = 1, 2, \ldots, N)$ is damped at each time step by the corresponding eigenvalue μ_s of the amplification matrix. Lawson and Morris (1978) noted that, to ensure that the higher frequency component $c_N\mathbf{v}_N$ is damped to zero faster than the lowest frequency component $c_1\mathbf{v}_1$, the inequality $|\mu_N| < |\mu_1|$ must be satisfied. This leads to

$$\frac{-1-\frac{1}{2}l\lambda_1}{1-\frac{1}{2}l\lambda_1} < \frac{1+\frac{1}{2}l\lambda_N}{1-\frac{1}{2}l\lambda_N} < \frac{1+\frac{1}{2}l\lambda_1}{1-\frac{1}{2}l\lambda_1} . \tag{8.35}$$

The right-hand side is trivially satisfied; the left-hand side gives $l^2\lambda_1\lambda_N < 4$.

It is seen from (8.12) that, for large values of N, $\lambda_1 \simeq -4h^{-2}[\pi^2/\{4(N+1)^2\}] = -\pi^2/(X_1-X_0)^2$ (since, for any small angle θ, $\sin\theta \simeq \theta$) and $\lambda_N \simeq -4/h^2$. Thus

$$r = \frac{l}{h} < \frac{X_1-X_0}{\pi} \text{ (approximately)} \tag{8.36}$$

which was clearly violated in the second numerical experiment of section 8.1 and explains the odd behaviour of U near the boundaries $x = 0$ and $x = 2$. The restriction (8.36) was not violated in the first experiment: this accounts for the smooth behaviour of U calculated using the Crank–Nicolson method over the whole of the range $0 \leqslant x \leqslant 2$ in Fig. 8.1.

8.2.3 Alternative definitions of stability

It has been remarked already that for a large number, N, of mesh points at each time level, the range of eigenvalues λ_s of the matrix A is large and, for large time steps, the magnitude of $l\lambda_s$ can also be large. The von Neumann necessary condition for stability, which requires

$$\left|\frac{1+(1-\theta)l\lambda_s}{1-\theta l\lambda_s}\right| \leqslant 1 , \tag{8.37}$$

is therefore a minimum requirement for stability (note that substituting for λ_s from (8.12) in (8.37) leads to (8.31)). Otherwise, some components of $\mathbf{U}$ might increase in magnitude while corresponding components of $\mathbf{u}$ decrease in magnitude. In the notation of Gourlay and Morris (1980)

$$R_\theta(l\lambda) = [1+(1-\theta)l\lambda]/(1-\theta l\lambda) \tag{8.38}$$

is called the *symbol* of the method (8.16).

The finite difference method arising from (8.16) is said to be A_0-*stable* if $|R_\theta(l\lambda)| \leqslant 1$ for all values of $p \geqslant 0$. It follows that A_0-stability guarantees unconditional stability in the conventional sense of perturbations in the initial data not being magnified for the same range of values of $p > 0$ as $t \to \infty$. It is left as an exercise to the reader to show that (8.16) is A_0-stable provided $\frac{1}{2} \leqslant \theta \leqslant 1$. In the terminology of Chapter 5 the family of methods arising from (8.16) to solve a system of first-order ordinary differential equations of the form (8.10) is *absolutely stable* for this same range of values of θ. This concept of absolute stability corresponds to the conventional concept of stability of linear finite difference methods for partial differential equations.

An A_0-stable method for which, additionally, $\lim_{l\lambda \to -\infty} R_\theta(l\lambda) = 0$ gives rise to a finite difference method which is L_0-*stable*. This property ensures that all components of the approximate solution $\mathbf{U}$ are damped as t increases. Rewriting $R_\theta(l\lambda)$ as

$$R_\theta(l\lambda) = [(1/l\lambda) + 1 - \theta]/[(1/l\lambda) - \theta]$$

it follows that (8.16) is L_0-stable provided $\theta = 1$. The only L_0-stable method of the family given by (8.16) is thus the fully implicit method (8.19) or (8.20). The Crank–Nicolson method (8.21) is not L_0-stable and the failure of this method to damp offending components of $\mathbf{U}$ is illustrated well by Fig. 8.2.

8.2.4 Convergence

A convergence analysis for parabolic equations is carried out in precisely the same way as for hyperbolic equations, by determining the restriction, if any, on the ratio p for which the local discretization error $z_m^n = u_m^n - U_m^n$, at the *fixed* mesh point (x_m, t_n), tends to zero as the mesh is refined by letting $h, l \to 0$ simultaneously. Following the arguments of Chapter 7 it may be shown, for instance, that the explicit method (8.18) is convergent provided $0 < p \leqslant \frac{1}{2}$; the fully implicit method (8.20) and the Crank–Nicolson method (8.22) require no restrictions to be placed on p for convergence.

8.3 LOCAL EXTRAPOLATIONS

The component of the principal part of the local truncation error due to the approximation of the space derivative in (8.6) by its central difference replacement (8.9) will not change. However, the component arising from the choice of θ in (8.16) can be improved by one or two powers of l by the process of *local extrapolation*.

The extrapolating procedure is carried out by, first of all, determining $\mathbf{U}^{(1)} = \mathbf{U}^{(1)}(t+2l)$ which is computed by applying (8.16) over two single time steps, then by determining $\mathbf{U}^{(2)} = \mathbf{U}^{(2)}(t+2l)$ which is computed by applying (8.16) over a double time step. Clearly

$$\begin{aligned}
\mathbf{U}^{(1)} &= (I-\theta lA)^{-1}[I+(1-\theta)lA]\mathbf{U}(t+l) \\
&= (I-\theta lA)^{-2}[I+(1-\theta)lA]^2\mathbf{U}(t) \\
&= [I+2lA+(1+2\theta)l^2A^2+(2\theta+2\theta^2)l^3A^3+(3\theta^2+2\theta^3)l^4A^4 \\
&\qquad +(4\theta^3+2\theta^4)l^5A^5+\ldots]\mathbf{U}(t) \qquad (8.39)
\end{aligned}$$

and

$$\begin{aligned}\mathbf{U}^{(2)} &= (I-2\theta lA)^{-1}[I+2(1-\theta)lA]\mathbf{U}(t)\\ &= [I+2lA+4\theta l^2A^2+8\theta^2l^3A^3+16\theta^3l^4A^4+32\theta^4l^5A^5+\ldots]\mathbf{U}(t)\ ,\end{aligned} \quad (8.40)$$

while the Maclaurin expansion of exp $(2lA)$ in (8.14) written over a time step $2l$ gives

$$\mathbf{U}^{(M)}(t+2l) = [I+2lA+2l^2A^2+\tfrac{4}{3}l^3A^3+\tfrac{2}{3}l^4A^4+\tfrac{4}{15}l^5A^5+\ldots]\mathbf{U}(t)\ . \quad (8.41)$$

Comparison of (8.39) and (8.40) with (8.41) verifies that $\mathbf{U}^{(1)}$ and $\mathbf{U}^{(2)}$ are each first-order accurate when $\theta \neq \frac{1}{2}$. The locally extrapolated value $\mathbf{U}^{(E)} = \mathbf{U}^{(E)}(t+2l)$ is determined form the second-order formula

$$\mathbf{U}^{(E)} = 2\mathbf{U}^{(1)} - \mathbf{U}^{(2)} + O(l^3) \quad (8.42)$$

when $\theta \neq \frac{1}{2}$ and from the fourth-order formula

$$\mathbf{U}^{(E)} = \tfrac{4}{3}\mathbf{U}^{(1)} - \tfrac{1}{3}\mathbf{U}^{(2)} + O(l^5) \quad (8.43)$$

when $\theta = \frac{1}{2}$ (note that the local extrapolation of the Crank–Nicolson method has produced two extra orders of accuracy).

The local extrapolation equations (8.42) and (8.43) may be written in the general form

$$\mathbf{U}^{(E)} = \left[\frac{2^q}{2^q-1}[R_\theta(lA)]^2 - \frac{1}{2^q-1}[R_\theta(2lA)]\right]\mathbf{U}(t)\ , \quad (8.44)$$

where q is the order of the method ($q = 1$ *for* $\theta \neq \frac{1}{2}$ and $q = 2$ for $\theta = \frac{1}{2}$), from which it follows that a perturbation of the initial conditions satisfies

$$\mathbf{Z}^{n+2} = \left[\frac{2^q}{2^q-1}[R_\theta(lA)]^2 - \frac{1}{2^q-1}[R_\theta(2lA)]\right]\mathbf{Z}^n\ . \quad (8.45)$$

The condition to be satisfied for A_0-stability is $|S_\theta(l\lambda_s)| \leqslant 1$, where $\lambda_s < 0$ is any eigenvalue of A and

$$S_\theta(l\lambda_s) = \frac{2^q}{2^q-1}[R_\theta(l\lambda_s)]^2 - \frac{1}{2^q-1}[R_\theta(2l\lambda_s)] \quad (8.46)$$

is the symbol of the locally extrapolated form of (8.16). For L_0-stability the additional condition is that $\lim_{l\lambda_s \to \infty} S_\theta(l\lambda_s) = 0$.

Writing $z = -l\lambda$, z is positive since all eigenvalues of A are negative. Then

$$R_\theta(-z) = [1-(1-\theta)z]/(1+\theta z) \tag{8.47}$$

and

$$S_\theta(-z) = \frac{2^q}{2^q-1}[R_\theta(-z)]^2 - \frac{1}{2^q-1}[R_\theta(-2z)] \; , \tag{8.48}$$

and (8.47), (8.48) may be used to analyse the locally extrapolated forms of the methods (8.17), (8.19), (8.21).

It is easy to see first of all that

$$S_0(-z) = 1-2z+2z^2 \tag{8.49}$$

and, taking $z = -l\lambda_s = 4p\sin^2[s\pi/\{2(N+1)\}]$ for any $s = 1,2,\ldots,N$, that $|S_0(-z)| \leqslant 1$ provided $0 < p \leqslant \frac{1}{4}$. Clearly, the local extrapolation of the explicit method (8.18) has not improved that method's stability properties.

The symbol for the extrapolated form of the fully implicit method (8.19) is given by

$$S_1(-z) = \frac{1+2z-z^2}{1+4z+5z^2+2z^3} \tag{8.50}$$

from which it may be verified that the locally extrapolated form of this method is L_0-stable. The curves for $R_1(-z)$ and $S_1(-z)$ are shown in Fig. 8.3.

Finally, the symbol for the extrapolated form of the Crank–Nicolson method (8.21) is given by

$$S_{\frac{1}{2}}(-z) = (3-\tfrac{9}{4}z^2+\tfrac{5}{4}z^3)/(3+6z+\tfrac{15}{4}z^2+\tfrac{3}{4}z^3) \tag{8.51}$$

from which it may be shown that $|S_{\frac{1}{2}}(-z)| \leqslant 1$ only for $0 \leqslant z \leqslant 6+4\sqrt{3}$. Then, taking $z = -l\lambda_s = 4p\sin^2[s\pi/\{2(N+1)\}]$ for any $s = 1,2,\ldots,N$ it may be shown that the locally extrapolated form of the Crank–Nicolson scheme is stable only for $0 < p \leqslant 3.23205$ (approximately). However, this range of values of p may lead to a value of l and subsequently a value of r which exceeds that permitted by (8.36) and the reader should be cautious about extrapolating the Crank–Nicolson method. The graphs of $R_{\frac{1}{2}}(-z)$ and $S_{\frac{1}{2}}(-z)$ are shown in Fig. 8.4.

Summarizing the preceding analyses, it appears that, of the family of methods (8.16), the locally extrapolated form of the fully implicit method is the most

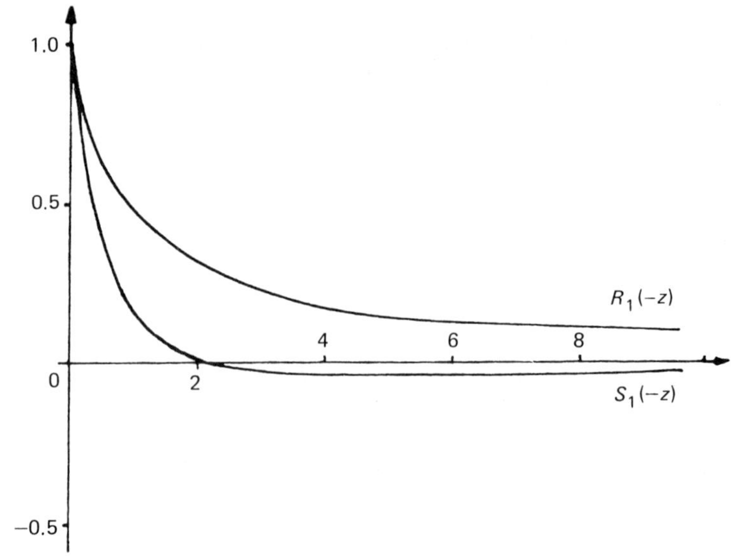

Fig. 8.3 — Amplification symbols $R_1(-z)$ and $S_1(-z)$.

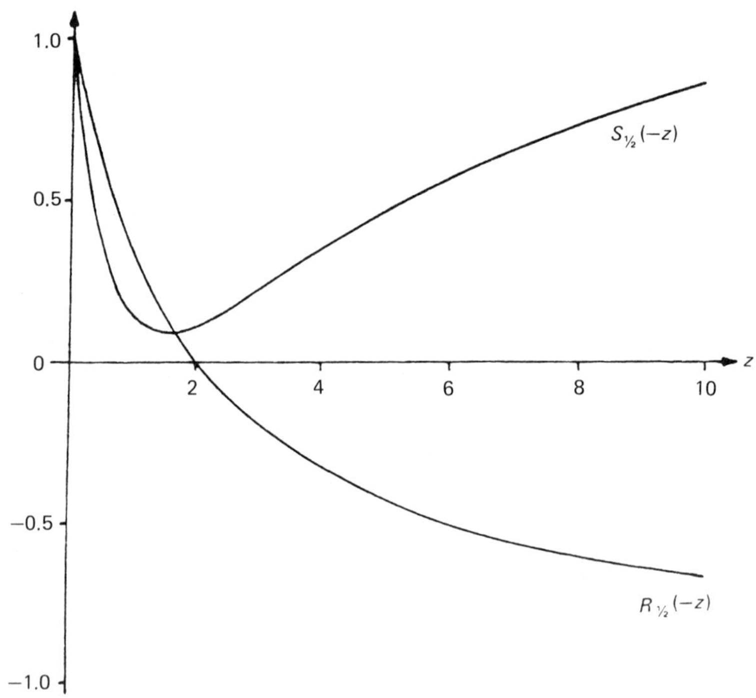

Fig. 8.4 — Amplification symbols $R_{\frac{1}{2}}(-z)$ and $S_{\frac{1}{2}}(-z)$.

attractive to the user. Generally speaking, a method which is not L_0-stable should not be extrapolated locally. Gourlay and Morris (1980) gave three- and four-stage local extrapolation procedures for the fully implicit method (8.19).

8.4 GLOBAL EXTRAPOLATIONS

Global extrapolation was dealt with in some detail in Chapter 5. It differs from local extrapolation in that the latter integrates from time t to, say, time $t+2l$ in two different ways; the two approximations to $\mathbf{u}(t+2l)$ are then combined linearly to give a better approximation $\mathbf{U}^{(E)}(t+2l)$ to $\mathbf{u}(t+2l)$. The procedure is repeated to integrate from $\mathbf{U}^{(E)}(t+2l)$ to determine $\mathbf{U}^{(E)}(t+4l)$, and so on until $\mathbf{U}^{(E)}(t+2kl)$ is found at some fixed time $T = t+2kl$ (k an integer).

Global extrapolation, on the other hand, integrates $2k$ times from time t using the time step l; starting again at time t, either $4k$ integrations on a finer grid with time step $\frac{1}{2}l$ or k integrations on a coarser grid with time step $2l$, are then carried out. A linear combination of the two sets of computed values at the times $t+jl$ ($j = 1,2,\ldots,2k$) should then give a more accurate approximation to $\mathbf{u}(t+jl)$. Global extrapolation formulas on two- and three-grids were given in Chapter 5.

8.5 PERCUTANEOUS DRUG ABSORPTION

8.5.1 Concentration in the skin in the transient stage

Postulating the skin to be a single barrier of thickness L with diffusion coefficient κ, the concentration $c(X, T)$ of a drug, applied as an ointment at the skin surface ($X = 0$) at time $T = 0$, at a point X in the skin ($0 < X < L$) at time $T > 0$, is given by *Fick's Second Law of Diffusion*, namely

$$\frac{\partial c}{\partial T} = \kappa \frac{\partial^2 c}{\partial X^2}\ ; \qquad 0 < X < L\ , \qquad T > 0\ . \tag{8.52}$$

Defining the dimensionless variables

$$t = \kappa T/L^2 \qquad \text{and} \qquad x = X/L \tag{8.53}$$

it is easy to show that the parabolic equation takes the form

$$\frac{\partial c}{\partial t} = \frac{\partial^2 c}{\partial x^2}\ ; \qquad 0 < x < 1\ , \qquad t > 0\ . \tag{8.54}$$

This modified mathematical model considers the skin to have uniform unit thickness and to have unit diffusion coefficient.

Assuming that the concentration of the uppermost epidermis ($x = 0$) is maintained at unity and that consequently the concentration at the skin–capillary boundary ($x = 1$) is r, the boundary conditions for the model are given by

$$c(0,t) = 1 \; ; \qquad t > 0 \tag{8.55}$$

$$\frac{\partial c}{\partial x}(1,t) + \omega c(1,t) = 0 \; ; \qquad t > 0 \tag{8.56}$$

where $\omega = (1-r)/r$; see Kubota and Ishizaki (1986a, 1986b). Initially, before application of the ointment, the concentration in the skin is zero and so the initial condition for the model is

$$c(x,0) = 0 \; ; \qquad 0 < x < 1 \; . \tag{8.57}$$

The initial-boundary value problem to be solved consists of the equations (8.54), (8.55), (8.56) and (8.57) and the method of solution follows closely that of section 8.2.

Subdividing the space interval into M parts each of width h so that $Mh = 1$, it is evident that the solution is sought at the mesh points $x_m = mh \, (m = 1,2,\ldots,M)$, $x_M = 1$ being the skin–capillary boundary. (The boundary condition at $x = 1$ is given in mixed function and derivative form by (8.56), in contrast to function form in section 8.2.)

Approximating the space derivative in (8.56) by its second-order central difference replacement gives $(C^n_{M+1} - C^n_{M-1})/(2h) + \omega C^n_M + O(h^2) = 0$ at time $t = nl$, so that

$$C^n_{M+1} = C^n_{M-1} - 2\omega h C^n_M + O(h^3) \; . \tag{8.58}$$

Note that, as in earlier sections, an upper case letter is used for the solution of an approximating difference scheme.

Applying (8.54) with (8.9), in which u is replaced by c, to all M mesh points x_m $(m = 1,2,\ldots,M)$ at time $t = nl \, (n = 0,1,2,\ldots)$, and using (8.58) when $\{(8.54), (8.9)\}$ are applied to (x_M, nl), leads to a linear system of first-order ordinary differential equations which has the form

$$\frac{\mathrm{d}\mathbf{C}(t)}{\mathrm{d}t} = A\mathbf{C}(t) + \mathbf{q} \; ; \qquad \mathbf{C}(0) = \mathbf{0} \; , \tag{8.59}$$

where A is now a matrix of order M given by

$$A = h^{-2} \begin{bmatrix} -2 & 1 & & & 0 \\ 1 & -2 & 1 & & \\ & \cdots & \cdots & \cdots & \\ & & 1 & -2 & 1 \\ 0 & & & 2 & -2(1+\omega h) \end{bmatrix} \tag{8.60}$$

and $\mathbf{q}$ is a vector of order M given by

$$\mathbf{q} = h^{-2}[1,0,\ldots,0]^T \tag{8.61}$$

(the first element in $\mathbf{q}$ arises from (8.55)).

It is left as an exercise to the reader to show that the solution of (8.59) is

$$\mathbf{C}(t) = -A^{-1}\mathbf{q} + A^{-1}\exp(tA)\mathbf{q} \tag{8.62}$$

and that $\mathbf{C}$ satisfies the recurrence relation

$$\mathbf{C}(t+l) = -A^{-1}\mathbf{q} + \exp(lA)\mathbf{C}(t) + A^{-1}\exp(lA)\mathbf{q}\ . \tag{8.63}$$

Introducing, next, the replacement (8.15) for $\exp(lA)$ leads to

$$(I - \theta lA)\mathbf{C}(t+l) = [I + (1-\theta)lA]\mathbf{C}(t) + l\mathbf{q} \tag{8.64}$$

which has local truncation error $O(l+h)$ if $\theta \neq \frac{1}{2}$ and error $O(l^2+h)$ if $\theta = \frac{1}{2}$. The loss of accuracy from $O(h^2)$ *to* $O(h)$ is due to the replacement of C^n_{M+1} by (8.58).

It is easy to show using Brauer's Theorem (Chapter 1) that all the eigenvalues λ_i ($i = 1,2,\ldots,M$) of A given in (8.60) are nonpositive and, following the arguments of subsection 8.2.3, it then follows that (8.64) is A_0-stable provided $\frac{1}{2} \leqslant \theta \leqslant 1$, and, furthermore, that the fully implicit method ($\theta = 1$) is L_0-stable.

Using $h = 0.1$ and $l = 0.0001$ the concentration distributions for $r = 0.001$ at times $t = 0.001, 0.01, 0.05, 0.1, 0.2, 0.4$, computed using the fully implicit L_0-stable method ($\theta = 1$), are depicted in Fig. 8.5. The analogous distributions for $r = 0.999$ at the same times and at the additional time $t = 1.0$ are depicted in Fig. 8.6. Comparisons with the theoretical findings of Kubota and Ishizaki (1986b) verify that the choice of an L_0-stable method was the correct one. This is in keeping with the related experiment reported in Example 5.18 of Chapter 5 where it was revealed that A_0-stable methods may induce oscillations in the solution whereas L_0-stable methods model decaying solutions more closely (see also Fig. 8.2 and the analysis of the A_0-stable Crank–Nicholson method ($\theta = \frac{1}{2}$) in subsection 8.2.2).

8.5.2 Concentration in the skin in the steady-state

Denoting by t^* the time at which a steady-state is reached, the concentration is given by

$$c(x,t^*) \equiv \phi(x) = 1-(1-r)x\ ; \tag{8.65}$$

see Kubota and Ishizaki (1985). Suppose that the ointment is removed from the skin surface when the steady-state has been reached: suppose further that this removal

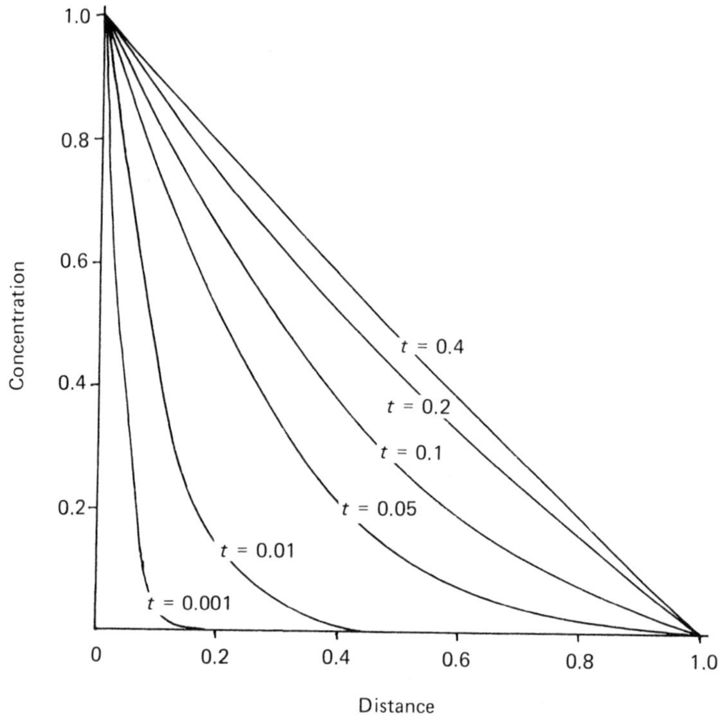

Fig. 8.5 — Concentration distributions for $r = 0.001$.

takes place at time $t = 0$. The concentration in the skin $c(x,t)$ at time $t > 0$ is modelled by the initial-boundary value problem

$$\frac{\partial c}{\partial t} = \frac{\partial^2 c}{\partial x^2} \; ; \qquad 0 < x < 1 \; , \qquad t > 0 \tag{8.66}$$

$$\frac{\partial c}{\partial x}(0,t) = 0 \; ; \qquad t > 0 \tag{8.67}$$

$$\frac{\partial c}{\partial x}(1,t) + \omega c(1,t) = 0 \; ; \qquad t > 0 \tag{8.68}$$

$$c(x,0) = \phi(x) \; ; \qquad 0 < x < 1 \; . \tag{8.69}$$

Proceeding as in subsection 8.5.1 it is clear that the solution is sought at the $M+1$ mesh points (mh, nl) $(m = 0, 1, \ldots, M; n = 1, 2, \ldots)$ at time $t = nl$. The derivatives in the boundary conditions (8.67) and (8.68) are replaced by their second-order

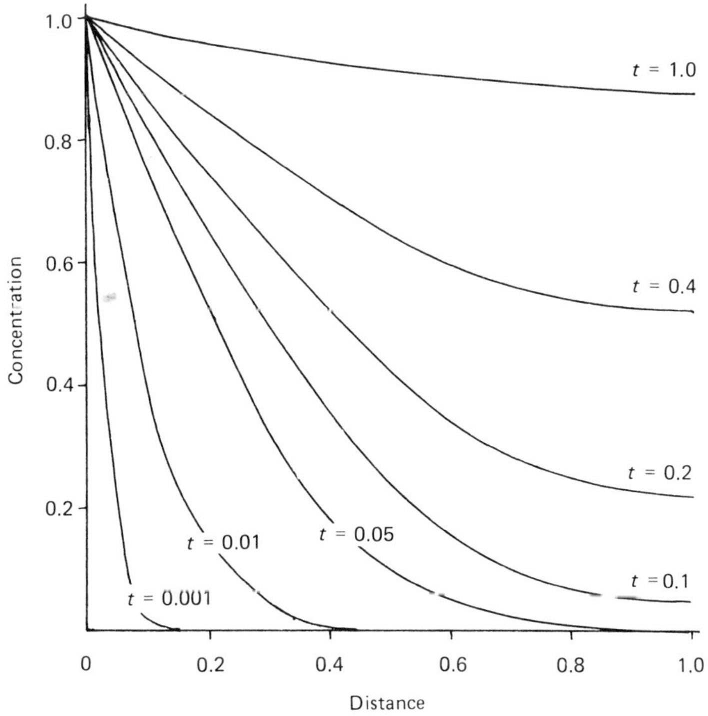

Fig. 8.6 — Concentration distributions for $r = 0.999$.

central difference approximants. This gives $C_{-1}^n = C_1^n + O(h^3)$ and $C_{M+1}^n = C_{M-1}^n - 2\omega h C_M^n + O(h^3)$; C_{-1}^n arises when {(8.66), (8.9)} are applied at point $(0, nh)$ and C_{M+1}^n arises at mesh point $(1, nh)$.

Making these applications leads in the, by now, familiar way to the system

$$\frac{d\mathbf{C}(t)}{dt} = A\mathbf{C}(t) \; ; \qquad \mathbf{C}(0) = \boldsymbol{\phi} \; , \tag{8.70}$$

the solution of which is

$$\mathbf{C}(t) = \exp(tA)\boldsymbol{\phi} \tag{8.71}$$

which satisfies

$$\mathbf{C}(t+l) = \exp(lA)\mathbf{C}(t) \; . \tag{8.72}$$

In (8.70), the matrix A is given by

$$A = h^{-2}\begin{bmatrix} -2 & 2 & & & 0 \\ 1 & -2 & 1 & & \\ & \cdots & \cdots & \cdots & \cdots \\ & & 1 & -2 & 1 \\ 0 & & & 2 & -2(1+\omega h) \end{bmatrix}; \qquad (8.73)$$

the eigenvalues of A are easily shown to be nonpositive by Brauer's Theorem. Replacing $\exp(lA)$ in (8.72) by (8.15) gives

$$(I-\theta lA)\mathbf{C}(t+l) = [I+(1-\theta)lA]\mathbf{C}(t) , \qquad (8.74)$$

which has $O(l+h)$ local truncation error if $\theta \neq \frac{1}{2}$ and $O(l^2+h)$ error if $\theta = \frac{1}{2}$. As before, (8.74) is A_0-stable if $\frac{1}{2} \leqslant \theta \leqslant 1$ with the case $\theta = 1$ being L_0-stable.

8.5.3 Alternative treatment of the steady-state case

Define a new function $g(x,t)$ by

$$g(x,t)+\phi(x) = c(x,t) \qquad (8.75)$$

with $\phi(x)$ given by (8.65). The equations of the mathematical model become

$$\frac{\partial g}{\partial t} = \frac{\partial^2 g}{\partial x^2} ; \qquad 0 < x < 1 , \qquad t > 0 \qquad (8.76)$$

$$\frac{\partial g}{\partial x}(0,t) = \frac{\omega}{\omega+1} ; \qquad t > 0 \qquad (8.77)$$

$$\frac{\partial g}{\partial x}(1,t)+\omega g(1,t) = 0 ; \qquad t > 0 \qquad (8.78)$$

$$g(x,0) = 0 ; \qquad 0 < x < 1 , \qquad (8.79)$$

and the concentration is sought at the same $M+1$ points as in the previous subsection. Replacing the derivatives in (8.77) and (8.78) by their second-order central difference approximants leads to $G^n_{-1} = G^n_1 - 2\omega h/(\omega+1) + O(h^3)$ and $G^n_{M+1} = G^n_{M-1} - 2\omega h G^n_M + O(h^3)$. The system obtained by applying (8.76) with (8.9) to all $M+1$ mesh points at a given time level is given by

$$\frac{\mathrm{d}\mathbf{G}(t)}{\mathrm{d}t} = A\mathbf{G}(t)+\mathbf{s} ; \qquad \mathbf{G}(0) = \mathbf{0} , \qquad (8.80)$$

where A is given by (8.73) and

$$\mathbf{s} = h^{-1}[-2\omega/(\omega+1), 0, \ldots, 0]^{\mathrm{T}} . \tag{8.81}$$

It follows from (8.62), (8.63) and (8.64) that

$$(I - \theta lA)\mathbf{G}(t+l) = [I + (1-\theta)lA]\mathbf{G}(t) + l\mathbf{s} \tag{8.82}$$

which is $O(l+h)$ if $\theta \neq \frac{1}{2}$ and $O(l^2+h)$ if $\theta = \frac{1}{2}$, and which is A_0-stable for $\frac{1}{2} \leqslant \theta \leqslant 1$ and L_0-stable when $\theta = 1$.

8.6 NEUROPHYSIOLOGICAL APPLICATIONS

8.6.1 Introduction

The function of the nervous system is to transmit information, and for this purpose it uses electrical impulses initiated and transmitted by individual nerve cells or neurons. One of the few mathematical models to give qualitatively accurate predictions of the conduction of electrical impulses in a nerve axon is in the pioneering paper of Hodgkin and Huxley (1952). In this particular paper the authors developed a system of nonlinear parabolic equations (now known as the HH equations) which describe the ionic and electrical events occurring during the transmission of an impulse along an axon, the filament carrying signals from the nerve cell body to other parts of the organism. These equations are based on the assumption that an axon behaves like a cylindrical electrical cable, with a conducting core and partially insulating sheath or membrane. The core and the surrounding medium are fluids which contain significant concentrations of charged metallic ions, mostly sodium and potassium. The currents which flow are due either to the capacitance of the membrane, or to the movement of ions (either through the membrane or longitudinally along the axon). A very important feature of the model is its ability to deal with changes in the permeability of the membrane to certain of these ions during an impulse.

The HH equations are given in full in the 1952 paper. A scale change allows them to take the form

$$\frac{\partial^2 u}{\partial x^2} = \frac{\partial u}{\partial t} + I(u, \mathbf{w}) \tag{8.83}$$

$$\frac{\partial \mathbf{w}}{\partial t} = P(u)\mathbf{w} + \mathbf{q}(u) . \tag{8.84}$$

In this nonlinear parabolic system the scalar $u = u(x,t)$ represents the electrical potential across the axon membrane (x is the distance along the axon, t is time, as usual), $\mathbf{w} = \mathbf{w}(x,t)$ is a three-dimensional vector which represents the permeability of the membrane to sodium and potassium (the principal ionic components of the membrane current), the matrix P and the vector $\mathbf{q}$ are each of order three and depend nonlinearly on u, and $I(u, \mathbf{w})$ is a scalar which depends linearly on u and nonlinearly on $\mathbf{w}$.

The left-hand side of (8.83) gives the net current across the membrane; this is expressed on the right-hand side as the sum of the capacitative term $\partial u/\partial t$ and the ionic term $I(u, \mathbf{w})$. Assuming the axon to be of infinite length, the system (8.83), (8.84) is solved on the half plane $-\infty < x < \infty$, $t \geqslant 0$ giving an initial value problem, or in the first quadrant $x \geqslant 0$, $t \geqslant 0$ giving an initial-boundary value problem.

Hastings (1975) notes that laboratory solutions of (8.83), (8.84) are primarily of two types, those which are independent of x and those which depend on the single quantity $x+ct$. The latter are *travelling waves* and c is called the *wave velocity*. It should be noted that travelling waves are not present in the solution of linear parabolic equations such as the simple heat equation (8.6).

8.6.2 A simpler model

A solution of the HH equations is difficult to obtain, either analytically or numerically, because of the nonlinear functions I, P and $\mathbf{q}$. A simpler formulation of the model, which retains most of the qualitative features of the work of Hodgkin and Huxley and is easier to solve numerically, is given by the FitzHugh–Nagumo equations

$$\frac{\partial u}{\partial t} = \frac{\partial_2 u}{\partial x_2} + u(1-u)(u-a) - w \ , \tag{8.85}$$

$$\frac{\partial w}{\partial t} = bu - cw \tag{8.86}$$

in which $0 < a < 1$ and $b > 0, c > 0$ are positive constants. The variable u continues to represent the potential across the membrane while the scalar w represents the permeability of the membrane to the single ionic component of the membrane current. The parameter b in (8.86) is small because w represents a pair of variables in the HH equations which change slowly relative to u. The reader is referred to the papers by FitzHugh (1961) and Nagumo *et al.* (1962) for the derivations of these equations.

Suppose that the solution is sought in some region $0 < x \leqslant X$, $t \geqslant 0$ of the (x, t)-plane and that the boundary and initial conditions are given by

$$U(0,t) = P(t) \ , \qquad w(0,t) = b\int_0^t P(s)\mathrm{e}^{-cs}\mathrm{d}s = Q(t) \ ; \qquad t \geqslant 0 \tag{8.87}$$

and

$$u(x,0) = w(x,0) = 0 \; ; \qquad x \geqslant 0 \; . \tag{8.88}$$

In (8.87), $P(t)$ represents the stimulus at $x = 0$ (and must not be confused with the function in (8.84)).

Discretizing $0 \leqslant x \leqslant X$ into M subintervals each of width h (so that $Mh = X$) and writing $t = t_n = nl$, the solution is sought at the mesh points (x_m, t_n) with $x_m = mh$ $(m = 1, 2, \ldots, M)$ and $n = 1, 2, \ldots$. In view of the necessity to find u and w at $x = X$, the central difference replacement (8.9) may not be applied at that point, though it may be applied at all other mesh points at a given time level. Instead, the *first*-order backward difference replacement

$$\frac{\partial^2 u(x,t)}{\partial t^2} = \frac{u(x,t) - 2u(x-h,t) + u(x-2h,t)}{h^2} + O(h) \tag{8.89}$$

may be used.

Applying (8.85) with (8.9), and then (8.86) to every mesh point with $x = x_m$ $(m = 1, 2, \ldots, M)$ at time level $t = nl$ $(n = 0, 1, 2, \ldots)$, leads to a nonlinear system of the form

$$\frac{d\mathbf{V}(t)}{dt} \equiv D\mathbf{V}(t) = A\mathbf{V}(t) + \boldsymbol{\phi}(\mathbf{V}) + \boldsymbol{\psi}(t) \; ; \qquad \mathbf{V}(0) = \mathbf{0} \tag{8.90}$$

where $\mathbf{V} = [\mathbf{U}^{\mathrm{T}}, \mathbf{W}^{\mathrm{T}}]^{\mathrm{T}}$ is the vector (of order $2M$) of approximate values of u and w, A is the $2M$-square matrix given by

$$A = \begin{bmatrix} A_1 & -I \\ \mathrm{bI} & -cI \end{bmatrix} \tag{8.91}$$

in which I is the identity matrix of order M and

$$A_1 = h^{-2} \begin{bmatrix} -2 & 1 & & & 0 \\ 1 & -2 & 1 & & \\ & \cdots & \cdots & \cdots & \\ & & 1 & -2 & 1 \\ 0 & & 1 & -2 & 1 \end{bmatrix} \tag{8.92}$$

is of order M,

$$\boldsymbol{\phi}(\mathbf{V}) = [\phi_1, \phi_2, \ldots, \phi_M, 0, 0, \ldots, 0]^{\mathrm{T}} \tag{8.93}$$

is of order $2M$ with

$$\phi_m = V_m(1-V_m)(V_m-a) \ , \tag{8.94}$$

$\psi(t)$ is of order $2M$ and is given by

$$\psi(t) = h^{-2}[P(t),0,0,\ldots,0]^{\mathrm{T}} \tag{8.95}$$

and $D = \operatorname{diag}\{\mathrm{d}/\mathrm{d}t\}$ is a diagonal differential operator, here of order $2M$.

Noting from Taylor's expansion that

$$\mathbf{V}(t+l) = \exp(lD)\mathbf{V}(t) \tag{8.96}$$

it follows from (8.15) that

$$(I-\theta lD)\mathbf{V}(t+l) = [I+(1-\theta)lD]\mathbf{V}(t) \ , \tag{8.97}$$

so that

$$\begin{aligned}(I-\theta lA)\mathbf{V}^{n+1} &- \theta l\boldsymbol{\phi}(\mathbf{V}^{n+1})-\theta l\boldsymbol{\psi}^{n+1}\\ &= [\mathrm{I}+(1-\theta)lA]\mathbf{V}^n+(1-\theta)l\boldsymbol{\phi}(\mathbf{V}^n)+(1-\theta)l\boldsymbol{\psi}^n \ ,\end{aligned} \tag{8.98}$$

in which I is now of order $2M$, gives a family of methods for solving (8.85), (8.86) with (8.87) and (8.88) which is $O(h+l)$ when $\theta \neq \frac{1}{2}$ and $O(h+l^2)$ when $\theta = \frac{1}{2}$ (note the lower accuracy in h because of (8.99)).

The solution vector $\mathbf{V}^{n+1} = \mathbf{V}((n+1)l)$ is obtained from (8.98) by solving a nonlinear algebraic system of the form

$$\mathbf{F}(\mathbf{V}^{n+1}) = \mathbf{0} \tag{8.99}$$

in which $\mathbf{F} = [f_1,f_2,\ldots,f_{2M}]^{\mathrm{T}}$ with

$$\begin{aligned}f_1 = {}&(1+2\theta r)U_1^{n+1}-\theta rU_2^{n+1}-\theta lU_1^{n+1}(1-U_1^{n+1})(U_1^{n+1}-a)+\theta lW_1^{n+1}-\theta rP(t+l)\\ &-[1+2(1-\theta)r]U_1^n-(1-\theta)rU_2^n-(1-\theta)lU_1^n(1-U_1^n)(U_1^n-a)+(1-\theta)lW_1^n-(1-\theta)rP(t) \ ,\end{aligned}$$

$$\begin{aligned}f_m = {}&-\theta rU_{m-1}^{n+1}+(1+2\theta r)U_m^{n+1}-\theta rU_{m+1}^{n+1}-\theta lU_m^{n+1}(1-U_m^{n+1})(U_m^{n+1}-a)+\theta rW_m^{n+1}\\ &-(1-\theta)rU_{m-1}^n-[1-2(1-\theta)r]U_m^n-(1-\theta)rU_{m+1}^n\\ &-(1-\theta)lU_m^n(1-U_m^n)(U_m^n-\mathrm{a})+(1-\theta)lW_m^n \ ; \qquad m = 2,3,\ldots,M-1 \ ,\end{aligned}$$

$$f_M = -\theta rU_{M-2}^{n+1}+2\theta rU_{M-1}^{n+1}+(1-\theta r)U_M^{n+1}-\theta lU_M^{n+1}(1-U_M^{n+1})(U_M^{n+1}-a)+\theta lW_M^{n+1}$$

$$-(1-\theta)rU_{M-2}^n+2(1-\theta)rU_{M-1}^n-[1+(1-\theta)r]U_M^n+(1-\theta)lW_M^n$$
$$-(1-\theta)lU_M^n(1-U_M^n)(U_M^n-a)\ .$$
$$f_j=-\theta lbU_j^{n+1}+(1+\theta lc)W_j^{n+1}-(1-\theta)lbU_j^n-[1-(1-\theta)lc]W_j^n\ ; \qquad j = M+1,\ldots,2M\ .$$

An iterative method, such as the Newton–Raphson method for a nonlinear algebraic system which was described in Chapter 3, should be used to obtain the solution.

8.7 A REACTION–DIFFUSION EQUATION IN GENETICS

The general one-dimensional reaction–diffusion equation is given by

$$\frac{\partial u}{\partial t} = \kappa\frac{\partial^2 u}{\partial x^2}+\phi(u)\ ; \qquad X_0 < x < X_1\ , \qquad t > 0,\ \kappa > 0 \tag{8.100}$$

in which $u = u(x,t)$. Boundary conditions associated with (8.100) have the form

$$u(X_0,t) = w_0\ , \qquad u(X_1,t) = w_1\ , \qquad t > 0 \tag{8.101}$$

and initial conditions are given by

$$u(x,0) = g(x)\ , \qquad X_0 \leqslant x \leqslant X_1\ . \tag{8.102}$$

It is assumed that the conductivity κ is constant, that $\phi(u)$ is sufficiently smooth for a unique solution to exist, and that $g(x)$ is continuous for $X_0 \leqslant x \leqslant X_1$; it is not assumed that $w_0 = g(X_0)$ nor that $w_1 = g(X_1)$.

The reaction–diffusion problem modelled by equations (8.100), (8.101) and (8.102) has applications in genetics and also in flame propagation, chemical reactions, the spread of infections and epidemics, and, as seen in section 8.6, in neurophysiology. A two- or three-dimensional form of (8.100) arises in nonlinear mathematical models of heat flow in the human body.

The problem is solved first of all by discretizing the interval $X_0 \leqslant x \leqslant X_1$ into $N+1$ subintervals each of width h so that $(N+1)h = X_1 - X_0$; the time increment is l as in previous sections. Replacing the space derivative by the central difference approximant (8.9) and applying (8.100) with (8.9) to all N internal mesh points at time level nl transforms the partial differential equation into the system of N nonlinear ordinary differential equations

$$\frac{\mathrm{d}\mathbf{U}(t)}{\mathrm{d}t} = D\mathbf{U}(t) - A\mathbf{U}(t)+\boldsymbol{\phi}(\mathbf{U})+\mathbf{b}\ ; \qquad \mathbf{U}(0) = \mathbf{g} \tag{8.103}$$

in which all vectors, the matrix A and the differential matrix $D = \operatorname{diag}\{\mathrm{d}/\mathrm{d}t\}$ are of order N with

$$A = \kappa h^{-2}\begin{bmatrix} -2 & 1 & & & 0 \\ 1 & -2 & 1 & & \\ & \cdots & \cdots & \cdots & \\ & & 1 & -2 & 1 \\ 0 & & & 1 & -2 \end{bmatrix} \tag{8.104}$$

$$\boldsymbol{\phi}(\mathbf{U}) = [\phi(U_1),\phi(U_2),\ldots,\phi(U_N)]^{\mathrm{T}} \tag{8.105}$$

and

$$\mathbf{b} = \kappa h^{-2}[w_0, 0, \ldots, 0, w_1]^{\mathrm{T}} \; . \tag{8.106}$$

The eigenvalues of A are given by

$$\lambda_s = -4\kappa h^{-2}[\sin^2 s\pi/\{2(N+1)\}] \; ; \qquad s = 1,2,\ldots,N \tag{8.107}$$

so that each $\lambda_s < 0$ since $\kappa > 0$.

Noting once more that $\mathbf{U}(t+l) = \exp(lD)\mathbf{U}(t)$ and using (8.15), it is easily seen that

$$(I - \theta l D)\mathbf{U}(t+l) = [I + (1-\theta)lD]\mathbf{U}(t) \tag{8.108}$$

from which if follows, using (8.103), that

$$(I - \theta l A)\mathbf{U}^{n+1} - \theta l \boldsymbol{\phi}^{n+1} = [I + (1-\theta)lA]\mathbf{U}^n + (1-\theta)l\boldsymbol{\phi}^n + \mathbf{b} \tag{8.109}$$

in which I is of order N and superscripts have been added in the usual way. Equation (8.109) describes a family of methods with $0 \leqslant \theta \leqslant 1$ which is $O(h^2 + l)$ accurate locally whenever $\theta \neq \frac{1}{2}$ and $O(h^2 + l^2)$ when $\theta = \frac{1}{2}$.

Manoranjan *et al.* (1984) solved this reaction–diffusion problem using finite element methods. These authors took $\kappa = 1$, $\phi(u) = u(1-u)(u-0.4)$, $X_1 = -X_0 = 2.48$, the boundary conditions to be

$$U(-2.48, t) = u(2.48, t) = 0.5 \; ; \qquad t > 0$$

and the initial conditions to be

$$u(x, 0) = x^2/94.4064 + 0.25 \; ; \qquad -2.48 \leqslant x \leqslant 2.48.$$

It was also shown in the paper that the solution converges to the steady-state unique function

$$u(x) = 1.2/[0.4 \cosh\{(0.4)^{\frac{1}{2}}x\} + 1.4] \; ; \qquad -2.48 \leqslant x \leqslant 2.48.$$

Taking $h = 0.0992$, which corresponds to $N = 49$ and 50 subintervals, and $l = 0.1$, the L_∞ errors using the second-order method with $\theta =$ and the finite element method of Manoranjan *et al.* (1984) are given in Table 8.1. The theoretical steady-

Table 8.1

Method	L_∞ error
$\theta = \frac{1}{2}$	0.164E–3
Manoranjan	0.497E–3

state solution at the point $x = 0$ where the maximum modulus error occurs is $u(0, \infty) = \frac{2}{3}$. Computational aspects involved in determining the solution were given in Chapter 3.

8.8 DIFFUSION–CONVECTION PROBLEMS

8.8.1 The linear case

One-dimensional diffusion–convection equations have important applications in convective heat transfer (in the arms and legs, for instance, where flowing blood makes a strong contribution to heat transfer) and other viscous flow problems.

The general diffusion–convection equation is given by (8.3). Without loss of generality, Re may be taken to be unity, so that the linear equation becomes

$$\frac{\partial u}{\partial t} = \frac{\partial^2 u}{\partial x^2} - \mu\frac{\partial u}{\partial x} \; ; \qquad 0 < x < X, t > 0 \; . \tag{8.110}$$

The parameter $\mu > 0$ is the convection parameter and may be large; it is constant in linear diffusion–convection problems. The initial conditions may be taken to be

$$u(x,0) = g(x) \; ; \qquad 0 \leqslant x \leqslant X \tag{8.111}$$

and the boundary conditions to be

$$u(0,t) = v \; ; \qquad > 0 \tag{8.112}$$

$$\frac{\partial u}{\partial x}(X,t) = 0 \; ; \qquad t > 0_{11} \tag{8.113}$$

with ω a constant.

In obtaining a numerical solution to the linear diffusion–convection problem, the interval $0 \leqslant x \leqslant X$ is divided into M subintervals each of width h so that $Mh = X$, and the time variable t is discretized in steps of length l as before. Retaining previous notations and replacing the space derivatives in (8.111) and (8.113) by (8.9) and $\partial u/\partial x = (2h)^{-1}\{u(x+h,t)-u(x-h,t)\}+O(h^2)$ produces a system of M first-order ordinary differential equations which is of the form

$$\frac{\mathrm{d}\mathbf{U}(t)}{\mathrm{d}t} = A\mathbf{U}(t)+\mathbf{b} \; ; \qquad \mathbf{U}(0) = \mathbf{g} \tag{8.114}$$

where

$$A = h^{-2}\begin{bmatrix} -2 & 1-\frac{1}{2}\mu h & & 0 \\ 1+\frac{1}{2}\mu h & -2 & 1-\frac{1}{2}\mu h & \\ \cdots & \cdots & \cdots & \\ & 1+\frac{1}{2}\mu h & -2 & 1-\frac{1}{2}\mu h \\ 0 & & 2 & -2 \end{bmatrix} \tag{8.115}$$

and

$$\mathbf{b} = h^{-2}[(1+\tfrac{1}{2}\mu h)\omega, 0, \ldots, 0]^{\mathrm{T}} \; . \tag{8.116}$$

(Note that $U_0(t) = \omega$ from (8.112) and $U_{N+1}(t) = U_{N-1}(t)$ from (8.113).)

It is easy to verify that the solution of (8.114) is

$$\mathbf{U}(t) = -A^{-1}\mathbf{b}+\exp(tA)\{\mathbf{g}+A^{-1}\mathbf{b}\} \tag{8.117}$$

which satisfies the recurrence relation

$$\mathbf{U}(t+l) = -A^{-1}\mathbf{b}+\exp(lA)\{\mathbf{U}(t)+A^{-1}\mathbf{b}\} \; . \tag{8.118}$$

Using (8.15) it is then easy to show that

$$(I-\theta lA)\mathbf{U}^{n+1} = [I+(1-\theta)lA]\mathbf{U}^n+l\mathbf{b} \tag{8.119}$$

is a family of methods, with $0 \leqslant \theta \leqslant 1$, for solving (8.110), (8.111), (8.112), (8.113).

The eigenvalues of the matrix A depend on μh and are different for the three cases $\mu h < 2$, $\mu h = 2$ and $\mu h > 2$. Obviously this has a bearing on stability analyses and the reader is referred to Twizell (1984, 1985) for a detailed account of stability. The local truncation error of each method arising from (8.119) has principal part of the form

$$\frac{1}{6}\mu h^2 \frac{\partial^3 u}{\partial x^3} + C_q l^{q-1} \frac{\partial^q u}{\partial t^q} \tag{8.120}$$

where, for $\theta \neq \frac{1}{2}$, $q = 2$, and for $\theta = \frac{1}{2}$, $q = 3$ with $C_3 = -\frac{1}{12}$.

8.8.2 The nonlinear case

In nonlinear diffusion–convection problems, the parabolic equation has the form

$$\frac{\partial u}{\partial t} = \frac{\partial^2 u}{\partial x^2} - \mu(u)\frac{\partial u}{\partial x} \; ; \qquad 0 < x < X \; , \qquad t > 0 \tag{8.121}$$

with initial conditions of the form

$$u(x,0) = g(x) \; ; \qquad 0 \leqslant x \leqslant X \tag{8.122}$$

and boundary conditions in the form

$$u(0,t) = \omega \; ; \qquad t > 0 \tag{8.123}$$

$$\frac{\partial u}{\partial x}(X,t) = 0 \; ; \qquad t > 0 \; . \tag{8.124}$$

Proceeding as in previous sections, it is easy to show that this parabolic problem may be transferred into one of solving the nonlinear system of N first-order ordinary differential equations given in vector form as

$$\frac{\mathrm{d}\mathbf{U}(t)}{\mathrm{d}t} \equiv D\mathbf{U}(t) = A\mathbf{U}(t) + \boldsymbol{\phi}(\mathbf{U}) + \mathbf{b} \tag{8.125}$$

where, now

$$A = h^{-2}\begin{bmatrix} -2 & 1 & & & 0 \\ 1 & -2 & 1 & & \\ & \cdots & \cdots & \cdots & \\ & & 1 & -2 & 1 \\ & & & 2 & -2 \end{bmatrix}, \tag{8.126}$$

$\boldsymbol{\phi}(\mathbf{U}^n) = [\phi_1^n, \phi_2^n, \ldots, \phi_N^n]^{\mathrm{T}}$ at time $t = nl$ with

$$\begin{aligned}\phi_m^n &= -(U_{m+1}^n - U_{m-1}^n)\mu(U_m^n)/2h \;; \qquad m = 1,2,\ldots,M-1 \;, \\ \phi_M^n &= 0\end{aligned} \tag{8.127}$$

(note: $U_0^n = \omega$ for $n = 1,2,\ldots$ and $U_0^0 = g(0)$) and

$$\mathbf{b} = h^{-1}[\omega,0,\ldots,0]^{\mathrm{T}} \;. \tag{8.128}$$

Using the Taylor relation $\mathbf{U}(t+l) = \exp(lD)\mathbf{U}(t)$ and introducing (8.15) once more, it is easy to show that

$$(I-\theta lA)\mathbf{U}^{n+1} - \theta l\boldsymbol{\phi}^{n+1} = [I+(1-\theta)lA]\mathbf{U}^n + (1-\theta)l\boldsymbol{\phi}^n + l\mathbf{b} \tag{8.129}$$

represents a family of methods, with $0 \leqslant \theta \leqslant 1$, for solving the nonlinear diffusion–convection problem. The local truncation errors are $O(h^2+l)$ whenever $\theta \neq \frac{1}{2}$ and $O(h^2+l^2)$ when $\theta = \frac{1}{2}$. The solution is obtained iteratively using a method such as the Newton–Raphson technique for solving a nonlinear algebraic system.

8.9 TWO SPACE DIMENSIONS

8.9.1 A recurrence relation

Consider the constant coefficient heat equation in two space variables given by

$$\frac{\partial u}{\partial t} = \frac{\partial^2 u}{\partial x^2} + \frac{\partial^2 u}{\partial y^2} \;; \qquad 0 < x,\, y < X,\, t > 0 \tag{8.130}$$

with homogeneous Dirichlet boundary conditions ($u \equiv 0$) on the boundary $\partial\Omega$ of the square Ω defined by the lines $x = 0$, $y = 0$, $x = X$, $y = X$, and initial conditions

$$u(x,y,0) = g(x,y) \;; \qquad (x,y) \in \Omega \times \partial\Omega \;. \tag{8.131}$$

It may be assumed that $g(x,y)$, which is a given continuous function of x and y, does not necesarily have the value zero for $(x,y) \in \partial\Omega$ so that discontinuities between initial conditions and boundary conditions may exist.

Both intervals $0 \leqslant x \leqslant X$ and $0 \leqslant y \leqslant X$ are divided into $N+1$ subintervals each of width h, so that $(N+1)h = X$, and the independent variable t is incremented in steps of length l. At each time level $t = nl$ ($n = 0,1,2,\ldots$) the square Ω together with its boundary $\partial\Omega$ have been superimposed by a square mesh of N^2 points within Ω and $N+2$ equally-spaced points along each side of $\partial\Omega$.

The solution $u(x,y,t)$ of (8.130) is sought at each point (kh,mh,nl) in the region $R = \Omega\times[t>0]$ where $k,m = 1,2,\ldots,N$ and $n = 1,2,\ldots$; for $k,m = 0$ and $N+1$ the boundary condition $u \equiv 0$ applies and for $n = 0$ the initial function g applies. The solution $u(kh,mh,nl)$ of (8.130) will be denoted by $u_{k,m}^n$, while the theoretical solution of an approximating difference scheme will be denoted by $U_{k,m}^n$. The elements $U^n_{k,m}$ will be ordered in rows parallel to the x-axis and written in vector form as

$$\mathbf{U}^n = [U^n_{1,1}, U^n_{2,1}, \ldots, U^n_{N,1} \;;\; U^n_{1,2}, U^n_{2,2}, \ldots, U^n_{N,2} \;; \ldots$$
$$\ldots ; U^n_{1,N}, U^n_{2,N}, \ldots, U^n_{N,N}]^{\mathrm{T}} \;. \quad (8.132)$$

The space derivatives in (8.130) may be approximated by the finite difference replacements

$$\frac{\partial^2 u}{\partial x^2} = h^{-1}\{u(x-h,y,t) - 2u(x,y,t) + u(x+h,y,t)\} + O(h^2) \quad (8.133)$$

and

$$\frac{\partial^2 u}{\partial y^2} = h^{-2}\{u(x,y-h,t) - 2u(x,y,t) + u(x,y+h,t)\} + O(h^2) \;. \quad (8.134)$$

Considering the time level $t = nl$, the differential equation (8.130), together with (8.133) and (8.134), is applied to all N interior mesh points of the square Ω.

These N^2 applications result in a system of N^2 linear, first-order ordinary differential equations of the form (8.10), in which the matrix A is now of order N^2 and may be *split* into constituents matrices B and C such that $A = B + C$. The matrix B arises from the use of (8.133) in (8.130); it is block diagonal with tridiagonal blocks and has the form

$$B = h^{-2}\begin{bmatrix} B_1 & & 0 \\ & \ddots \; B_1 & \\ 0 & & \ddots \; B_1 \end{bmatrix} \quad (8.135)$$

where B_1 is the tridiagonal matrix of order N given by

$$B_1 = \begin{bmatrix} -2 & 1 & & & 0 \\ 1 & -2 & 1 & & \\ & \cdots & \cdots & \cdots & \\ 0 & & & -2 & 1 \end{bmatrix} . \quad (8.136)$$

The matrix C arises from the use of (8.134) in (8.130); it is block tridiagonal with diagonal blocks and has the form

$$C = h^{-2}\begin{bmatrix} -2I & I & & & 0 \\ I & -2I & I & & \\ & \cdots & \cdots & \cdots & \\ & & I & -2I & I \\ 0 & & & I & -2I \end{bmatrix}, \tag{8.137}$$

where I is the identity matrix of order N. The N^2 eigenvalues of A are real and negative and are given by

$$\lambda_{i,j} = -4h^{-2}\left(\sin^2\frac{i\pi}{2(N+1)} + \sin^2\frac{j\pi}{2(N+1)}\right) ; \qquad i,j = 1,2,\ldots,N . \tag{8.138}$$

It is noted that the matrices B and C may commute, depending on the splitting, but in all that follows in this section, the more general case of noncommutative B and C will be considered.

Solving the system of ordinary differential equations, subject to the initial condition $\mathbf{U}(0) = \mathbf{g}$, gives

$$\mathbf{U}(t) = \exp[t(B+C)]\mathbf{g}$$

which satisfies the recurrence relation

$$\mathbf{U}(t+l) = \exp[l(B+C)]\mathbf{U}(t) ; \qquad t = 0, l, 2l, \ldots . \tag{8.139}$$

It is this relation, which is the two-dimensional equivalent of (8.14), which will yield a family of method for solving the two-dimensional heat equation.

8.9.2 An explicit method and its analysis

Using (8.15) in (8.139) with $\theta = 0$ gives

$$\mathbf{U}(t+l) = [I + l(B+C)]\mathbf{U}(t) \tag{8.140}$$

which, when applied to the mesh point (kh, mh, nl) in R, gives

$$U_{k,m}^{n+1} = U_{k,m}^{n} + \frac{l}{h^2}\left(U_{k-1,m}^{n} - 2U_{k,m}^{n} + U_{k+1,m}^{n}\right) + \frac{l}{h^2}\left(U_{k,m-1}^{n} - 2U_{k,m}^{n} + U_{k,m+1}^{n}\right) . \tag{8.141}$$

Gathering terms leads to the six-point, two-level, explicit scheme

$$U_{k,m}^{n+1} = (1-4p)U_{k,m}^{n} + p\left(U_{k,m-1}^{n} + U_{k-1,m}^{n} + U_{k,m+1}^{n} + U_{k+1,m}^{n}\right). \tag{8.142}$$

The local truncation error of (8.142) is given by

$$\begin{aligned} L(x,y,t) &= \frac{u(x,y,t+l)-u(x,y,t)}{l} - \frac{u(x-h,y,t)-2u(x,y,t)+u(x+h,y,t)}{h^2} \\ &\quad - \frac{u(x,y-h,t)-2u(x,y,t)+u(x,y+h,t)}{h^2} - \left(\frac{\partial u}{\partial t} - \frac{\partial^2 u}{\partial x^2}\right) \end{aligned} \tag{8.143}$$

which follows from an obvious rearrangement of (8.141). Expanding all terms in (8.143) about $u(x,y,t)$ shows that the principal part of $L(x,y,t)$ is

$$\frac{1}{2}l\frac{\partial^2 u}{\partial t^2} - \frac{1}{12}h^2\left(\frac{\partial^2 u}{\partial x^4} + \frac{\partial^2 u}{\partial y^4}\right). \tag{8.144}$$

The first term in (8.144) is the same as the analogous term in (8.25) while the second is a simple extension of its analogous term in (8.25). Clearly, the local truncation error $L(x,y,z,t)$ can be written down for this particular method in three space dimensions.

Like the one-dimensional explicit method ($\theta = 0$) given by (8.18), equation (8.142) enjoys only conditional stability. The amplification factor $R_\theta(l\lambda)$ must not exceed unity in modulus for stability in the conventional sense. This leads to $|1+l\lambda_{i,j}| \leqslant 1$ for all $i,j = 1,2,\ldots,N$ from which it is easy to show that the finite difference scheme is stable provided $-l\lambda_{i,j} \leqslant 2$. Referring to (8.138) from which it may be deduced that $|\lambda_{i,j}| < 8h^{-2}$, it follows that the von Neumann necessary condition for the stability of (8.142) is $p \leqslant \frac{1}{4}$.

It may be shown that the explicit scheme development in this way for problems in three space dimensions is stable provided $p \leqslant \frac{1}{6}$.

8.9.3 A fully implicit method

Returning to the recurrence relation (8.139), a fully implicit method may be developed by first writing (8.139) in the form

$$\mathbf{U}(t+l) = \exp(lC)\exp(lB)\mathbf{U}(t) \tag{8.145}$$

which has error $O(l^2)$ in time. Equation (8.145) may be approximated further also with error $O(l^2)$ by applying (8.15) with $\theta = 1$ to $\exp(lC)$ and $\exp(lB)$ and writing (8.145) in the *split form*

$$(I - lB)\mathbf{U}^* = \mathbf{U}(t) \tag{8.146}$$

$$(I - lC)\mathbf{U}(t + l) = \mathbf{U}^* \; . \tag{8.147}$$

This splitting has the disadvantage of introducing an additional, intermediate vector $\mathbf{U}^*$ which must be stored. However, $\mathbf{U}^*$ can easily be found from (8.146) using a tridiagonal linear solver; the solution $\mathbf{U}(t + l)$ is then found from (8.147) using the special algorithm given in Chapter 1.

The fully implicit method described by (8.146) and (8.147) is first-order accurate in time and is L_0-stable. The accuracy in time may be improved by the process of local extrapolation as in Lawson and Morris (1978), the algorithm being written over two single time steps as

$$(I - lB)\mathbf{U}^* = \mathbf{U}(t) \; ,$$

$$(I - lC)\mathbf{U}(t + l) = \mathbf{U}^* \; ,$$

$$(I - lC)\mathbf{U}^\dagger = \mathbf{U}(t + l) \; ,$$

$$(I - lB)\mathbf{U}^{(0)}(t + 2l) = \mathbf{U}^\dagger \; . \tag{8.148}$$

Eliminating the intermediate vectors $\mathbf{U}^*$, $\mathbf{U}(t + l)$, $\mathbf{U}^\dagger$ in (8.148) gives

$$\mathbf{U}^{(0)}(t + 2l) = (I - lB)^{-1}(I - lC)^{-1}(I - lC)^{-1}(I - lB)^{-1}\mathbf{U}(t) \tag{8.149}$$

and expanding the matrix inverses leads to

$$\mathbf{U}^{(0)}(t + 2l) = [I + 2l(B + C) + l^2(3B^2 + 3C^2 + 2BC + 2CB) + \ldots]\mathbf{U}(t) \; . \tag{8.150}$$

The Maclaurin expansion of $\exp[2l(B + C)]$ in (8.139) written over a double time step $2l$ gives

$$\mathbf{U}^{(\mathrm{M})}(t + 2l) = [I + 2l(B + C) + 2l^2(B^2 + C^2 + BC + CB) + \ldots]\mathbf{U}(t) \tag{8.151}$$

so that (8.148) is verified to be first-order accurate.

Lawson and Morris (1978) proceed by writing the recurrence relation (8.139) as

$$\mathbf{U}(t + l) = \exp(lB)\exp(lC)\mathbf{U}(t) \tag{8.152}$$

in addition to (8.145). Equations (8.152) and (8.145) are then written over a double time step $2l$ and the exponential matrices by (8.15) with $\theta = 1$, giving

$$\begin{aligned} \mathbf{U}^{(1)}(t+2l) &= (I-2lB)^{-1}(I-2lC)^{-1}\mathbf{U}(t) \\ &= [I+2l(B+C)+4l^2(B^2+BC+C^2)+\ldots]\mathbf{U}(t) \end{aligned} \tag{8.153}$$

and

$$\begin{aligned} \mathbf{U}^{(2)}(t+2l) &= (I-2lC)^{-1}(I-2lB)^{-1}\mathbf{U}(t) \\ &= [I+2l(B+C)+4l^2(B^2+CB+C^2)+\ldots]\mathbf{U}(t)\ . \end{aligned} \tag{8.154}$$

Comparing (8.153) and (8.154) with (8.151) it is seen that, in addition to $\mathbf{U}^{(0)}(t+2l)$, the approximations $\mathbf{U}^{(1)}(t+2l)$ and $\mathbf{U}^{(2)}(t+2l)$ are first-order accurate in time.

Lawson and Morris (1978) then point out that the local extrapolation

$$\mathbf{U}^{(E)}(t+2l) = 2\mathbf{U}^{(0)} - \tfrac{1}{2}(\mathbf{U}^{(1)}+\mathbf{U}^{(2)}) \tag{8.155}$$

is second-order accurate in time, the first-order approximations (8.149), (8.153) and (8.154) having been combined linearly to give higher-order accuracy. Lawson and Morris state that this extrapolated method is L_0-stable.

8.9.4 The Peaceman–Rachford method

This method is derived from the recurrence relation (8.139) by writing it to $O(l^2)$ error as

$$\mathbf{U}(t+l) = \exp(\tfrac{1}{2}lB)\exp(lC)\exp(\tfrac{1}{2}lB)\mathbf{U}(t)\ . \tag{8.156}$$

Then, by using (8.15) with $\theta = 1, \tfrac{1}{2}, 0$, respectively, to replace the three exponential terms in (8.156), (8.156) becomes the second-order method

$$\begin{aligned} (I-\tfrac{1}{2}lC)\mathbf{U}^* &= (I+\tfrac{1}{2}lB)\mathbf{U}(t)\ , \\ (I-\tfrac{1}{2}lB)\mathbf{U}(t+l) &= (\mathrm{I}+\tfrac{1}{2}lC)\mathbf{U}^*\ , \end{aligned} \tag{8.157}$$

where $\mathbf{U}^*$ is an intermediate vector. This A_0-stable method is the *Peaceman–Rachford method*.

Lawson and Morris (1977) give an algorithm which, they claim, gives a saving of 30–40 per cent in computational arithmetic on the implementation of the Peaceman–Rachford method. The additional storage requirement of this efficient implementation is one vector of order N^2.

8.9.5 A numerical example

In order to illustrate the behaviours of the L_0-stable and A_0-stable methods (8.155) and (8.157), each is used to solve the following model problem which is due to

Lawson and Morris (1978):

$$\frac{\partial u}{\partial t} = \frac{\partial^2 u}{\partial x^2} + \frac{\partial^2 u}{\partial y^2} \ ; \qquad 0 < x,y < 2 \ , t > 0$$

subject to the initial conditions

$$u(x,y,0) = \sin(\tfrac{1}{2}\pi y) \ ; \qquad 0 \leqslant x,y \leqslant 0$$

and the boundary conditions

$$u(x,y,t) = 0 \ ; \qquad x = 0 \ , y = 0 \ , x = 2 \ , y = 2 \ .$$

The initial distribution is shown in Fig. 8.7 (note the discontinuity between the

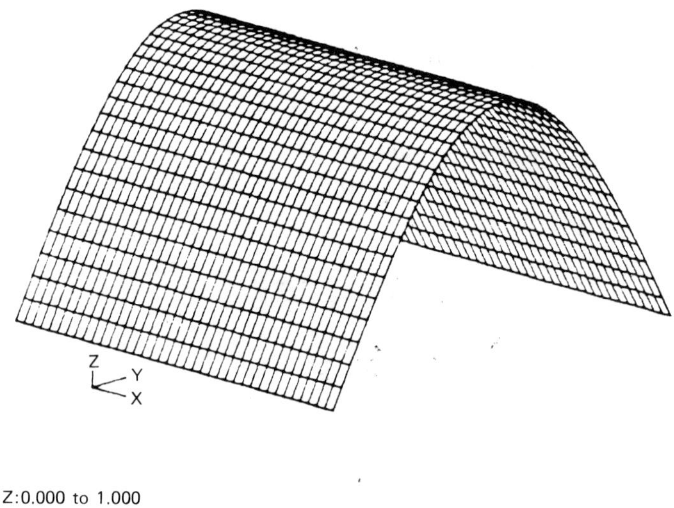

Fig. 8.7 — Initial distribution for model problem with two space dimensions.

boundary conditions and initial conditions for $y = 0$ and $y = 2$), and the theoretical solution

$$u(x,y,t) = \sin\tfrac{1}{2}\pi y \sum_{k=1}^{\infty} [1 - (-1)^k] \frac{2}{k\pi} \sin(\tfrac{1}{2}k\pi x) \exp[-\tfrac{1}{4}\pi^2(k^2+1)t] \qquad (8.158)$$

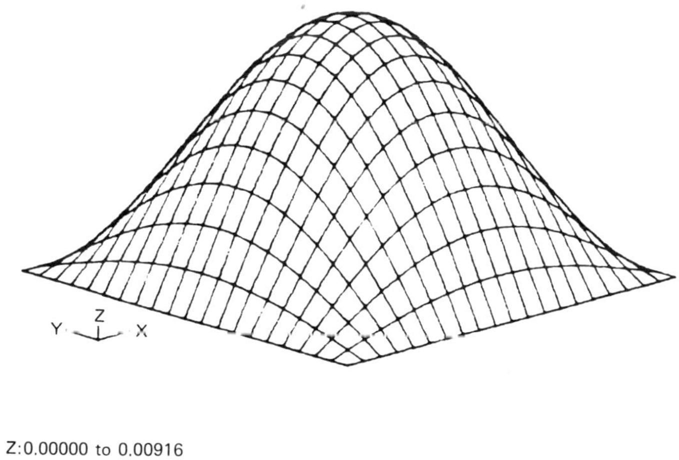

Fig. 8.8 — Theoretical solution at time $t = 1.0$ for the model problem with two space dimensions.

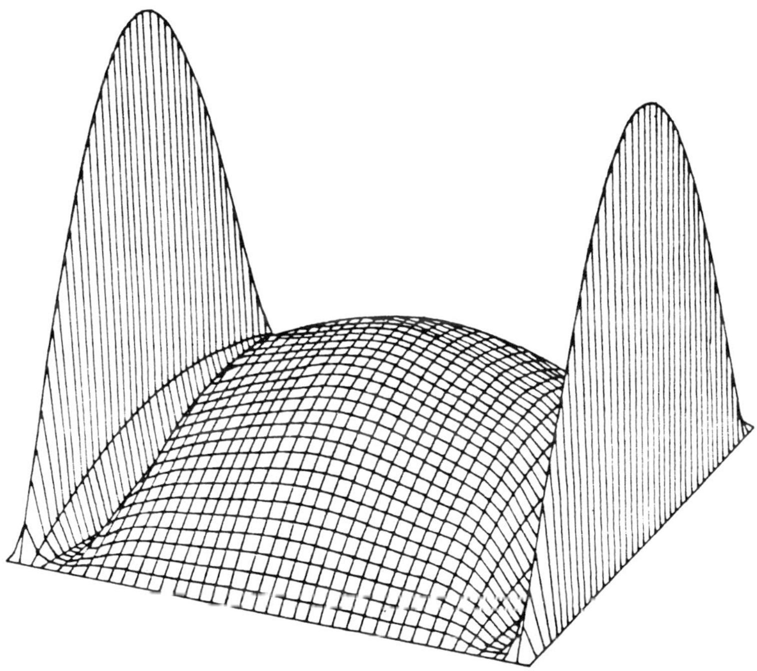

Fig. 8.9 — Computed solution at time $t = 1.0$ using the Peaceman–Rachford method with $h = 0.05$ and $l = 0.1$.

is depicted at time $t = 1.0$ in Fig. 8.8. The maximum value of u at time $t = 1.0$ occurs for $x = 1$ and $y = 1$ and has the approximate value 0.00916.

Following Lawson and Morris (1978), the solution is computed at time $t = 1.0$ using $h = 0.05$ and $l = 0.1$; the ratios $p = l/h^2$ and $r = l/h$ therefore have the values $p = 40$ and $r = 2$. The distribution for $t = 1.0$ is shown in Fig. 8.9 for the A_0-stable Peaceman–Rachford method and in Fig. 8.10 for the L_0-stable locally extrapolated method (8.155).

It has been necessary to scale down the interior peak in Fig. 8.9. In fact, the Peaceman–Rachford method, like the Crank–Nicolson method (8.21) in one space variable, is very accurate away from the boundary of the region, and if Figs. 8.8 and 8.9 had been drawn to the same scale, the interior peaks would have been almost coincident. The peaks near the boundary of Fig. 8.9, however, would then have been about twice their height.

The presence of these peaks near the boundaries $y = 0$ and $y = 2$ of Fig. 8.9 shows that the time step $l = 0.1$ is too large for use with a space step of $h = 0.05$ when the Peaceman–Rachford method is used to solve the model problem which has a decaying solution as time increases. On the other hand, it is evident from Fig. 8.10 that the L_0-stable locally extrapolated method (8.155), which is also second-order accurate, does not suffer from this restriction and the computed solution is a smooth representation of the theoretical solution (8.158).

The maximum errors for this experiment using these two methods, are included in Table 8.2. The maximum errors for the Peaceman–Rachford method (8.157) are those near the boundaries $y = 0$ and $y = 2$ and, because of symmetry, these errors are the same; the maximum error for the locally extrapolated method (8.155) occurs at the point where $x = 1, y = 1$

The same model problem can be used to show that the Peaceman–Rachford method gives good results when the ratio $r = l/h$ is not too large. The maximum errors at time $t = 1.0$ for $h = 0.05$, as before, and l decreased to 0.025 are also given in Table 8.2. For this experiment, the maximum errors for both methods occur at the point where $x = 1$ and $y = 1$, showing that the discontinuities between boundary conditions and initial conditions have been smoothed out by both methods. The superior results of the Peaceman–Rachford method over those of the L_0-stable method (8.155) are due to the more favourable time error constant, $C_3 = -\frac{1}{12}$, of the former method over that of the latter, which has $C_3 = \frac{4}{3}$.

8.10 HEAT FLOW IN THE HUMAN BODY

8.10.1 Introduction

The regulation of temperature within the human body is dependent upon a number of factors. These are: generation of heat by metabolic reactions (basal or due to external work); the geometry of the body; insulation due to layers of fat and skin; conduction of heat in tissue; convection of heat by flowing blood, counter-current heat exchange between large arteries and veins; heat loss due to respiration, sweating, and shivering; loss of heat at the surface due to convection, evaporation, and radiation; and environmental conditions.

The most realistic models of the human body consider the individual elements

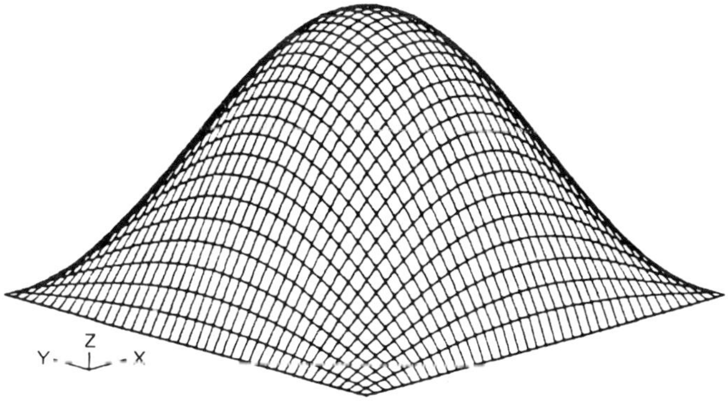

Fig. 8.10 — Computed solution at time $t = 1.0$ using the locally extrapolated method (8.155).

Table 8.2

Method	Maximum errors ($h = 0.05$)	
	$l = 0.1$	$l = 0.025$
(8.155)	0.34E–3	0.46E–4
(8.157)	0.23E–1	0.33E–4

(head, torso, arms and legs) to consist of four concentric layers of different tissue type, namely core, muscle, fat and skin with each layer of tissue having different thermal properties. However, the thermal conductivities of the core and muscle are so nearly equal that they are frequently considered to be just one tissue called the *core*: likewise, the thermal conductivities of the fat and skin are approximately the same and they are treated as one tissue called the *insulating layer*. The overall ratio of the thickness of the core and insulating layer is about 6.7:1. The torso, arms and legs are usually treated as circular cylinders while the head is sometimes treated as a sphere and sometimes as a cylinder.

8.10.2 The equations of a mathematical model

Mathematical models of heat flow in the human body may be one of two types: steady-state models which will be discussed in Chapter 9 or time-dependent models which will be discussed in this subsection. The simple model to be outlined will consider the torso only and longitudinal heat flow will be neglected so that only a cross-section of the torso (which will be assumed to be circular) need be taken into account.

Five heat exchanges will be balanced in the model: the differential equation which does this will be given in each case, the symbols being listed in section 8.10.4. The exchanges are as follows:

(1) heat stored in the tissues

= heat transferred by conduction between tissue elements
+ heat transferred from tissues to blood stream
+ metabolic heat generation,

$$(\rho c)_i \frac{\partial u}{\partial t} = \kappa_i \left\{ \frac{\partial^2 u}{\partial r^2} + \frac{1}{r}\frac{\partial u}{\partial r} + \frac{1}{r^2}\frac{\partial^2 u}{\partial \theta^2} \right\} + Q_{ci}(v-u) + h_i \ , \tag{8.159}$$

where the subscript $i = 1,2$ refers to the core or insulating layer, respectively;

(2) rate of accumulation of thermal energy by the blood

= rate at which it is supplied

$$(MC)_i \frac{\mathrm{d}v}{\mathrm{d}t} = Q_{ai}(v-u) \ ; \tag{8.160}$$

(3) rate of heat transferred by the torso to the environment

= rate at which it is gained by the environment at the skin surface

$$-\kappa_2 \frac{\partial u}{\partial r} = H\{u - E(\theta)\} + S(u) \ , \tag{8.161}$$

where

$$S(u) = \gamma\gamma_{\mathrm{T}}(2^{u-u^*})/(10\Delta) \ ; \tag{8.162}$$

(4) discontinuities in thermal conductivity at the core/insulating layer interface but flux equality must be retained,

$$-\kappa_1 \frac{\partial u}{\partial r} = -\kappa_2 \frac{\partial u}{\partial r} \ ; \tag{8.163}$$

(5) there is no heat flow at the centre of the torso

$$\left.\frac{\partial u}{\partial r}\right|_{r=0} = 0 \ . \tag{8.164}$$

The terms in (8.162) are from the body's heat control mechanism which was contained in the papers by Stolwijk and Hardy (1966) and Stolwijk (1979); the reader is referred to these papers and to the Ph.D. thesis of Smith (1981).

8.10.3 The numerical solution

The space derivatives in (8.159) are replaced by

$$\frac{\partial^2 u}{\partial r^2} = (\delta r)^{-2}\{u(r-\delta r,\theta,t) - 2u(r,\theta,t) + u(r+\delta r,\theta,t)\} + O((\delta r)^2) \ , \tag{8.165}$$

$$\frac{\partial u}{\partial r} = (2\delta r)^{-1}\{u(r+\delta r,\theta,t) - u(r-\delta r,\theta,t) + O((\delta r)^2) \tag{8.166}$$

and

$$\frac{\partial^2 u}{\partial \theta^2} = (\delta\theta)^{-2}\{u(r,\theta-\delta\theta,t) - 2u(r,\theta,t) + u(r,\theta+\delta\theta,t)\} + O((\delta\theta)^2) \ , \tag{8.167}$$

that in (8.161) and that on the left-hand side of (8.163) by

$$\frac{\partial u}{\partial r} = (\delta r)^{-1}\{u(r,\theta,t) - u(r-\delta r,\theta,t)\} + O(\delta r) \tag{8.168}$$

that on the right-hand side of (8.163) by

$$\frac{\partial u}{\partial r} = (\delta r)^{-1}\{u(r+\delta r,\theta,t) - u(r,\theta,t)\} + O(\delta r) \tag{8.169}$$

and that in (8.164) by (8.168) with $r = 0$.

In (8.165)–(8.169), δr and $\delta\theta$ are increments in r and θ associated with the discretization of the (r,θ)-plane. This discretization takes the form of K concentric circles, one coincident with the skin surface and one coincident with the core/insulating layer interface, together with M radii, one coincident with the r-axis. The notation used in section 8.9 can then be adapted in an obvious way, the vector $\mathbf{U}$ having $KM+2$ elements (the temperature at the centre, the temperature at the other KM points of the discretization, and the temperature of the blood).

Applying the differential equations and the appropriate replacements for the space derivatives to the $KM+1$ points of the discretization and to the blood gives a system of ordinary differential equations of the form

$$\frac{\mathrm{d}\mathbf{U}}{\mathrm{d}t} = A\mathbf{U} + \boldsymbol{\phi}(\mathbf{U}) + \mathbf{b} \tag{8.170}$$

where the square matrix A, and the vectors $\mathbf{U}$, $\boldsymbol{\phi}$, $\mathbf{b}$ are all of order $KM+2$ with A depending on the space discretization and physical constants, $\boldsymbol{\phi}$ arising because of the nonlinear term in (8.161), and the vector $\mathbf{b}$ depending on physical constants. The solution of equations of the form (8.170) has already been discussed in sections 8.6 and 8.7.

8.10.4 Nomenclature

$u(r,\theta,t)$: temperature of tissue at time t at point (r,θ),
ρc : product of density and specific heat of tissue,
κ : conductivity of tissue,
Q_c : product of mass flow rate and specific heat of blood entering capillary beds per unit volume,
v : temperature of blood,
h : metabolic heat generation per unit volume,
MC : product of mass and specific heat of blood,
Q_a : product of mass flow rate and specific heat of blood entering the arterial pool,
$E = E(\theta)$: environmental temperature distribution around the torso,
H : effective heat transfer coefficient of convection and radiation at environment temperature E,
$S(u)$: heat loss due to sweating at the skin surface,
γ : sweating command of body's heat control mechanism,
γ_T : proportion of γ to the torso,
u^* : set-point or threshold temperature for the body's thermoreceptors in the torso,
Δ : area of surface of torso.

Values of the constants described above may be located in Smith and Twizell (1985).

PROBLEMS

1. The constant coefficient heat equation in one space dimension is given by

$$\text{(A)} \qquad \frac{\partial u}{\partial t} = \frac{\partial^2 u}{\partial x^2}\,; \qquad 0 < x < X\,, \qquad t > 0$$

with boundary conditions $u(0,t) = u(X,t) = 0$, $t > 0$, and initial conditions $u(x,0) = g(x), 0 \leqslant x \leqslant \mathrm{X}$, where $g(x)$ is a given continuous function of x. Obtain the equivalent first order system

$$\frac{d\mathbf{U}(t)}{dt} = A\mathbf{U}(t) ,$$

carefully defining the vector **U** and the matrix A.

Show that the solution of this system satisfies the relation

$$\text{(B)} \qquad \mathbf{U}(t+l) = \exp(lA)\mathbf{U}(t)$$

where l is a convenient time step.

The matrix exponential function $\exp(lA)$ in (B) is to be replaced by

$$L_l = [I - l(1-\theta)A]^{-1}[I + l\theta A] ,$$

where I is the identity matrix of order N and θ is a parameter, and the vectors

$$\mathbf{U}^{(1)}(t+2l) = L_l^2\mathbf{U}(t) \; ; \qquad \mathbf{U}^{(2)}(t+2l) = L_{2l}\mathbf{U}(t)$$
$$\mathbf{V}(t+2l) = \alpha\mathbf{U}^{(1)}(t+2l) - (\alpha-1)\mathbf{U}^{(2)}(t+2l)$$

are to be computed. Verify that

$$\mathbf{U}^{(1)}(t+2l) = [I + 2lA + l^2(3-2\theta)A^2 + 2l^3(1-\theta)(2-\theta)A^3 + \ldots]\mathbf{U}(t) ,$$

$$\mathbf{U}^{(2)}(t+2l) = [I + 2lA + 4l^2(1-\theta)A^2 + 8l^3(1-\theta)^2A^3 + \ldots]\mathbf{U}(t)$$

and find the value of the parameter α if $\mathbf{V}(t+2l)$ is to be a second order approximation in time to $\mathbf{U}(t+2l)$.

Given that the matrix A has negative real eigenvalues, write down the amplification symbol for $\mathbf{V}(t+2l)$ when $\theta = 0$ and determine whether the method is A_0-stable, L_0-stable, or otherwise.

2. The matrix exponential function in question 1 is to be replaced by the approximation

$$L_l = (I - lA)^{-1} ,$$

where I is the identity matrix order N, and the vectors

$$\mathbf{U}^{(1)}(t+3l) = L_l^3\mathbf{U}(t) \; ; \quad \mathbf{U}^{(2)}(t+3l) = L_{2l}L_l\mathbf{U}(t) \; ; \quad \mathbf{U}^{(3)}(t+3l) = L_{3l}\mathbf{U}(t) \; ;$$

$$\mathbf{V}(t+3l) = \alpha\mathbf{U}^{(1)} + \beta\mathbf{U}^{(2)} + (1-\alpha-\beta)\mathbf{U}^{(3)}$$

are to be computed. Find values of the parameters α and β if $\mathbf{V}(t+3l)$ is to be a second order approximation in time to $\mathbf{U}(t+3l)$.

Given that the matrix A has negative real eigenvalues, write down the amplification symbol for $\mathbf{V}(t+3l)$ and determine whether the method is A_0-stable, L_0-stable, or otherwise.

[N.B.: The Maclaurin expansion of $\mathbf{U}(t+3l)$ is

$$\mathbf{U}(t+3l) = \left(I+3lA+\frac{9}{2}l^2A^2+\frac{9}{2}l^3A^3+\frac{27}{8}l^4A^4+\ldots\right)\mathbf{U}(t) \text{ .]}$$

3. The function $u(x,t)$ satisfies the differential equation

$$\frac{\partial u}{\partial t}+\frac{\partial u}{\partial x} = \frac{\partial^2 u}{\partial x^2}\ ; \qquad 0 < x < 1\ ,\ t > 0\ ,$$

with initial conditions $u(x,0) = 0 (0 \leqslant x \leqslant 1)$ and boundary conditions $u(0,t) = \pi t$, $\partial u/\partial x\,(1,t) = 0$ for $t > 0$. The interval $0 \leqslant x \leqslant 1$ is discretized into N subintervals each of width h and the time variable is discretized in steps of length l so that $t = nl$ $(n = 0,1,2,\ldots)$.

Obtain the finite difference scheme

$$U_m^{n+1} = (p+\tfrac{1}{2}r)_{m-1}^{n}+(1-2p)U_m^n+(p-\tfrac{1}{2}r)U_{m+1}^n\ ;$$
$$n = 0,\ 1,\ \ldots,\ m = 1,\ 2,\ \ldots,\ N-1$$

for this problem, where $p = l/h^2$, $r = l/h$ and $U_m^n \equiv U(mh,nl)$. Find the local truncation error for the scheme and show that its amplification factor ξ satisfies the criterion $|\xi|^2 \leqslant 1+0(l)$ provided $p \leqslant \frac{1}{2}$. Comment on the practical inadequacy of this criterion.

Modify the scheme for use with $m = N$.

4. The function $u(x,t)$ satisfies the simple heat equation

$$\frac{\partial u}{\partial t} = \frac{\partial^2 u}{\partial x^2}\ ; \qquad 0 < x < X\ ,\ t > 0$$

with boundary conditions $u(0,t) = u(X,t) = 0$ for $t > 0$ and initial conditions $u(x,0) = g(x)$ for $0 \leqslant x \leqslant X$.

The solution of the problem is to be determined using the finite difference method

(A) $U_m^{n+1} = (1-2p+3p^2)U_m^n+p(1-2p)(U_{m-1}^n+U_{m+1}^n)+\tfrac{1}{2}p^2(U_{m-2}^n+$
$+\,U_{m+2}^n)$

where $0 < p = l/h^2$, $l > 0$ and $h > 0$ being the increments in t and x respectively and U_m^n is the theoretical solution of (A) at the point (mh,nl).

(i) Find the local truncation error of (A).
(ii) Writing the difference between the theoretical and computed solutions of (A) at the point (mh,nl), as $e^{\alpha nl}e^{i\beta mh}$, where α is complex and β is real, show that the amplication factor $e^{\alpha l}$ satisfies the equation

$$e^{\alpha l} = 1 - 2p + 3p^2 + 2p(1-2p)\cos\beta h + p^2\cos 2\beta h\ .$$

Hence show that (A) is stable provided $p \leqslant \frac{1}{2}$.
(Note: $\cos 2\beta h = 1 - 8\sin^2\frac{1}{2}\beta h + 8\sin^4\frac{1}{2}\beta h$.)

5. The finite difference scheme

$$U_m^{n+1} = U_m^{n-1} + 2p(U_{m-1}^u - U_m^{n+1} - U_m^{n-1} + U_{m-1}^n)$$

was proposed by du fort and Frankel for solving the one dimensional heat equation

$$\frac{\partial u}{\partial t} = \frac{\partial^2 u}{\partial x^2}\ ; \qquad 0 < x < X\ , t > 0$$

where $p = l/h^2$, h and l being the increments in x and t; boundary conditions are specified at $x = 0, x = 1\ (t > 0)$ and initial conditions are specified at $t = 0 (0 \leqslant x \leqslant 1)$. Consider the consistency of the scheme in the two cases

(a) $\dfrac{l}{h} \to 0$ as $l \to 0$,

(b) $\dfrac{l}{h} \to$ C, some constant, as $l \to 0$.

6. The space derivative in the parabolic problem

$$\frac{\partial u}{\partial t} = u\frac{\partial^2 u}{\partial x^2}\ ; \qquad 0 < x < X\ , \qquad t < 0$$

$$u(x,0) = g(x)\ ,\ 0 \leqslant x \leqslant X\ ; \qquad u(0,t) = \alpha, u(X,t) = \beta,\ t > 0$$

is replaced by a second order central difference approximant. Show that the problem is transformed into one of solving the nonlinear initial value problem

$$D\mathbf{U}(t) \equiv \frac{d\mathbf{U}(t)}{dt} = G(\mathbf{U})A\mathbf{U}(t) + \mathbf{b}(\mathbf{U}) \; ; \qquad \mathbf{U}(0) = \mathbf{g} \; ,$$

where $D = \operatorname{diag}\{d/dt\}$, carefully specifying the vectors $\mathbf{U} = \mathbf{U}(t), \mathbf{b}(\mathbf{U})$ and $\mathbf{g}$ and the matrices $G(\mathbf{U})$ and A.

Show that

$$(I - lD)\mathbf{U}^{n+1} = \mathbf{U}^n$$

is a first order method (in time) for computing $\mathbf{U}^{n+1} = \mathbf{U}(t+l)$, where $t = nl$ ($n = 0, 1, 2, \ldots$) and l is a time step. Show further that the solution may be obtained by solving a nonlinear algebraic system of the form

$$\mathbf{F}(\mathbf{U}^{n+1}) = \mathbf{0}$$

carefully specifying $\mathbf{F}(\mathbf{U}^{n+1})$.

Indicate why this method is to be preferred to the second order method

$$(I - \tfrac{1}{2}lD)\mathbf{U}^{n+1} = (I + \tfrac{1}{2}lD)\mathbf{U}^n$$

for solving parabolic problems where the differential equation is

$$\frac{\partial u}{\partial t} = \frac{u}{t}\frac{\partial^2 u}{\partial x^2} \; ; \qquad 0 < x < X \qquad t > 0 \; .$$

7. The non-linear cubic Schrödinger equation

$$\text{(A)} \qquad i\frac{\partial u}{\partial t} + \frac{\partial^2 u}{\partial x^2} + u|u|^2 = 0 \; ; \qquad L_0 < x < L_1, t > 0, i = +\sqrt{-1}$$

where $u(x,t) = v(x,t) + iw(x,t)$ is a complex field, models the evolution of a weakly nonlinear, strongly dispersive, almost monochromatic wave. The initial conditions associated with (A) are

$$\text{(B)} \qquad u(x,0) = g(x) = g_R(x) + ig_I(x) \; ; \qquad L_0 \leqslant x \leqslant L_1 \; ,$$

where $g_R(x)$ and $g_I(x)$ are continuous, and the boundary conditions are given by

$$\text{(C)} \qquad \frac{\partial u}{\partial x} = 0 \; ; \qquad x = L_0 \text{ and } x = L_1 \; , \qquad t > 0 \; .$$

Show that {(A), (B)} is equivalent to the nonlinear system

$$\text{(D)} \qquad \frac{\partial \mathbf{u}}{\partial t} + A\frac{\partial^2 \mathbf{u}}{\partial x^2} + \mathbf{f}(\mathbf{u}) = \mathbf{0} \ ; \qquad \mathbf{u}(0) = \mathbf{g} \ ,$$

where $\mathbf{u} = \mathbf{u}(t) = [v(t), w(t)]^T$, T denoting transpose, and give the precise forms of A, $\mathbf{f}(\mathbf{u})$ and $\mathbf{g}$.

Replacing the boundary conditions and the space derivatives in (D) by second order finite difference approximants, show that {(D), (C)} may be transformed to the system of nonlinear ordinary differential equations given by

$$\text{(E)} \qquad \frac{d\boldsymbol{\alpha}}{dt} + S\boldsymbol{\alpha} + \mathbf{F}(\boldsymbol{\alpha}) = \mathbf{0} \ ; \qquad \boldsymbol{\alpha}(0) = \mathbf{g}$$

where $\boldsymbol{\alpha} = \boldsymbol{\alpha}(t) = [V(t), W(t)]^T$, V and W being the approximating values of v and w, and give the precise forms of S (in terms of A) and $\mathbf{F}(\boldsymbol{\alpha})$.

Develop an algorithm in *PECE* mode, based on the Euler–modified Euler combination, for solving (E).

Estimate the maximum value of l which may be used if the algorithm is to be stable.

Indicate *briefly* how the stability of the algorithm may be improved.

8. Given the parabolic equation

$$\text{(A)} \qquad \frac{\partial u}{\partial t} = \frac{\partial^2 u}{\partial x^2} + \frac{\partial^2 u}{\partial y^2} \ ; \qquad 0 < x, y < L \ , t > 0$$

in two space dimensions with initial conditions

$$u(x,y,0) = g(x,y) \ ; \qquad 0 \leqslant x,y \leqslant L$$

and boundary conditions

$$u(x,y,t) = 0 \ ; \qquad (x,y) \in \partial\Omega \ , t > 0 \ ,$$

where $\partial\Omega$ is the boundary of the square Ω having sides $x = 0, y = 0, x = L, y = L$.

Using the usual second order central difference replacements of the space derivatives, with a step size of h in both x and y, show that an approximate solution of (A) may be found by solving a system of linear ordinary differential equations of the form

$$\text{(B)} \qquad \frac{d\mathbf{U}(t)}{dt} = A\mathbf{U}(t)\ ; \qquad \mathbf{U}(0) = \mathbf{g}\ .$$

Given that the matrix A may be split into the form $A = B + C$, where B and C are related to the replacements of $\partial^2 u/\partial x^2$ *and* $\partial^2 u/\partial y^2$ respectively, write down the forms of B and C.

Show that the solution of (B) satisfies the recurrence relation

$$\mathbf{U}(t+kl) = \exp[kl(B+C)]\mathbf{U}(t)\ ; \qquad k = 0,1,2,\ldots$$

The matrices L_k ($k = 1,2,3$) are defined by

$$L_k = (I - klB)^{-1}(I - klC)^{-1}\ ,$$

where l is an increment in t. Determine α, β, γ if

$$\mathbf{U}^{(E)}(t+3l) = \alpha\mathbf{U}^{(0)} + \beta\mathbf{U}^{(1)} + \gamma\mathbf{U}^{(2)} + (1-\alpha-\beta-\gamma)\mathbf{U}^{(3)}$$

is to be a third order approximation to the solution of (A) or (B) at time $t + 3l$, where

$$\mathbf{U}^{(0)} = L_1^3\mathbf{U}(t),\ \mathbf{U}^{(1)} = L_2 L_1\mathbf{U}(t),\ \mathbf{U}^{(2)} = L_1 L_2\mathbf{U}(t),\ \mathbf{U}^{(3)} = L_3\mathbf{U}(t)\ .$$

9

Elliptic equations

9.1 INTRODUCTION

Elliptic partial differential equations are usually associated with steady-state or equilibrium problems in which there is more than one independent variable. In this chapter, problems with two independent space variables x and y will be considered. The domain of integration will be some region R in the (x,y)-plane bounded by a closed curve ∂R. Together with the elliptic equation in R, boundary conditions will be specified everywhere on ∂R. Problems for which the boundary conditions are given as function values of the dependent variable u are known as *Dirichlet* problems; those for which the boundary conditions are given as the value of the outward normal derivative are known as *Neumann* problems; and problems which contain a mixture of the two types of boundary conditions are known as *Robin* problems.

Elliptic equations are one of the three classes of the general partial differential equation

$$a\frac{\partial^2 u}{\partial x^2} + b\frac{\partial^2 u}{\partial x \partial y} + c\frac{\partial^2 u}{\partial y^2} = e \tag{9.1}$$

in which a,b,c,e are functions of $x,y,u,\partial u/\partial x,\partial u/\partial y$ but not of the second-order derivatives; for elliptic equations, $b^2 - 4ac < 0$. The simplest elliptic form of (9.1) has $a = c = 1$ and $b = e = 0$, giving $b^2 - 4ac = -4 < 0$; equation (9.1) is thus

$$\frac{\partial^2 u}{\partial x^2} + \frac{\partial^2 u}{\partial y^2} = 0; \quad (x,y) \in R \tag{9.2}$$

which is known as *Laplace's equation*. Writing $a = c = 1$, $b = 0$ and $e = F(x,y)$ gives *Poisson's equation*

$$\frac{\partial^2 u}{\partial x^2}+\frac{\partial^2 u}{\partial y^2} = F(x,y); \quad (x,y)\in R. \tag{9.3}$$

(The reader is refered to Appendix A for details of the classification of (9.1).)

The simplest region R is the unit square bounded by the lines $x=0$, $y=0$, $x=1$, $y=1$ in the (x,y)-plane: these lines form ∂R on which the value of u is specified. A square mesh is now superimposed on R and the value of u calculated at each point of the mesh. Usually the mesh consists of equally spaced straight lines parallel to the x,y axes. Of course, in cases where ∂R is rectangular or irregular, it may not be possible to have such a convenient grid; examples of such cases will be met later in the chapter.

Returning to the square boundary, suppose that N lines are drawn parallel to each axis. Then $R\cup\partial R$ has been discretized into N^2 mesh points in R together with $4N+4$ mesh points on ∂R. Defining h to be the constant distance between adjacent mesh lines, it follows that $(N+1)h=1$ and that the (x,y) coordinates of the mesh points are $x_k=kh$ and $y_m=mh$ where $k,m = 0,1,2,\ldots,N,N+1$.

The theoretical solution of the ellipic equation at the mesh point (kh,mh) is $u(kh,mh)$, which will be denoted by $u_{k,m}$ while $U_{k,m}$ will be used to denote the theoretical solution of an approximating numerical method.

9.2 THE DIRICHLET PROBLEM ON A SQUARE REGION

In subsequent sections, more difficult examples of the three types of elliptic problem will be encountered (indeed, all three types of boundary condition were met in Chapter 8). In the present section, however, the reader will see how to treat Dirichlet boundary conditions when ∂R is closed.

Consider, then, Poisson's equation (9.3) in the unit square R together with Dirichlet boundary conditions given by

$$u = \begin{cases} f_0(x) \text{ along } y = 0 \text{ with } 0 \leqslant x \leqslant 1, \\ f_1(x) \text{ along } y = 1 \text{ with } 0 \leqslant x \leqslant 1, \\ g_0(y) \text{ along } x = 0 \text{ with } 0 \leqslant y \leqslant 1, \\ g_1(y) \text{ along } x = 1 \text{ with } 0 \leqslant y \leqslant 1. \end{cases} \tag{9.4}$$

Replacing the derivatives in (9.2) by their second-order approximants

$$\frac{\partial^2 u}{\partial x^2} = h^{-2}\{u(x-h,y)-2u(x,y)+u(x+h,y)\}+O(h^2) \tag{9.5}$$

and

$$\frac{\partial^2 u}{\partial y^2} = h^{-2}\{u(x,y-h)-2u(x,y)+u(x,y+h)\}+O(h^2) \tag{9.6}$$

leads to the five-point finite difference replacement for Poisson's equation given by

$$U_{k,m-1} + U_{k-1,m} - 4U_{k,m} + U_{k+1,m} + U_{k,m+1} = h^2 F_{k,m} \tag{9.7}$$

where $F_{k,m} = F(x_k, y_m)$.

The local truncation error $L(x,y)$ of (9.7) is given by

$$L(x,y) = \frac{u(x-h,y) - 2u(x,y) + u(x+h,y)}{h^2} + \frac{u(x,y-h) - 2u(x,y) + u(x,y+h)}{h^2} - F(x,y)$$
$$- \left(\frac{\partial^2 u}{\partial x^2} + \frac{\partial^2 u}{\partial y^2} - F(x,y)\right). \tag{9.8}$$

Expanding about $u(x,t)$ using Taylor's series it may be shown that

$$L(x,y) = \frac{1}{12}h^2\left(\frac{\partial^4 u}{\partial x^4} + \frac{\partial^4 u}{\partial y^4}\right) + \frac{1}{360}h^4\left(\frac{\partial^6 u}{\partial x^6} + \frac{\partial^6 u}{\partial y^6}\right) + \ldots. \tag{9.9}$$

The principal part of $L(x,y)$ is clearly $O(h^2)$ and (9.7) is thus a second-order replacement of (9.2).

The numerical method (9.7) is implemented by applying it to all N^2 interior mesh points in the order

$$(x_1, y_m), (x_2, y_m), \ldots, (x_N, y_m)$$

for each $m = 1, 2, \ldots, N$. This gives a linear system of N^2 equations in the function values $U_{k,m}$ $(k,m = 1, 2, \ldots, N)$. It is convenient to write this system in matrix–vector form as

$$A\mathbf{U} = \mathbf{b} + h^2\mathbf{F} \tag{9.10}$$

where A is an N^2-square matrix which may be written in block tridiagonal form as

$$A = \begin{bmatrix} B & I & & \\ I & B & I & \\ & \cdots & \cdots & \cdots \\ & I & B & I \\ & & I & B \end{bmatrix} \tag{9.11}$$

In (9.11), I is the identity matrix of order N, and B is the tridiagonal matrix

$$B = \begin{bmatrix} -4 & 1 & & & 0 \\ 1 & -4 & 1 & & \\ & \cdots & \cdots & \cdots & \\ & & 1 & -4 & 1 \\ 0 & & & 1 & -4 \end{bmatrix} \tag{9.12}$$

The vector $\mathbf{U}$ is of order N^2 and is given by

$$\mathbf{U} = [U_{1,1}, U_{2,1}, \ldots, U_{N,1}; U_{1,2}, U_{2,2}, \ldots, U_{N,2}; \ldots \\ \ldots; U_{1,N}, U_{2,N}, \ldots, U_{N,N}]^{\mathrm{T}}, \tag{9.13}$$

and the vector $\mathbf{b}$ is also of order N^2 and is the vector of boundary conditions; it is given by

$$\mathbf{b} = [-f_0(x_1) - g_0(y_1), -f_0(x_2), \ldots, -f_0(x_{N-1}), -f_0(x_N) - g_1(y_1); \\ -g_0(y_2), 0, \ldots, 0, -g_1(y_2); \ldots; -g_0(y_{N-1}), 0, \ldots, 0, -g_1(y_{N-1}); \\ -f_1(x_1) - g_0(y_N), -f_1(x_2), \ldots, -f_1(x_{N-1}), -f_1(x_N) - g_1(y_N)]^{\mathrm{T}}. \tag{9.14}$$

It is noted that, in (9.14), there are $4N-4$ nonzero elements corresponding to the number of interior mesh points adjacent to the boundary lines. The vector $\mathbf{F}$ in (9.10) is also of order N^2 and is given by

$$\mathbf{F} = [F_{1,1}, F_{2,1}, \ldots, F_{N,1}; F_{1,2}, F_{2,2}. \ldots, F_{N,2}; \ldots \\ \ldots; F_{1,N}, F_{2,N}, \ldots, F_{N,N}]^{\mathrm{T}}. \tag{9.15}$$

The eigenvalues of the matrix A are given by

$$\lambda_{i,j} = -4\left(\sin^2\frac{i\pi}{2(N+1)} + \sin^2\frac{j\pi}{2(N+1)}\right); \quad i,j = 1,2,\ldots,N \tag{9.16}$$

and so the solution obtained for (9.3) is unique. The solution can be obtained using one of the iterative methods named in Chapter 1. The matrix A is sparse and an iterative method will require less storage, as only the nonzero elements, not more than five in each row, need be stored.

The reader will note that Laplace's equation (9.2) can be solved in the same way, the only difference being that $\mathbf{F} \equiv \mathbf{0}$ in (9.10).

The approach to be followed in solving Neumann and Robin problems is similar to that just described for Dirichlet problems, the added complication being that derivatives in boundary conditions must be replaced by finite difference approximants (see Twizell (1984)).

9.3 ALTERNATING DIRECTION IMPLICIT (ADI) METHODS

9.3.1 Introduction

Alternating Direction Implicit (*ADI*) methods were introduced by Peaceman and Rachford (1955) for the numerical solution of elliptic and parabolic equations (see subsection 8.9.4 of Chapter 8). In the case of elliptic equations the method uses an iterative procedure which if it converges at all does so to the solution of the partial differential equation.

In solving Laplace's equation (9.2), for example, over the mesh superimposed on the unit square described in section 9.1, one of the second-order derivatives, say $\partial^2 u/\partial y^2$, is approximated at the rth iterate ($r=0,1,2,\ldots$) by its finite difference replacement (9.6) and the other second-order derivative by the replacement (9.5) at an intermediate iterate $r+\frac{1}{2}$. In keeping with Peaceman and Rachford (1955), this suggests the iteration scheme

$$2U_{k,m}^{(r+\frac{1}{2})} - 2U_{k,m}^{(r)} = [U_{k-1,m}^{(r+\frac{1}{2})} - 2U_{k,m}^{(r+\frac{1}{2})} + U_{k+1,m}^{(r+\frac{1}{2})} + U_{k,m-1}^{(r)} - 2U_{k,m}^{(r)} + U_{k,m+1}^{(r)}]/h^2 + O(h^2) \tag{9.17}$$

at the point (kh,mh), where $U_{k,m}^{(r)}$ denotes the value of $U_{k,m}$ at the rth iterate.

In proceeding from the intermediate iterate $r+\frac{1}{2}$ to the iterate $r+1$, $\partial^2 u/\partial y^2$ is now replaced by (9.6) at the iterate $r+1$, and $\partial^2 u/\partial x^2$ is again replaced at the intermediate iterate by (9.5). At the mesh point (kh,mh)this suggests the iteration scheme

$$2U_{k,m}^{(r+1)} - 2U_{k,m}^{(r+\frac{1}{2})} = [U_{k-1,m}^{(r+\frac{1}{2})} - 2U_{k,m}^{(r+\frac{1}{2})} + U_{k+1,m}^{(r+\frac{1}{2})} + U_{k,m-1}^{(r+1)} - 2U_{k,m}^{(r+1)} + U_{k,m+1}^{(r+1)}]/h^2 + O(h^2) \tag{9.18}$$

The value of $U_{k,m}^{(r+1)}$ is found in terms of $U_{k,m}^{(r)}$ by eliminating the intermediate iterates from (9.17) and (9.18). The resulting iteration procedure is said to converge if

$$U_{k,m}^{(r+1)} = U_{k,m}^{(r)} = U_{k,m}$$

for some r and if the $U_{k,m}$ $(k,m=1,2,\ldots,N)$ satisfy (9.7) with $F_{k,m}\equiv 0$.

9.3.2 Formulation for the Dirichlet problem

For the Dirichlet problem, the solution $\mathbf{U}$ is computed by solving the system

$$A\mathbf{U} = \mathbf{b} \tag{9.19}$$

where the matrix A is given by (9.11) and the vector $\mathbf{b}$ by (9.14). The matrix B in (9.11) may be written as

$$B = -C - 2I \tag{9.20}$$

where

$$C = \begin{bmatrix} 2 & -1 & & & 0 \\ -1 & 2 & -1 & & \\ & \cdots & \cdots & \cdots & \\ & & -1 & 2 & -1 \\ 0 & & & -1 & 2 \end{bmatrix} \tag{9.21}$$

is of order N and I is the identity matrix of order N.

It is now easy to show that

$$A = -D - E \tag{9.22}$$

where

$$D = \begin{bmatrix} C & & & \\ & C & & \\ & & \cdots & \\ & & & C \end{bmatrix} \tag{9.23}$$

is block diagonal and

$$E = \begin{bmatrix} 2I & -I & & & \\ -I & 2I & -I & & \\ & \cdots & \cdots & \cdots & \\ & & -I & 2I & -I \\ & & & -I & 2I \end{bmatrix} \tag{9.24}$$

is block tridiagonal.

Using this splitting of the matrix A and applying the pair of equations (9.17) and (9.18) to every mesh point in R (the interior of the square $0 < x, y < 1$), the resulting

ADI method for the Dirichlet problem is given by

$$(2h^2I+D)\mathbf{U}^{(r+\frac{1}{2})} = (2h^2I-E)\mathbf{U}^{(r)}-\mathbf{b} \tag{9.25}$$

and

$$(2h^2I+E)\mathbf{U}^{(r+1)} = (2h^2I-D)\mathbf{U}^{(r+\frac{1}{2})}-\mathbf{b} \tag{9.26}$$

where $\mathbf{U}^{(r)}$ denotes $\mathbf{U}$ at the rth iterate.

Eliminating the intermediate vector $\mathbf{U}^{(r+\frac{1}{2})}$ from (9.25) and (9.26) gives an equation of thc form

$$\mathbf{U}^{(r+1)} = F\mathbf{U}^{(r)}+\mathbf{c} \tag{9.27}$$

where F is the matrix of order N^2 given by

$$F = (2h^2I+E)^{-1}(2h^2I+D)^{-1}(2h^2I-D)(2h^2I-E) \tag{9.28}$$

and $\mathbf{c}$ is the vector of order N given by

$$\mathbf{c} = -(2h^2I+E)^{-1}[(2h^2I+D)^{-1}(2h^2I-D)+I]\mathbf{b}. \tag{9.29}$$

It is easy to see that the matrices D and E given by (9.23) and (9.24) commute. They therefore have a common system of orthonormal eigenvectors (Forsythe and Moler, 1967) and their eigenvalues λ_i, μ_j $(i,j=1,2,\ldots,N)$ are given, respectively, by

$$\lambda_i = 4\sin^2\{i\pi/2(N+1)\} \quad \text{and} \quad \mu_j = 4\sin^2\{j\pi/2(N+1)\} \tag{9.30}$$

with $i,j=1,2,\ldots,N$.

The eigenvalues of the matrix F are therefore

$$\frac{(2h^2-\lambda_i)(2h^2-\mu_j)}{(2h^2-\lambda_i)(2h^2+\mu_j)}; \quad i,j=1,2,\ldots,N \tag{9.31}$$

and since each λ_i, μ_j satisfies $0<\lambda_i\mu_j<4$, it is easy to show that the spectral norm of F satisfies

$$\|F\| < 1. \tag{9.32}$$

It was seen in Chapter 1 that an iterative procedure of the form (9.27) is convergent when (9.32) is satisfied and so this ADI formulation for the Dirichlet form of Laplace's equation is convergent to $(D+E)\mathbf{U} = -\mathbf{b}$ which is equivalent to (9.19) since $A = -D-E$.

Details of ADI methods for the Neumann and Robin problems associated with Laplace's equation are given in Twizell (1984).

9.4 IRREGULAR BOUNDARIES: THE OTOLITH MEMBRANE

Situated within the nonauditory part of the inner ear are two sacs called the utricle and saccule which respond to the acceleration due to gravity and to linear acceleration. These sacs are membranous and are approximately ellipsoidal in shape. They are located within a cavity in the vestibule and each contains a sense organ called the macula, a plaque of hair cells covered by a thin layer of gelatine called the *otolith membrane*. Crystals of calcium carbonate called the otoconia adhere to the otolith; these crystals are much denser than the substance of the otolith.

The utricle and saccule have primary rôles in the stability and equilibrium of the body, stabilizing the eyes during movements of the head and initiating compensatory muscle movements to adjust the position of the body. The macula lies in an approximately horizontal plane when the head is in the normal upright position, and disturbance of the head causes gravitational forces to act on the macula. The resulting displacement of the macula causes shearing forces in the hair cells, thus distorting the otolith membrane and displacing the otoconia. Responding movements of the head for stabilizing the eyes and restoring balance can therefore be estimated by computing the displacements of points of the otolith membrane.

In compiling a mathematical model of the otolith membrane it may be assumed that the material of the membrane is gelatinous with Poisson ratio ν in the range $0.3 \leqslant \nu \leqslant 0.8$; usually, $\nu = 0.5$. Young's modulus E for the membrane may be taken to be 200000 dynes/cm^2 and its density may be taken to be 0.903 gm/cm^3. Typically, the maximum length of the membrane is 0.256 cm and its maximum breadth is 0.192 cm.

The boundaries of most, if not all, organs in the human body are not regular and a discussion of curved and/or irregular boundaries is appropriate. By way of illustration, the geometry of the otolith membrane will now be taken to be that discussed in the works by Hudetz (1973), Twizell and Curran (1977) and Twizell (1980), where the shape of the boundary ∂R is as shown in Fig. 9.1.

The portion ACB in Fig. 9.1 is given by the ellipse

$$\frac{x^2}{0.009216} + \frac{y^2}{0.016384} = 1 \tag{9.33}$$

and the portion ADB by the polynomial

$$x = -0.084 - 5.0y + 2.1701y^2 + 54.2535y^3. \tag{9.34}$$

The membrane is discretized by superimposing a square grid of length $h = 0.032$ cm on R producing 30 interior nodes as in Fig. 9.1.

It is known from the small displacement theory of elasticity that the displacements u, v in the x, y directions of the points of the membrane, following a rotation of

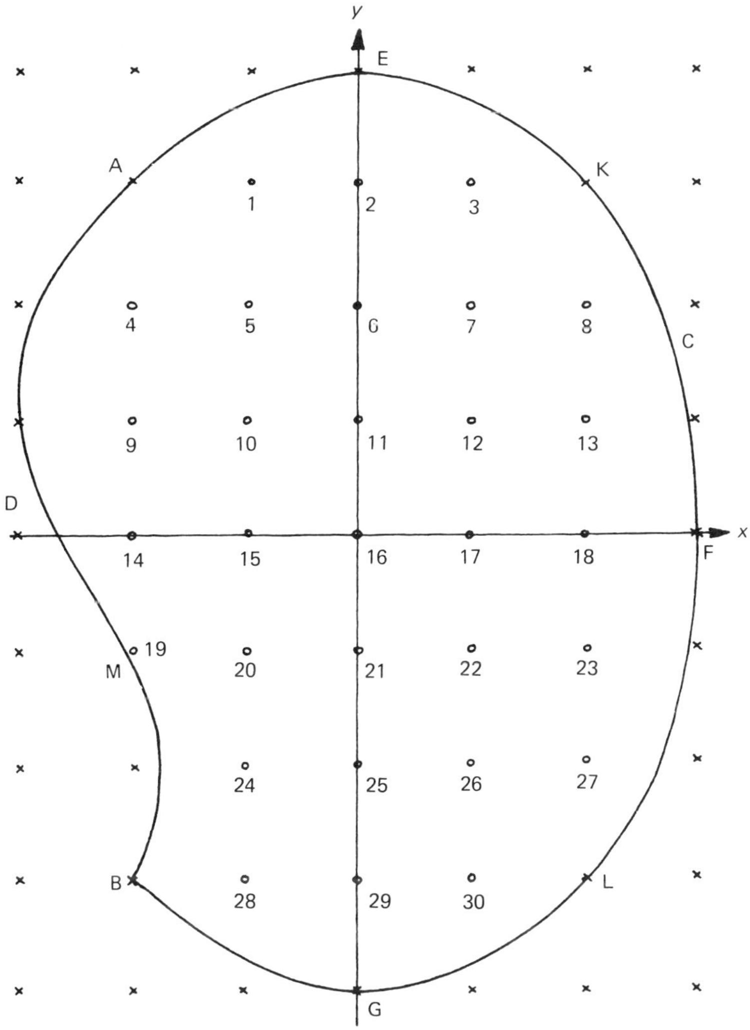

Fig. 9.1 — The model otolith membrane.

the membrane, may be modelled by a system of elliptic partial differential equations of the form

$$\frac{\partial^2 u}{\partial x^2}+\frac{1-\nu}{2}\frac{\partial^2 u}{\partial y^2}+\frac{1+\nu}{2}\frac{\partial^2 v}{\partial x \partial y} = -\frac{1-\nu^2}{E}F^{(x)} \tag{9.35}$$

and

$$\frac{\partial^2 v}{\partial y^2}+\frac{1-\nu}{2}\frac{\partial^2 v}{\partial x^2}+\frac{1+\nu}{2}\frac{\partial^2 u}{\partial x \partial y} = -\frac{1-\nu^2}{E}F^{(y)} \tag{9.36}$$

in R together with the Dirichlet boundary conditions

$$u(x,y) = v(x,y) = 0; \quad (x,y)\in\partial R \tag{9.37}$$

where ∂R is the boundary of the membrane.

The terms $F^{(x)}$ and $F^{(y)}$ in (9.36) and (9.37) are the x and y components of the graviational force and are given by

$$F^{(x)} = \rho g\{\cos\psi(\sin\alpha\sin\gamma + \cos\alpha\sin\beta\cos\gamma) + \sin\psi(\sin\alpha\cos\gamma - \cos\alpha\sin\beta\sin\gamma)\} \tag{9.38}$$

and

$$F^{(y)} = \rho g\{-\sin\psi(\sin\alpha\sin\gamma + \cos\alpha\sin\beta\cos\gamma) + \cos\psi(\sin\alpha\cos\gamma - \cos\alpha\sin\beta\sin\gamma)\} \tag{9.39}$$

where ψ is the azimuth angle and α, β, γ are the angles of rotation of the membrane about the x, y, z axes (the z-axis is perpendicular to the x,y-plane and acts upwards through the centre of the ellipse (9.33)).

The term g in (9.38) and (9.39) is the acceleration due to gravity, and by changing the value of g, the right-hand sides of the elliptic equations and consequently their solutions are changed. In this way graviational conditions in locations other than on the surface of the Earth can be modelled (see Twizell (1980)).

The coordinates of the 30 mesh points in Fig. 9.1 are given by (kh,mh) with $k = -2, -1, 0, 1, 2$ *and* $m = -3, -2, \ldots, 2, 3$. The reader should note that this discretization leads to three points of the mesh on ∂R (the points E, F, G in Fig. 9.1) and five other points very close to the boundary (the points A, K, L, B, M in Fig. 9.1).

As in previous sections, $U_{k,m}$ and $V_{k,m}$ will be used to denote the theoretical solution of a difference scheme at the mesh point (kh,mh) where the theoretical solutions of (9.35) and (9.36) are $u(kh,mh)$ and $v(kh,mh)$, respectively.

The next step is to approximate the derivatives in (9.35) and (9.36) by the finite difference replacements

$$\frac{\partial^2 u}{\partial x^2} = h^{-2}\{u(x-h,y) - 2u(x,y) + u(x+h,y)\} + O(h^2), \tag{9.40}$$

$$\frac{\partial^2 u}{\partial y^2} = h^{-2}\{u(x,y-h) - 2u(x,y) + u(x,y+h)\} + O(h^2), \tag{9.41}$$

$$\frac{\partial^2 u}{\partial x\partial y} = \tfrac{1}{4}h^{-2}\{u(x-h,y-h) - u(x+h,y-h) - u(x-h,y+h) + u(x+h,y+h)\} + O(h^2), \tag{9.42}$$

$$\frac{\partial^2 v}{\partial x^2} = h^{-2}\{v(x-h,y) - 2v(x,y) + v(x+h,y)\} + O(h^2), \tag{9.43}$$

$$\frac{\partial^2 v}{\partial y^2} = h^{-2}\{v(x,y-h) - 2v(x,y) + v(x,y+h)\} + O(h^2), \tag{9.44}$$

$$\frac{\partial^2 v}{\partial x \partial y} = \tfrac{1}{4}h^{-2}\{v(x-h,y-h) - v(x+h,y-h) - v(x-h,y+h) + v(x+h,y+h)\} + O(h^2). \tag{9.45}$$

Substituting (9.40), (9.41) and (9.45) into (9.35), and (9.44), (9.43) and (9.42) into (9.36) gives the difference schemes

$$\begin{aligned}
&\frac{1-\nu}{2h^2}U_{k,m-1} + \frac{1}{h^2}U_{k-1,m} - \frac{2}{h^2}\left(1+\frac{1-\nu}{2}\right)U_{k,m} + \frac{1}{h^2}U_{k+1,m} \\
&+\frac{1-\nu}{2h^2}U_{k,m+1} + \frac{1+\nu}{8h^2}(V_{k-1,m-1} - V_{k+1,m-1} - V_{k-1,m+1} + V_{k+1,m+1}) \\
&= -\frac{1-\nu}{E}F^{(x)}
\end{aligned} \tag{9.46}$$

and

$$\begin{aligned}
&\frac{1+\nu}{8h^2}(U_{k-1,m-1} - U_{k+1,m-1} - U_{k-1,m+1} + U_{k+1,m+1}) \\
&+\frac{1}{h^2}V_{k,m-1} + \frac{1-\nu}{2h^2}V_{k-1,m} - \frac{2}{h^2}\left(1+\frac{1-\nu}{2}\right)V_{k,m} + \frac{1-\nu}{2h^2}V_{k+1,m} \\
&+\frac{1}{h^2}V_{k,m+1} = -\frac{1-\nu^2}{E}F^{(y)}.
\end{aligned} \tag{9.47}$$

Proceeding as in Chapters 7 and 8 and as in section 9.2 of the present chapter, it may be shown that the local truncation error associated with (9.46) has principal part

$$\frac{h^2}{12}\frac{\partial^4 u}{\partial x^4} + \frac{(1-\nu)h^2}{24}\frac{\partial^4 u}{\partial y^4} + \frac{(1+\nu)h^2}{12}\left(\frac{\partial^4 v}{\partial x^3 \partial y} + \frac{\partial^4 v}{\partial x \partial y^3}\right). \tag{9.48}$$

It is then easy to see, by interchanging x with y and u with v, that the local truncation error of (9.47) has principal part

$$\frac{(1+\nu)h^2}{12}\left(\frac{\partial^4 u}{\partial x^3 \partial y} + \frac{\partial^4 u}{\partial x \partial y^3}\right) + \frac{(1-\nu)h^2}{24}\frac{\partial^4 v}{\partial x^4} + \frac{h^2}{12}\frac{\partial^4 v}{\partial y^4}. \tag{9.49}$$

It follows from (9.48) and (9.49) that (9.46) and (9.47) are consistent with (9.35) and (9.36).

To find the solution, equations (9.46) and (9.47) are each applied to the 30 interior nodes in Fig. 9.1, giving a linear system of 60 equations in 60 unknown values of U and V. Applying (9.46) to all 30 grid points followed by (9.47) to all 30 grid points ensures that the system may be written in block form as

$$\begin{bmatrix} P & Q \\ Q & S \end{bmatrix} \begin{bmatrix} \mathbf{U} \\ \mathbf{V} \end{bmatrix} = \begin{bmatrix} \mathbf{F}^{(x)} \\ \mathbf{F}^{(y)} \end{bmatrix} \tag{9.50}$$

where P, Q, S are square matrices of order 30, $\mathbf{U}$ and $\mathbf{V}$ are the vectors of values of U and V at the 30 interior mesh points, and $\mathbf{F}^{(x)}$ and $\mathbf{F}^{(y)}$ are the vectors which have all elements equal to $-(1-\nu^2)F^{(x)}/E$ and $-(1-\nu^2)F^{(y)}/E$ respectively.

When (9.46) and (9.47) are applied to grid points adjacent to ∂R, grid points on or outside ∂R must be used and function values outside R must be estimated in terms of function values at internal nodes. This is easily achieved by linear Lagrangian interpolation which was described in Chapter 2.

All interior mesh points adjacent to ∂R are in one or more of three types of position relative to the exterior mesh points adjacent to ∂R. These three types of position are depicted in Fig 9.2 where 'a' is the interior mesh point, 'b' is the boundary point (where $u = \upsilon = 0$) and 'c' is the exterior mesh point.

Denoting, temporarily, the values of x, y, u, υ at the mesh point a by $x_a, y_a, u_a, \upsilon_a$, with similar notations at the points b and c, linear Lagrangian interpolation in situation (i) of Fig. 9.2 gives

$$u_x = \frac{x - x_c}{x_a - x_c} u_a + \frac{x - x_a}{x_c - x_a} u_c$$

and

$$\upsilon_x = \frac{x - x_c}{x_a - x_c} \upsilon_a + \frac{x - x_a}{x_c - x_a} \upsilon_c .$$

Writing $x = x_b$ and noting that $u_b = \upsilon_b = 0$ leads to

$$u_c = \frac{x_b - x_c}{x_b - x_a} u_a \quad \text{and} \quad \upsilon_c = \frac{x_b - x_c}{x_b - x_a} \upsilon_a , \tag{9.51}$$

where the coordinates x_a, x_c are known multiples of h. The y-coordinate of b in situation (i) of Fig. 9.2 is the same as the y-coordinate of a and c, and x_b can thus be found from (9.33) or (9.34), taking care to attach the appropriate sign after taking the square root in (9.33); u_c and υ_c can then be written in terms of u_a and υ_a using (9.51).

It is easy to see that in situation (ii) of Fig. 9.2 the corresponding values of u_c and υ_c are given by

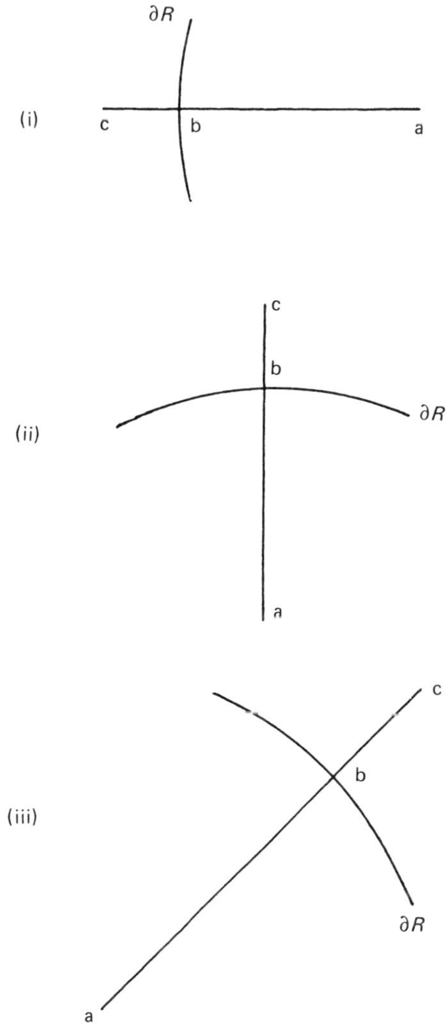

Fig. 9.2 — Three types of position of mesh points adjacent to ∂R.

$$u_c = \frac{y_b - y_c}{y_b - y_a} u_a \quad \text{and} \quad v_c = \frac{y_b - y_c}{y_b - y_a} v_a . \tag{9.53}$$

The y-coordinate of the boundary point in situation (ii) is determined in the same way as the x-coordinate in situation (i) when a and c are on opposite sides of the ellipse (9.33). The mesh point numbered 19 in Fig. 9.1, however, is separated from the exterior point $(-2h, -2h)$ by the cubic polynomial (9.34). The x-coordinate of the corresponding boundary point is $x_b = -2h$; the value of y_b is thus found by solving the equation

$$f(y) \equiv 54.2535y^3 + 2.1701y^2 - 0.5y - 0.084 + 2h = 0. \tag{9.53}$$

Taking $y_b^{(0)} = -2h = -0.064$, the Newton–Raphson method, which converges quadratically, was used in Chapter 3 to estimate y_b which is then used in (9.52) to give u_c and v_c at the point $(-2h, -2h)$ in terms of the values of u and v at point numbered 19 in Fig. 9.1, which has coordinates (2h, – h).

Situation (iii) in Fig. 9.2 is much more difficult to deal with in that neither the x-coordinate nor the y-coordinate of the boundary point b is the same as the corresponding coordinate of a or c. It is known, however, that the boundary point lies at the intersection of a straight line of the form

$$y = x + rh \quad \text{or} \quad y = -x + rh, \tag{9.54}$$

where r is some integer which can easily be determined by inspection of Fig. 9.1, and a curve of one of the forms

$$\begin{aligned} x &= \tfrac{3}{4}(0.016384 - y^2)^{\frac{1}{2}}, \quad x = -\tfrac{3}{4}(0.016384 - y^2)^{\frac{1}{2}}, \\ x &= -0.084 - 0.05y + 2.1701y^2 + 54.2535y^3. \end{aligned} \tag{9.55}$$

The choice of line from (9.54) and curve from (9.55) clearly depends on the positions of a and c.

To determine the value of x_b, the appropriate form of y from (9.54) is substituted into the appropriate equation from (9.55). This results in some one-point function of the form $x = \phi(x)$ which may be solved iteratively to give the value of x_b by writing

$$x_b^{(s+1)} = \phi(x_b^{(s)}); \quad s = 0, 1, 2, \ldots. \tag{9.56}$$

Taking $x_b^{(0)} = \frac{1}{2}(x_a + x_c)$ as starting value, the one-point iteration formula (9.56) will give linear convergence to x_b (see Chapter 3) which is then used in (9.51) to give u_c and v_c.

All information required for the compilation of the linear system (9.50) is now known and this system may be solved using one of the methods described in Chapter 1.

The overall displacement at the point (kh, mh) in the membrane is given by

$$(U_{k,m}^2 + V_{k,m}^2)^{\frac{1}{2}}$$

and is inclined at an angle

$$\tan^{-1}(V_{k,m}/U_{k,m})$$

to the positive x-axis.

9.5 NONRECTANGULAR COORDINATES

Occasionally, it is useful to have a finite difference approximation to Poisson's equation on a circular region (see, for example, section 8.10 of Chapter 8). Poisson's equation is polar coordinates is given by

$$\frac{\partial^2 u}{\partial r^2}+\frac{1}{r}\frac{\partial u}{\partial r}+\frac{1}{r^2}\frac{\partial^2 u}{\partial \theta^2} = F(r,\theta); \quad (r,\theta)\in R \tag{9.57}$$

and the associated Dirichlet boundary conditions are expressed in the form

$$u(a,\theta) = f(\theta); \quad (a,\theta)\in\partial R. \tag{9.58}$$

In equation (9.57), $0 \leqslant r < a$ and in (9.58), $0 \leqslant \theta < 2\pi$; R is the interior of the circle with centre $(0,0)$ and radius a; and ∂R is the boundary of R.

The region R is discretized at $KM+1$ points, including the centre, by drawing K concentric circles within ∂R and M radii in such a way that one radius is coincident with $\theta=0$; the circles (including ∂R) are the same distance δr apart, and adjacent radii make an angle $2\pi/M$ with each other as in Fig. 9.3.

The Dirichlet problem requires the solution at each of the $KM+1$ internal mesh points which have coordinates $(k\delta r, m\delta\theta)$ with $k=0,1,\ldots,K$ and $m=1,2,\ldots,M$. Clearly, $\delta r = a/K$ and $\delta\theta = 2\pi/M$. The notation $U_{k,m}$ will be used to denote the theoretical solution of some numerical method at the mesh point $(k\delta r, m\delta\theta)$ where the solution of Poisson's equation is $u_{k,m} \equiv u(k\delta r, m\delta\theta)$, $k=0$ referring to the centre of the circle and $k=K+1$ to points on ∂R. Similarly, $F_{k,m} = F(k\delta r, m\delta\theta)$.

Poisson's equation (9.57) canot be applied at the origin where $r=0$. The value of U at the origin will, however, be required in the application of (9.57) to the M mesh points $(\delta r, m\delta\theta)$, $m = 1,2,\ldots,M$. An estimate of U at the origin can be obtained by referring back to the Cartesian form of (9.57) given by (9.3). Replacing (9.3) by the five-point formula (9.7) gives

$$U(\delta r,0) + U(\delta r,\pi/2) + U(\delta r,\pi) + U(\delta r,3\pi/2) - 4U(0,0) = F(0,0). \tag{9.59}$$

Rotation of the axes through an angle $\delta\theta$ leads to a similar equation: repetition through $M-2$ further rotations of $\delta\theta$ and addition of the M equations of the form (9.59) leads to

$$U_{0,0} = \sum_{m=1}^{M} U_{1,m}/M - \tfrac{1}{4}F_{0,0}. \tag{9.60}$$

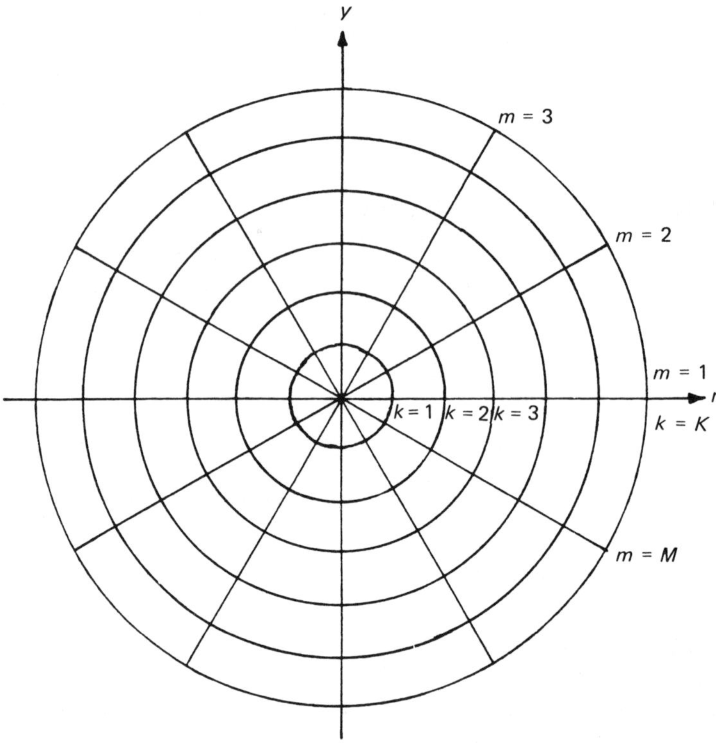

Fig. 9.3 —
Discretization of circular region with K concentric circles and M radial lines.

The three derivatives in (9.57) may now be approximated by the finite difference replacements

$$\frac{\partial^2 u}{\partial r^2} = (u_{k-1,m} - 2u_{k,m} + u_{k+1,m})/(\delta r)^2 + O((\delta r)^2), \tag{9.61}$$

$$\frac{\partial u}{\partial r} = (u_{k-1,m} - u_{k-1,m})/(2\delta r) + O((\delta r)^2), \tag{9.62}$$

$$\frac{\partial^2 u}{\partial \theta^2} = (u_{k,m-1} - 2u_{k,m} + u_{k,m+1})/(\delta\theta)^2 + \mathrm{O}((\delta\theta)^2), \tag{9.63}$$

Substitution of (9.61), (9.62), (9.63) into (9.57) gives, when applied to the mesh point $(k\delta r, m\delta\theta)$, a difference equation which simplifies to

$$\frac{M^2}{4\pi^2 k^2} U_{k,m-1} + \left(1 - \frac{1}{2k}\right) U_{k-1,m} - 2\left(1 + \frac{M^2}{4\pi^2 k^2}\right) U_{k,m}$$

$$+\left(1+\frac{1}{2k}\right)U_{k+1,m}+\frac{M^2}{4\pi^2k^2}U_{k,m+1} \;=\; \frac{a^2}{k^2}F_{k,m}. \tag{9.64}$$

The local truncation error of (9.64) is $O((\delta r)^2+(\delta\theta)^2)$.

Equation (9.64) is applied to the KM mesh points $(k\delta r, m\delta\theta)$, $k=1,2,\ldots,K$ for each $m=1,2,\ldots,M$. This leads to a linear system of the form

$$A\mathbf{U} \;=\; \mathbf{b}+\frac{a^2}{K^2}\mathbf{F}, \tag{9.65}$$

where A, $\mathbf{U}$, $\mathbf{b}$ and $\mathbf{F}$ depend on the order in which (9.64) is applied to the mesh points. A full listing of the elements of each is given in Twizell (1984) for the case when the points are ordered along consecutive radii.

9.6 FINITE ELEMENTS

9.6.1 Introduction

The finite element method evolved during attempts by engineers and applied mathematicians to solve some equations of structural engineering and elasticity. In the early stages of the development of finite element methods, engineers were concerned with the implementation of algorithms to get good results, while mathematicians were more concerned with making the algorithms respectable by thorough analyses. Finite element methods have subsequently enjoyed enormous success in civil and mechanical engineering, where it is often necessary to solve systems of elliptic partial differential equations, and there is no doubt that this success has been fuelled by close cooperation between engineers and methematicians.

The growth in the use of finite element methods has occurred simultaneously with the phenomenal growth in the use of high-speed digital computers. This in turn has been accompanied by greatly increased efforts by *practical* numerical analysts to develop, publish and market computer software for would-be users of finite element methods. It is not surprising, then, that the research literature relating to finite element methods has grown at an unprecedented rate in the past two decades. The number of textbooks on the market has also grown, and research workers who need, or wish, to use the finite element method can confidently expect to find a text suited to their needs (for instance, Strang and Fix (1973), Davies (1980) and Wait and Mitchell (1986)).

Prerequisites for the use of finite element methods include no more than a knowledge of vector spaces and advanced calculus, and the ability to write computer programs in a high-level scientific language. It is not intended to give here a rigorous analysis of any of the finite element methods in common use, rather to show how such methods can be used.

There has been much debate on the relative merits and shortcomings of finite element and finite difference methods: it is reassuring to know that two powerful methods are available from which to choose. It was seen in section 9.4 that great care must be taken when using finite difference methods to solve elliptic equations over regions with irregular boundaries. It is perhaps in such problems, where flexibility of geometry is important, that finite element methods have an advantage over finite difference techniques.

In using finite element methods to solve elliptic equations, the usual approach is to express the problem in the equivalent variational form, then to subdivide the region of integration into subregions (the elements). With two space dimensions these are usually triangles or quadrilaterals with straight or curved sides. The finite element approach has the advantage that it is usually possible to discretize the region in such a way that a very close approximation to the shape of the boundary is retained. It was seen in section 9.4 that this is possible with finite difference methods, too, provided interpolation is used.

The next step in the finite element solution is to choose an approximation to each dependent variable in the elliptic system, then to find the combination of values of the approximating functions within the elements which minimizes the equivalent variational problem. The approximating functions are often polynomials of low degree and the combination linear, though the text by Wachspress (1975) discusses in detail the use of rational approximating functions. These were also used by Smith and Twizell (1980) in a study of steady-state heat distribution in the human torso. Choosing the linear combination leads to a system of equations, the unknowns of which are the values of the approximating functions at the vertices of the elements.

9.6.2 Variational formulations

The linear equation $Tu=f$, where T is a linear operator, is related to the quadractic functional (see Appendix A)

$$I(U) = (TU,U)-2(f,U) \tag{9.66}$$

in that $I(U)$ is minimized at $U=u$ only if its derivative vanishes there. The necessary condition for the vanishing of this derivative is known as the *Euler–Lagrange equation* which is of course $Tu=f$. The problem of inverting T is therefore equivalent to minimizing I in that they have the same solution u.

Such problems can thus be solved in the *operational form* $Tu=f$ or in the *variational form* of minimizing $I(U)$. The operational form is used in a finite difference solution of an elliptic equation and the variational formulation in a finite element solution.

It may be helpful to the first-time reader to consider the easy example in which T and f are real constants, with $T>0$, and U is a real variable. The functional $I(U)$ becomes

$$I(U) = TU^2-2fU_i$$

and it follows immediately that the necessary condition ($\mathrm{d}I/\mathrm{d}U=0$) for $I(U)$ to have a minimum value at $U=u$ is $Tu=f$.

A more difficult example is that in which $\mathbf{U} = [U_1, U_2, \ldots, U_N]^{\mathrm{T}}$, $\mathbf{f}$ =

$[f_1, f_2, \ldots, f_N]^{\mathrm{T}}$ and $\mathbf{u} = [u_1, u_2, \ldots, u_N]^{\mathrm{T}}$ are vectors and A is a symmetric, positive definite matrix of order N. The quantity $(A\mathbf{U}, \mathbf{U})$ is then the inner product of the vectors $A\mathbf{U}$ and $\mathbf{U}$, and $I(\mathbf{U})$ becomes

$$I(\mathbf{U}) = \sum_{i,j} A_{i,j} U_j U_i - 2 \sum f_i U_i .$$

Noting that $A_{j,i} = A_{i,j}$ since A is symmetric, the necessary conditions for $I(\mathbf{U})$ to have a minimum value are

$$\frac{\partial I}{\partial U_i} = 2 \sum_i A_{i,j} U_j - 2f_i = 0; \quad i = 1, 2, \ldots, N.$$

These necessary conditions form a linear system of N equations, the unknowns of which are the elements of $\mathbf{U}$. If $\mathbf{U} = \mathbf{u}$ is the solution of this system, $\mathbf{u}$ satisfies the equation

$$A\mathbf{u} = \mathbf{f}$$

which is the Euler–Lagrange equation for the problem.

9.6.3 Finite element methods, basis functions, and different elements

There are four commonly used methods of finite element solution. The first is the *Ritz method* in which the functional

$$I(U) = \int_0^1 (A(tU), U)\,\mathrm{d}t - (f, U) \tag{9.67}$$

is considered, where A is some differential operator. $I(U)$, defined by (9.67), has a minimum value at $U = u$ when the Euler–Lagrange equation

$$Au = f \tag{9.68}$$

is satisfied. In the case when A is a linear operator, $I(U)$ becomes

$$I(U) = \tfrac{1}{2}(AU, U) - (f, U). \tag{9.69}$$

The Ritz method entails minimizing $I(U)$ using the approximation

$$U = \sum_{i=1}^{m} \alpha_i \phi_i(x,y) \tag{9.70}$$

where the ϕ_i are approximations to u. Clearly in N-dimensional space, $\phi = \phi(x_1,x_2,\ldots,x_N)$. The minimization of $I(U)$ involves differentiating (9.70) with respect to each of the parameters α_i $(i = 1,2,\ldots,m)$ and solving the system

$$\frac{\partial}{\partial \alpha_i}[I(U)] \equiv (AU,\phi_i) - (f,\phi_i) = 0.$$

This approach may be used when A is nonlinear.

The *Galerkin method* does not use the *classical solution* which satisfies (9.68). It uses instead the *weak solution* which satisfies one of the equations

$$(Au,U) = (F,U) \quad \text{for all } U \in H_1 \tag{9.71}$$

or

$$(u,A^*U) = (f,U) \quad \text{for all } U \in H_2 \tag{9.72}$$

where A^* is the adjoint of A (see section 1 of Appendix A).

In (9.71) the space H_1 contains all functions which are measurable (Appendix A) and which vanish on ∂R; such functions are known as *admissible functions*. In (9.72) the space H_2 contains only those admissible functions for which A^*U is measurable; clearly $H_2 \subset H_1$. It is clear that if the operator A is of order $2s$ then (9.71) requires u to have measurable derivatives of order $2s$.

The Galerkin solution is derived from (9.71) or (9.72) by integrating by parts a total of s times. It is written

$$a(u,U) = (f,U) \quad \text{for all } U \subset H_3$$

where H_3 satisfies $H_2 \subset H_3 \subset H_1$, and $a(u,U)$ is the *energy inner product*; the equivalent norm is the *energy norm*.

For the Laplace operator, $a(u,U)$ takes the form

$$a(u,U) = \iint_R \left[\left(\frac{\partial u}{\partial x}\right)\left(\frac{\partial U}{\partial x}\right) + \left(\frac{\partial u}{\partial y}\right)\left(\frac{\partial U}{\partial y}\right)\right] \mathrm{d}x\,\mathrm{d}y.$$

The third method is the *Method of Collocation* which bears a strong resemblance to the Galerkin method. The coefficients α_i $(i = 1,2,\ldots,m)$ in (9.70) are chosen in such a way that the differential equation is satisfied exactly at certain points, known as *collocation points*. The main disadvantage of the method of collocation is that the approximating functions ϕ_i $(i = 1,2,\ldots,m)$ must be of degree $2s$ for a differential equation of degree $2s$. On the other hand, there are no inner products to integrate as in the Ritz and Galerkin methods and consequently the matrix of the resulting linear system is more sparse.

The *Method of Least Squares* is appropriate for use when the operator A is non-self-adjoint. The least-squares solution is that which minimizes

$$(AU-f, AU-f) + K\langle U-g, U-g\rangle$$

where g is the inhomogeneous boundary condition on ∂R. This is equivalent to minimizing the functional

$$I(U) = \iint_R (AU-f)^2 \mathrm{d}x\,\mathrm{d}y + K\int_{\partial R} (U-g)^2 \mathrm{d}s$$

where K is some constant. One important property of the method is that the approximating functions U do not have to satisfy the boundary conditions. However, higher continuity requirements are required for the least-squares method: in two space dimensions, for example, $U \in C^{1,1}$ for the least-squares method but $U \in C^{0,0}$ for the Ritz method (the notation $U \in C^{1,1}$ means U and its first-order derivatives are continuous in R).

Obviously, an important step in the numerical solution of partial differential equations using any of these four finite element methods is the choice of the approximate function, which are also known as *basis functions*. The choice of basis function is influenced by the shape of the finite elements into which R has been subdivided: in two space dimensions these are usually triangles, quadrilaterals or rectangles. Away from the boundary ∂R the elements are usually congruent, so facilitating the computation of the solution, while adjacent to the boundary the elements may be irregular.

Within each element, whatever its shape, the basis function depends on the value of the function and, in certain cases, on the derivative of the function, at certain *nodes* of the element. These nodes may be the vertices of the element, points on its sides, or its centroid. In the case of elements adjacent to the boundary, the basis functions may have to satisfy the boundary conditions. In addition, the basis functions must have continuous derivatives of all orders up to and including $s-1$ along the sides of the elements if the function $I(U)$ to be minimized over R contains derivatives of order s. This is known as the *conforming condition* or *condition of compatibility* and, if it is met, integration over R may be carried out over the elements individually since the sth derivatives in $I(U)$ give rise only to step functions across the sides of adjacent elements.

Basis functions for use with triangles, quadrilaterals, parallelograms, rectangles, trapezia, and curved elements are contained in detail in the paper by Mitchell and Phillips (1972).

9.6.4 Formulations for the three types of elliptic problem

The Dirichlet, Neumann and Robin-type boundary conditions associated with elliptic equations were outlined in section 9.1 and treated further, in the Dirichlet case, in section 9.2 for the Poisson problem (9.3) where it is assumed that $u(x,y)\in C^{2,2}(R)$ and that $F(x,y)$ is continuous in R. Equation (9.3) is of the form $Tu=F$ with $T=\partial^2/\partial x^2+\partial^2/\partial y^2$.

Consider the minimization of the functional

$$I(U) = \iint_R \{U_x^2+U_y^2+2F(x,y)U\}\,\mathrm{d}x\,\mathrm{d}y \tag{9.73}$$

with Dirichlet boundary condition $u=0$ everywhere on ∂R. In (9.73), the notations $U_x\equiv\partial U/\partial x$, $U_y=\partial U/\partial_y$ are introduced; similarly $U_{xx}\equiv\partial^2U/\partial x^2$, $U_{xy}\equiv\partial^2Uy\partial x\partial y$ and $U_{yy}\equiv\partial^2U/\partial y^2$ with similar notations for the derivatives of other functions.

Suppose $U=u$ minimizes (9.73); then the necessary condition for this minimum is

$$\delta I = \frac{\mathrm{d}}{\mathrm{d}\alpha}[I(u+\alpha\eta)]|_{\alpha=0} = 0.$$

where $\eta=\eta(x,y)\in C^{2,2}(R)$ and $\eta=0$ on ∂R. Therefore

$$\delta I = \iint_R \{2u_x\eta_x+2u_y\eta_y+2F\eta\}\,\mathrm{d}x\,\mathrm{d}y = 0.$$

Using Green's Theorem,

$$\begin{aligned}\iint_R (u_x\eta_x+u_y\eta_y)\,\mathrm{d}x\,\mathrm{d}y &= \iint_R\left\{\frac{\partial}{\partial x}(\eta u_x)+\frac{\partial}{\partial y}(\eta u_y)\right\}\\ &\quad -\iint_R (u_{xx}+u_{yy})\eta\,\mathrm{d}x\,\mathrm{d}y\\ &= \int_{\partial R}(u_x\eta\,\mathrm{d}y-u_y\eta\,\mathrm{d}x)-\iint_R(u_{xx}+u_{yy})\eta\,\mathrm{d}x\,\mathrm{d}y,\\ &= -\iint_R(u_{xx}+u_{yy})\eta\,\mathrm{d}x\,\mathrm{d}y\end{aligned}$$

since $\eta=0$ on ∂R. Therefore,

$$\delta I = -2\iint_R \{u_{xx}+u_{yy}-F(x,y)\}\eta\,\mathrm{d}x\,\mathrm{d}y$$

for all functions $\eta(x,y)$, and clearly $\delta I=0$ when

$$\frac{\partial^2 u}{\partial x^2}+\frac{\partial^2 u}{\partial y^2} = F(x,y)$$

which is Poisson's equation (9.3). Hence the solution of (9.73) is the solution of Poisson's equation with Dirichlet boundary conditions. By showing that $U=u$ gives the global minimum of $I(U)$ in (9.73) it may be shown that Poisson's equation (9.3) is also sufficient (see Twizell (1984)).

Equation (9.73) is an example of the functional

$$I(U) = \iint_R \phi(x,y,u,u_x,u_y)\,\mathrm{d}x\,\mathrm{d}y \tag{9.74}$$

where $u(x,y)$ takes prescribed values on ∂R. The necessary condition for $I(U)$ given by (9.74) to have an extremum at $U=u$ is that u must satisfy the Euler Lagrange equation

$$\frac{\partial}{\partial x}\phi_{u_x}+\frac{\partial}{\partial y}\phi_{u_y}-\phi_u = 0. \tag{9.75}$$

This condition is not usually sufficient for the functional to have a maximum or minimum value: it is a necessary condition.

Occasionally, the differential equation

$$\frac{\partial^2 u}{\partial x^2}+\frac{\partial^2 u}{\partial y^2}+q(x,y)u = F(x,y)$$

in R arises with natural boundary condition $\partial u/\partial n=0$ (the Neumann problem), where $\partial u/\partial n$ is the outward normal derivative. The corresponding functional $I(U)$ is given by

$$I(U) = \iint_R (U_x^2+U_y^2-qU^2+2FU)\,\mathrm{d}x\,\mathrm{d}y.$$

Finally, consider Poisson's equation (9.3) together with the mixed boundary condition $\partial u/\partial n+q(x,y)u=g(x,y)$ on ∂R (the Robin problem). The corresponding functional $I(U)$ is given by

$$I(U) = \iint_R (U_x^2 + U_y^2 - 2FU)\,dx\,dy + \int_{\partial R} (qU^2 - 2gU)\,ds,$$

where s is an element of ∂R.

9.6.5 A worked example

The aim of this subsection is to present to the reader the details of a finite element solution of an elliptic problem. To this end, consider Laplace's equation (9.2) where R is some general region, together with boundary conditions

$$u = f(x,y) \tag{9.76}$$

on the boundary ∂R of R.

It was seen in subsection 9.6.4 that the solution of (9.2) is equivalent to minimizing the functional

$$I(U) = \iint_R (U_x^2 + U_y^2)\,dx\,dy \tag{9.77}$$

over all functions $U(x,y)$ satisfying (9.76).

The next step is to subdivide the region R into finite elements. Choosing triangles results in the approximation of a general region R by a polygonal region. A general region is partially triangulated in Fig. 9.4.

On each triangle T_i of the polygonal region, the solution of (9.2) is approximated by

$$U(x,y) = p + qx + ry; \tag{9.78}$$

this gives $C^{0,0}$ continuity over the polygonal region. The terms p, q, r are constants and are unique to T_i.

Suppose triangle T_i has vertices (x_j, y_j), $j = 1,2,3$; its area A_i is then given by

$$\begin{aligned} A_i &= \tfrac{1}{2}\begin{vmatrix} 1 & x_1 & y_1 \\ 1 & x_2 & y_2 \\ 1 & x_3 & y_3 \end{vmatrix} \\ &= \tfrac{1}{2}[(x_2 - x_1)(y_3 - y_1) - (x_3 - x_1)(y_2 - y_1)]. \end{aligned} \tag{9.79}$$

At each nodal vertex the function U is given by

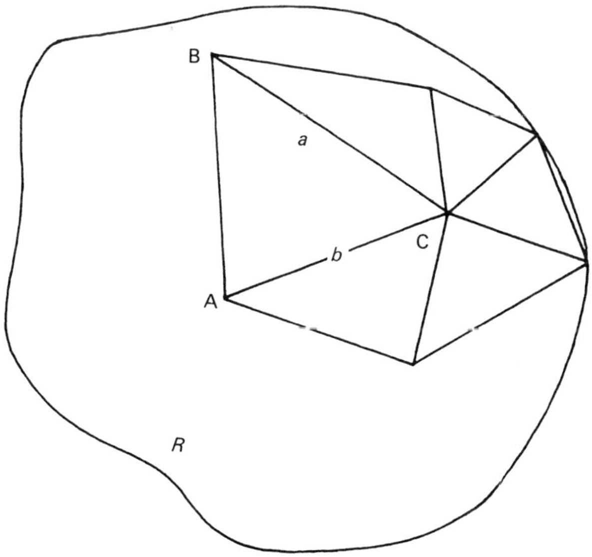

Fig. 9.4 — Partial triangulation of a general region R

$$U_j = p + qx_j + ry_j \quad (j = 1, 2, 3). \tag{9.80}$$

This leads to a system of equations written in matrix form as

$$\begin{bmatrix} U_1 \\ U_2 \\ U_3 \end{bmatrix} = \begin{bmatrix} 1 & x_1 & y_1 \\ 1 & x_2 & y_2 \\ 1 & x_3 & y_3 \end{bmatrix} \begin{bmatrix} p \\ q \\ r \end{bmatrix}. \tag{9.81}$$

Equation (9.81) is easily inverted to give

$$\begin{bmatrix} p \\ q \\ r \end{bmatrix} = \frac{1}{2A_i} \begin{bmatrix} x_2y_3 - x_3y_2 & x_3y_1 - x_1y_3 & x_1y_2 - x_2y_1 \\ y_2 - y_3 & y_3 - y_1 & y_1 - y_2 \\ x_3 - x_2 & x_1 - x_3 & x_2 - x_1 \end{bmatrix} \begin{bmatrix} U_1 \\ U_2 \\ U_3 \end{bmatrix}. \tag{9.82}$$

Consider now equation (9.77) with U given by (9.78) so that $U_x = q$ and $U_y = r$. Then

$$I(U) = \iint_R (q^2 + r^2)\,\mathrm{d}x\,\mathrm{d}y = (q^2 + r^2)\iint_R \mathrm{d}x\,\mathrm{d}y = (q^2 + r^2)A_i$$

so that

$$\frac{\partial}{\partial U_j}I(U) = 2A_i\left[q\frac{\partial q}{\partial U_j}+r\frac{\partial r}{\partial U_j}\right]$$
$$= \psi_j, \tag{9.83}$$

say, for $j=1,2,3$. The necessary conditions for $I(U)$ to have a minimum value are that

$$\psi_j = 0; \quad j=1,2,3 \tag{9.84}$$

simultaneously.

It is easy to show, using q and r from (9.82), that

$$2\left[q\frac{\partial q}{\partial U_1}+r\frac{\partial r}{\partial U_1}\right] = [\{(x_3-x_2)^2+(y_2-y_3)^2\}U_1 + \{(x_3-x_2)(x_1-x_3)+(y_2-y_3)(y_3-y_1)\}U_2 + \{(x_3-x_2)(x_2-x_1)+(y_2-y_3)(y_1-y_2)\}U_3]/(2A_i^2). \tag{9.85}$$

In any triangle with sides a, b, c opposite the vertices A, B, C, respectively (see Fig. 9.4), it is well known that

$$\mathbf{a} = (x_2-x_3)\mathbf{i}+(y_2-y_3)\mathbf{j},$$
$$\mathbf{b} = (x_3-x_1)\mathbf{i}+(y_3-y_1)\mathbf{j},$$
$$\mathbf{c} = (x_1-x_2)\mathbf{i}+(y_1-y_2)\mathbf{j},$$

where **i**, **j** are unit vectors in the positive x, y directions, respectively, and that

$$\mathbf{a}.\mathbf{a} = a^2 = (x_2-x_3)^2+(y_2-y_3)^2,$$
$$\mathbf{a}.\mathbf{b} = -ab\cos\mathrm{C} = (x_2-x_3)(x_3-x_1)+(y_2-y_3)(y_3-y_1),$$
$$\mathbf{a}.\mathbf{c} = -ac\cos\mathrm{B} = (x_2-x_3)(x_1-x_2)+(y_2-y_3)(y_1-y_2).$$

It follows, then, that

$$q\frac{\partial q}{\partial U_1}+r\frac{\partial r}{\partial U_1} = [a^2U_1-(ab\cos\mathrm{C})\,U_2-(ac\cos\mathrm{B})\,U_3]/(4A_i^2) \tag{9.86}$$

and similarly that

$$q\frac{\partial q}{\partial U_2}+r\frac{\partial r}{\partial U_2} = [-(ba\cos\mathrm{C})\,U_1+b^2U_2-(bc\cos\mathrm{A})\,U_3]/(4A_i^2) \tag{9.87}$$

$$q\frac{\partial q}{\partial U_3}+r\frac{\partial r}{\partial U_3} = [-(ca\cos B)\,U_1-(bc\cos A)\,U_2+c^2U_3]/(4A_i^2) \qquad (9.88)$$

It is also known that for the triangle ABC of Fig. 9.4,

$$A_i = \tfrac{1}{2}bc\sin A = \tfrac{1}{2}ca\sin B = \tfrac{1}{2}ab\sin C. \qquad (9.89)$$

The equations in (9.89) give values of ab, bc, ac for use in (9.86), (9.87), (9.88). Using (9.89), a^2, b^2, c^2 can also be found in terms of the area A_i and the angles A, B, C as follows:

$$\begin{aligned} a^2 &= \frac{abac}{bc} = \frac{2A_i}{\sin C}\cdot\frac{2A_i}{\sin B}\cdot\frac{\sin A}{2A_i} \\ &= 2A_i\frac{\sin A}{\sin B\sin C} = 2A_i\frac{\sin(B+C)}{\sin B\sin C} \\ &= 2A_i\frac{\sin B\cos C+\cos B\sin C}{\sin B\sin C} \end{aligned}$$

so that

$$a^2 = 2A_i(\cot B+\cot C). \qquad (9.90)$$

Similarly, $b^2 = 2A_i(\cot C+\cot A)$ and $c^2 = 2A_i(\cot A+\cot B)$ and substituting for a^2, b^2, c^2, ab, bc, ca in (9.86), 9.87) and (9.88) gives

$$\begin{aligned} q\frac{\partial q}{\partial U_1}+r\frac{\partial r}{\partial U_1} &= [(\cot B+\cot C)\,U_1-(\cot C)\,U_2-(\cot B)\,U_3]/(2A_i), \\ q\frac{\partial q}{\partial U_2}+r\frac{\partial r}{\partial U_2} &= [-(\cot C)\,U_1+(\cot C+\cot A)\,U_2-(\cot A)\,U_3]/(2A_i), \\ q\frac{\partial q}{\partial U_3}+r\frac{\partial r}{\partial U_3} &= [-(\cot B)\,U_1-(\cot A)\,U_2+(\cot A+\cot B)\,U_3]/(2A_i), \end{aligned}$$

so that

$$\begin{bmatrix} \psi_1 \\ \psi_2 \\ \psi_3 \end{bmatrix} = \begin{bmatrix} \cot B + \cot C & -\cot C & -\cot B \\ -\cot C & \cot C + \cot A & -\cot A \\ -\cot B & -\cot A & \cot A + \cot B \end{bmatrix} \begin{bmatrix} U_1 \\ U_2 \\ U_3 \end{bmatrix}. \tag{9.91}$$

Clearly, the terms ψ_1, ψ_2, ψ_3 are dependent on the geometry of the triangle T_i. It was noted earlier that $I(U)$ is minimized when each $\psi_j = 0$ simultaneously $(j = 1, 2, 3)$, and the values of U_1, U_2, U_3 which satisfy (9.84) are easily determined for triangle T_i from (9.91).

As an example, suppose the region R is the unit square bounded by the lines $x = 0, x = 1, y = 0, y = 1$ as in Fig. 9.5(a), where each triangle is one of two isosceles forms (b) or (c) in which $A = 90°$ and $B = C = 45°$. For this triangulation, (9.91) becomes;

$$\begin{bmatrix} \psi_1 \\ \psi_2 \\ \psi_3 \end{bmatrix} = \begin{bmatrix} 2 & -1 & -1 & U_1 \\ -1 & 1 & 0 & U_2 \\ -1 & 0 & 0 & U_3 \end{bmatrix}. \tag{9.92}$$

Each interior node of Fig. 9.5(a) has six common triangles; for the node with coordinates (kh, mh), $h = \frac{1}{4}$, the six triangles are shown in Fig. 9.5(d).

Applying $\psi_1 = 0$ to triangles 1 and 4 in Fig. 9.5(d), $\psi_2 = 0$ to triangles 2 and 5, and $\psi_3 = 0$ to triangles 3 and 6, and adding, gives

$$\begin{aligned} &(2U_{k,m} - U_{k+1,m} - U_{k,m-1}) + (-U_{k+1,m} + U_{k,m}) \\ &\quad + (-U_{k,m+1} + U_{k,m}) + (2U_{k,m} - U_{k,m+1} - U_{k-1,m}) \\ &\quad + (-U_{k-1,m} + U_{k,m}) + (-U_{k,m-1} + U_{k,m}) = 0. \end{aligned}$$

Gathering terms and dividing by -2 gives

$$U_{k,m-1} + U_{k-1,m} - 4U_{k,m} + U_{k+1,m} + U_{k,m+1} = 0 \tag{9.93}$$

which is the five-point difference scheme (with the right-hand side put equal to zero) derived in section 9.2. The finite difference and finite element approaches have produced the same formula for the solution of Laplace's equation in a square region with Dirichlet boundary conditions.

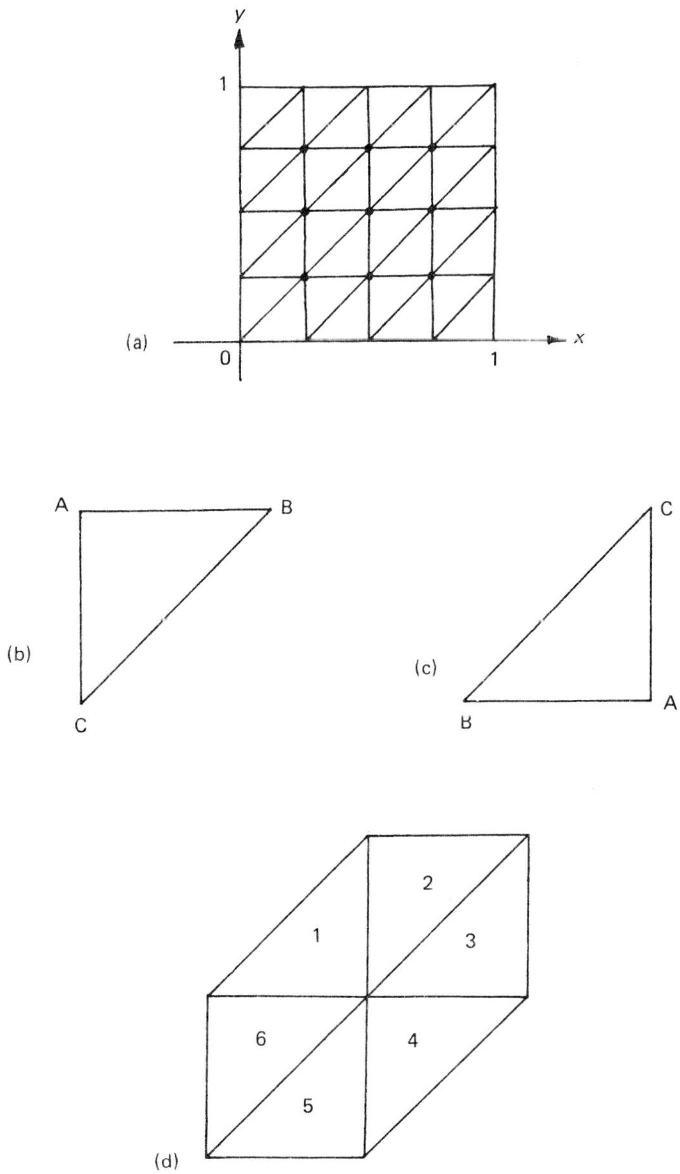

Fig. 9.5 — Triangulation of unit square for model problem: (a) triangulation of the square, (b) Type I isosceles triangle, (c) Type II isosceles triangle, (d) the six common triangles of any interior node of the square.

9.6.6 Finite elements and the otolith membrane

The displacements u and v of the membrane in the x and y directions, respectively, are modelled by the system of elliptic equations given by (9.35) and (9.36) with

$u = v = 0$ everywhere on the boundary ∂R of the region R occupied by the membrane. The functional $I(U)$ for this problem is given by

$$I(U) = \iint_R W\,\mathrm{d}x\mathrm{d}y,$$

where

$$W = \tfrac{1}{2}(1-\nu)(U_x^2 + V_y^2) + \tfrac{1}{2}\nu(U_x + V_y)^2 + \tfrac{1}{4}(1-\nu)(U_y + V_x)^2 - \frac{(1-\nu^2)}{E}F^{(x)}U - \frac{(1-\nu^2)}{E}F^{(y)}V, \tag{9.94}$$

for the system of Euler–Lagrange equations

$$\frac{\partial W}{\partial U} - \frac{\partial}{\partial x}\left(\frac{\partial W}{\partial U_x}\right) - \frac{\partial}{\partial y}\left(\frac{\partial W}{\partial U_y}\right) = 0, \tag{9.95}$$

$$\frac{\partial W}{\partial V} - \frac{\partial}{\partial x}\left(\frac{\partial W}{\partial V_x}\right) - \frac{\partial}{\partial y}\left(\frac{\partial W}{\partial V_y}\right) = 0 \tag{9.96}$$

reduces to the system of elliptic equations (9.35), (9.36).

A detailed description of a triangulation of R, the choice of basis function (linear polynomials in x and y), the minimization of the integral, and of a series of numerical experiments are given in the paper by Twizell and Curran (1977).

9.6.7 Finite elements and heat distribution in the human torso

A brief description of the human thermoregulatory system, which is very difficult to model mathematically, was given in section 8.10 of Chapter 8. The more sophisticated mathematical models of heat flow in the human torso consider the torso to consist of four concentric layers of different types of tissue, namely core, muscle, fat and skin (see Kuznetz (1979)), the first two being described as the core and the other two the insulating layer.

The physical structure of the torso lends itself well to the spatial discretization necessary for the finite element method, in that an arbitrarily close approximation to the boundary (the skin surface) may be obtained. Smith and Twizell (1980) used elements with curved sides and rational basis functions, but triangle elements with linear basis functions give very acceptable results also (Twizell and Smith, 1982).

The steady-state temperature $u = u(x,y)$ at points in tissue type s ($s = 1,2,3,4$ refer to the core, muscle, fat, skin, respectively) is modelled by the elliptic equation

$$\kappa_s\left(\frac{\partial^2 u}{\partial x^2} + \frac{\partial^2 u}{\partial y^2}\right) + q_s + (\dot{m}C)_s(v - u) = 0 \tag{9.97}$$

in which κ represents the thermal conductivity of the tissue, q represents heat generation rate, $\dot{m}$ represents blood flow rate, C represents the specific heat of the blood and v represents its temperature.

The elliptic equation (9.97) is subject to a number of boundary conditions. The first is at the skin surface and is given by

$$\kappa_4 \frac{\partial u}{\partial n} + H(u - E) = 0 \tag{9.98}$$

where n is the direction of the outward normal to the skin surface, H is the effective heat transfer coefficient of convection, evaporation and radiation at the skin surface, and E is the environmental temperature adjacent to the skin (see Wissler (1961)). The remaining boundary conditions associated with (9.97) occur at the core/muscle, muscle/fat and fat/skin interfaces and have the forms

$$\begin{aligned} \kappa_s \frac{\partial u_s}{\partial n} &= \kappa_{s+1} \frac{\partial u_{s+1}}{\partial n} \\ u_s &= u_{s+1} \end{aligned} \tag{9.99}$$

for $s - 1, 2, 3$. The equations in (9.99) are flux equality equations and express the discontinuity in conductivity and the continuity in temperature at the interfaces.

The solution of (9.97), (9.98), (9.99) may be found by minimizing the integral

$$I(U,V) = \sum_{s=1}^{4} \iint_{R_s} W_s \mathrm{d}x\mathrm{d}y + H \int_S (\tfrac{1}{2}U^2 - EU)\mathrm{d}s \tag{9.100}$$

where S is the skin surface and

$$W_s = \tfrac{1}{2}\kappa_s(U_x^2 + U_y^2) + \tfrac{1}{2}(\dot{m}C)_s U^2 - q_s U - (\dot{m}C)_s VU.$$

A detailed description of one triangulation, the minimization of $I(U,V)$ in (9.100) using linear basis functions, and a number of numerical experiments are reported in Twizell and Smith (1982).

PROBLEMS

1. The function $u(x,y)$ satisfies the Poisson equation

$$\Delta[u(x,y)] \equiv \frac{\partial^2 u}{\partial x^2} + \frac{\partial^2 u}{\partial y^2} = x^4 - 6x^2y^2 + y^4$$

in a rectangular region Ω with boundary $\partial\Omega$. The region $\overline{\Omega} \equiv \Omega \cup \partial\Omega$ is covered with a square mesh of length h so that all parts of $\partial\Omega$ are mesh lines. The set of all mesh points in Ω is Ω_h. At each point $P \equiv (x,y) \in \Omega_h$ the differential form $\Delta[U(P)]$ is approximated by the nine-point finite difference replacement $\Delta_h[U(P)]$, where

$$\Delta_h[U(P)] = \frac{1}{6h^2}\{u(x+h,y+h) + u(x-h,y+h) + u(x-h,y-h) \\ + u(x+h,y-h) + 4[u(x+h,y) + u(x,y+h) + u(x-h,y) + \\ u(x,y-h)] - 20u(x,y)\}.$$

Show that the leading term of the truncation error of this difference replacement has the form $Kh^2\Delta^2 u$, where

$$\Delta^2 \equiv \frac{\partial^4}{\partial x^4} + \frac{2\partial^4}{\partial x^2 \partial y^2} + \frac{\partial^4}{\partial y^4},$$

and evaluate K. Hence evaluate this leading term for the case when the difference replacement is applied to the above Poisson equation at a point $P \in \Omega_h$, indicating the significance of your result.

2. The function $u(x,y)$ satisfies the Laplace equation (9.2) in a rectangular region Ω with boundary $\partial\Omega$. The region $\bar{\Omega} \equiv \Omega \cup \partial\Omega$ is covered with a square mesh of length h, so that all parts of $\partial\Omega$ are mesh lines. The set of mesh points in Ω and $\partial\Omega$ are respectively Ω_h and $\partial\Omega_h$. At each point $P \in \Omega_h$ the differential operator is approximated by a difference operator. For any point $P \equiv (x,y) \in \Omega_h$ such that the eight neighbouring points $(x \pm h, y)$, $(x \pm 2h, y)$, $(x, y \pm h)$, $(x, y \pm 2h)$ all lie in $\Omega_h \cup \partial\Omega_h$, devise a nine-point finite difference replacement for (9.2) which has truncation error with leading term

$$Kh^4\left(\frac{\partial^6 u}{\partial x^6} + \frac{\partial^6 u}{\partial y^6}\right),$$

and evaluate K.

Appendix A
Background mathematics

A.1 ELEMENTARY CALCULUS

Fundamental to the study of calculus are the concepts of *limit*, *continuity*, *convergence of a sequence to a limit* and *differentiability* which are now defined.

Definition A.1 Let $f = f(x)$ be a function defined on a set S of real numbers; f is said to have the *limit* L at a, written

$$\lim_{x \to a} f(x) = L \ ,$$

if, given any real number $\varepsilon > 0$, there exists a number $\delta > 0$ such that $|f(x) - L| < \varepsilon$, whenever $x \in S$ and $0 < |x - a| < \delta$.

This definition is illustrated in Fig. A.1.

Definition A.2 Let $f = f(x)$ be a function defined on a set S of real numbers and let $a \subset S$; f is said to be *continuous* at a if

$$\lim_{x \to a} f(x) = f(a) \ .$$

Definition A.3 The function in Definition 1.2 is said to be *continuous on S* if it is continuous at each number in S.

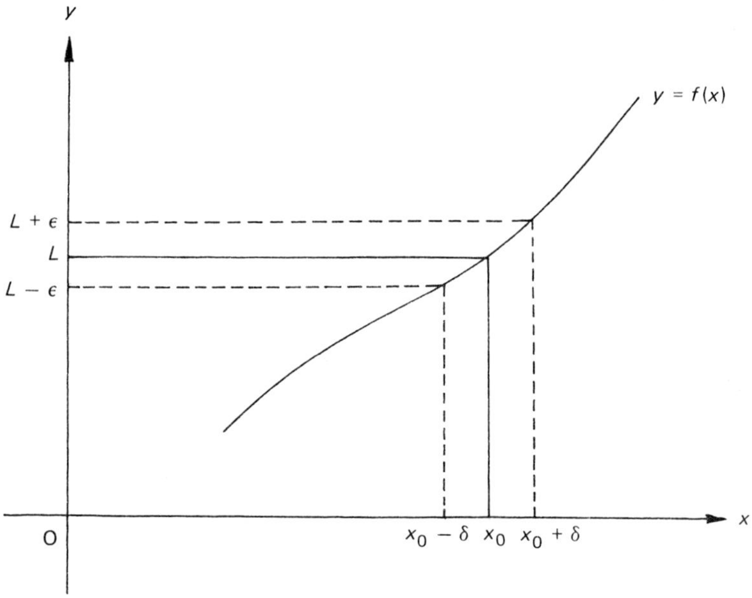

Fig. A.1 — Geometric illustration of a limit.

The notation $C(S)$ is used to denote the set of all functions continuous on S, and the function referred to in Definition A.3 may be denoted by $f(x) \in C(S)$. When S is some open interval (a,b) of the real number line, the notation $f(x) \in C(a,b)$ is used; clearly $f(x) \in C[a,b]$ refers to the closed interval $[a,b]$.

Definition A.4 Let $x_1,x_2,x_3,\ldots$ be an infinite sequence of real (or complex) numbers. The sequence is said to *converge* to a number x called the *limit of the sequence* if, for any $\varepsilon > 0$, there exists a positive integer $N(\varepsilon)$ such that $n > N(\varepsilon)$ implies $|x_n - x| < \varepsilon$. The notation

$$\lim_{n \to \infty} x_n = x \text{ or } x_n \to x \text{ as } n \to \infty$$

means that the sequence $x_1,x_2,x_3,\ldots$ converges to the limiting value x.

Theorem A.1 Let $f = f(x)$ be defined on a set S of real numbers and let $a \in S$. Then the following statements are equivalent:

(a) $f(x)$ is continuous at a;
(b) if any sequence $x_1,x_2,x_3,\ldots$ in S converges to a, then

$$\lim_{n\to\infty} f(x_n) = f(a) \ .$$

This theorem relates the concepts of convergence and continuity.

Definition A.5 Suppose $f = f(x)$ is a function defined in an open interval containing a point a. Then f is said to be *differentiable* at a if

$$\lim_{x\to a} \frac{f(x) - f(a)}{x - a}$$

exists.

When this limit does exist it is called the *derivative* of $f(x)$ at $x = a$ and is denoted by $f'(a)$.

Definition A.6 A function that has a derivative at each number in a set S is said to be differentiable on the set.

Theorem A.2 If a function $f = f(x)$ is differentiable at $x = a$, then f is continuous at a.

Theorem A.3 (*Rolle's Theorem*) Suppose $f \in C[a,\ b]$ and f is differentiable on (a,b). If $f(a) = f(b) = 0$, then a number c exists, with $a < c < b$, such that $f'(c) = 0$.

This theorem is illustrated in Fig. A.2; Theorem A.4 is illustrated in Fig. A.3.

Theorem A.4 (*The Mean Value Theorem*) If $f \in C[a,b]$ and f is differentiable on (a,b), then a number c exists, with $a < c < b$, such that

$$f'(c) = \frac{f(b) - f(a)}{b - a} \ .$$

Theorems A.3 and A.4 are important in deriving methods for error estimation. Their proofs can be found in any elementary calculus textbook. Theorem A.5 which follows has an application in estimating the error in the Lagrange interpolating polynomial (see section 2.3 of Chapter 2).

Theorem A.5 (*Generalized Rolle's Theorem*) Let $f = f(x) \in C[a,b]$, be n times differentiable on (a,b). If $f(x)$ vanishes at the $n + 1$ distinct points $x_0, x_1, \ldots, x_n$ in

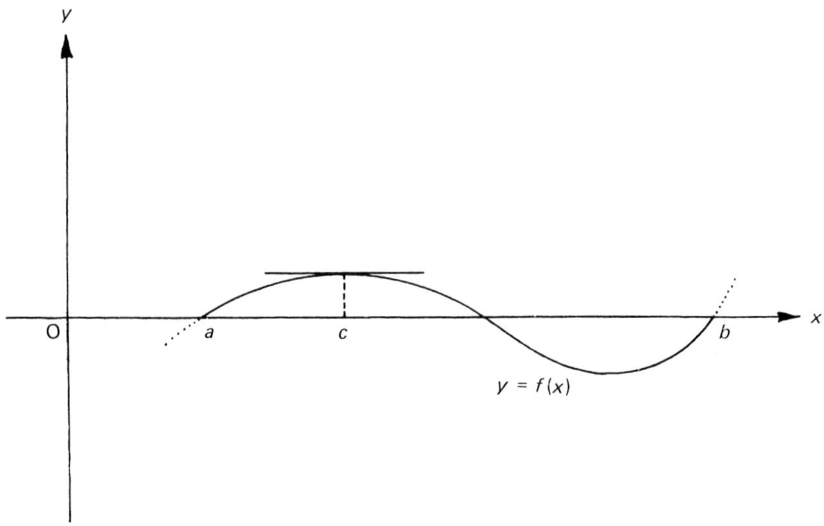

Fig. A.2 — Geometric illustration of Rolle's Theorem.

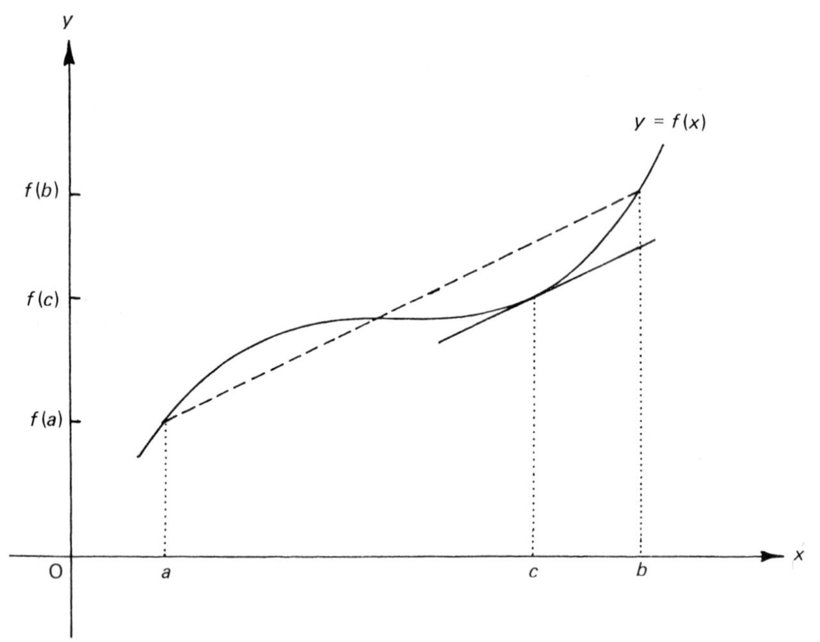

Fig. A.3 — Geometric illustration of the Mean Value Theorem.

$[a,b]$ then a number c exists, with $a<c<b$, such that $f^{(n)}(c)=0$.

Theorem A.6 (*Intermediate Value Theorem*) If $f(x)\in C[a,b]$ and k is any number between $f(a)$ and $f(b)$, then there exists a number c, with $a<c<b$, such that $f(c)=k$ (see Fig. A.4).

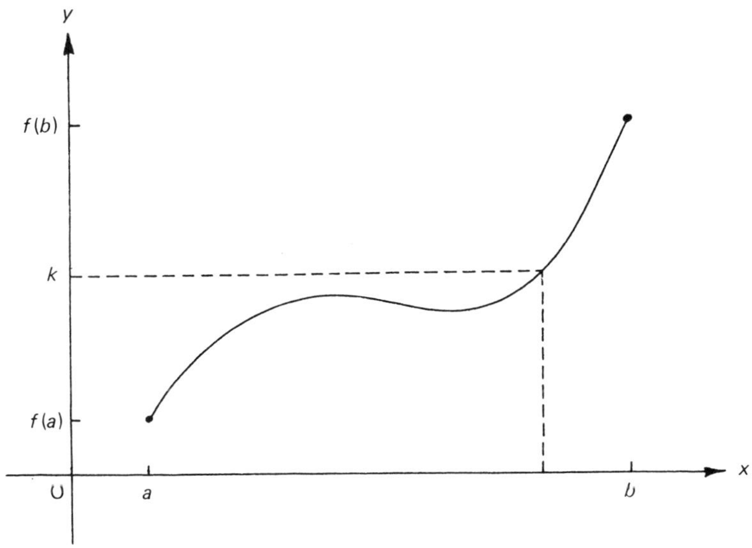

Fig. A.4 — Geometric illustration of the Intermediate Value Theorem.

A.2 SOME EXPANSIONS

The key formula in expressing the derivatives which appear in differential equations in terms of function values is *Taylor's formula*. Suppose, first of all, that $u=u(x)$ is a function of one independent variable x. If this function is continuous and has $n+1$ continuous derivatives on some closed interval I containing the point $x=a$ (that is, $u(x)\in C^{n+1}(I)$) then, for each $x\in I$,

$$u(x) = u(a)+\frac{(x-a)}{1!}u'(a)+\frac{(x-a)^2}{2!}u''(a)+\ldots+\frac{(x-a)^n}{n!}u^{(n)}(a)+R_n(x,a)\ . \tag{A.1}$$

This is known as Taylor's formula and the term $R_n(x,a)$ is usually referred to as the *remainder term*. The remainder term is usually written in one of the following forms, where r depends on n and x:

(1) *The Lagrange form*

$$R_n(x,a) = \frac{(x-a)^{n+1}}{(n+1)!} u^{(n+1)}(r) \ , \qquad r \text{ between } a \text{ and } x \ ;$$

(2) *The Cauchy form*

$$R_n(x,a) = \frac{(x-a)(x-r)^n}{n!} u^{(n+1)}(r) \ , \qquad r \text{ between } a \text{ and } x \ ;$$

(3) *The integral form*

$$R_n(x,a) = \int_a^x \frac{(x-t)^n}{n!} u^{(n+1)}(t) \, \mathrm{d}t \ .$$

The polynomial $P_n(x,a)$ given by

$$P_n(x,a) = u(a) + \frac{(x-a)}{1!} u'(a) + \frac{(x-a)^2}{2!} u''(a) + \ldots + \frac{(x-a)^n}{n!} u^{(n)}(a) \tag{A.2}$$

is known as the *Taylor polynomial* of degree n of $u(x)$ at a, and the coefficients $u(a)$, $u'(a)/1$, $u''(a)/2!, \ldots$ are called the *Taylor coefficients* of $u(x)$ at a. Hence

$$u(x) = P_n(x,a) + R_n(x,a) \ .$$

The Taylor polynomial $P_n(x,a)$ is therefore an approximation to $u(x)$, the error in this approximation being $R_n(x,a)$.

In the special case $a = 0$, Taylor's formula (A.1) becomes

$$u(x) = u(0) + \frac{x}{1!} u'(0) + \frac{x^2}{2!} u''(0) + \ldots + \frac{x^n}{n!} u^{(n)}(0) + R_n(x) \ . \tag{A.3}$$

In this form, the formula is known as *Maclaurin's formula* and the remainder term $R_n(x) \equiv R_n(x,0)$ takes one of the following forms, where r depends on n and x:

(1) *The Lagrange form*

$$R_n(x) = \frac{x^{n+1}}{(n+1)!} u^{(n+1)}(r) \ , \qquad r \text{ between } 0 \text{ and } x \ ;$$

(2) *The Cauchy form*

$$R_n(x) = \frac{x(x-r)^n}{n!} u^{(n+1)}(r) , \qquad r \text{ between 0 and } x ;$$

(3) *The integral form*

$$R_n(x) = \int_0^x \frac{(x-t)}{n!} u^{(n+1)}(t) \, dt .$$

Often Taylor's formula is used to express $u(x)$ as a power series. The series

$$u(x) = u(a) + \frac{(x-a)}{1!} u'(a) + \frac{(x-a)^2}{2!} u''(a) + \ldots = \sum_{k=0}^{\infty} \frac{(x-a)^k}{k!} u^{(k)}(a) \tag{A.4}$$

is called the *Taylor series* of $u(x)$ at a or the *Taylor expansion* of $u(x)$ in powers of $x-a$. Similarly, the series

$$u(x) = u(0) + \frac{x}{1!} u'(0) + \frac{x^2}{2!} u''(0) + \ldots = \sum_{k=0}^{\infty} \frac{x^k}{k!} u^{(k)}(0) \tag{A.5}$$

is called the *Maclaurin series* or the *Maclaurin expansion* of $u(x)$.

In the numerical solution of ordinary differential equations (see Chapters 5 and 6) it is more convenient to consider the Taylor series in a different form. Replacing x and a (simultaneously) in (A.4) by $x+h$ and x, where $h > 0$ is called an *increment* in x, gives

$$u(x+h) = u(x) + \frac{h}{1!} u'(x) + \frac{h^2}{2!} u''(x) + \frac{h^3}{3!} u'''(x) + \frac{h^4}{4!} u^{(iv)}(x) + \ldots . \tag{A.6}$$

Taking only the first two terms of the right-hand side of (A.6) gives

$$u(x+h) = u(x) + hu'(x) + O(h^2) , \tag{A.7}$$

where the expression $O(h^2)$ indicates that the error $R_2(x)$ has *principal part* proportional to h^2. This terminology is formalized as follows:

Definition A.7 Let $f(x)$ and $g(x)$ be given functions of x. If there exists a positive constant K such that $f(x) \leqslant Kg(x)$ for all x, then $f(x) = O(g(x))$ where the letter O stands for 'order'. If instead there is a positive constant M such that $f(x) \geqslant Mg(x)$ for all x, then $f(x) = \Omega(g(x))$. If $f(x) = O(g(x))$ and $f(x) = \Omega(g(x))$ are both valid, then

$f(x) = \Theta(g(x))$.

Returning to (A.7) and dividing by h leads to

$$\frac{\mathrm{d}u}{\mathrm{d}x} = \frac{u(x+h)-u(x)}{h} + O(h) \tag{A.8}$$

which is called a *first-order* (because of the $O(h)$ error), *forward difference* approximant to $\mathrm{d}u/\mathrm{d}x$.

The Taylor expansion for $u(x-h)$ is found by replacing h by $-h$ in (A.6) to give

$$u(x-h) = u(x) - \frac{h}{1!}u'(x) + \frac{h^2}{2!}u''(x) - \frac{h^3}{3!}u'''(x) + \frac{h^4}{4!}u^{(\mathrm{iv})}(x) - \ldots \tag{A.9}$$

from which it is easy to derive the *first-order*, *backward difference* approximant to $\mathrm{d}u/\mathrm{d}x$ given by

$$\frac{\mathrm{d}u}{\mathrm{d}x} = \frac{u(x)-u(x-h)}{h} + O(h)\ . \tag{A.10}$$

Taking terms up to and including h^3 on the right-hand sides of (A.6) and (A.9) and subtracting gives

$$u(x+h) - u(x-h) = 2hu'(x) + O(h^3)$$

so that

$$\frac{\mathrm{d}u}{\mathrm{d}x} = \frac{u(x+h)-u(x-h)}{2h} + O(h^2)\ . \tag{A.11}$$

This is a *second-order* (because of the $O(h^2)$ error), *central difference* approximant to $\mathrm{d}u/\mathrm{d}x$.

Taking terms up to and including h^4 on the left-hand sides of (A.6) and (A.9) and adding gives

$$u(x+h) + u(x-h) = 2u(x) + h^2u''(x) + O(h^4)$$

which leads to

$$\frac{\mathrm{d}^2u}{\mathrm{d}x^2} = \frac{u(x-h)-2u(x)+u(x-h)}{h^2} + O(h^2)\ . \tag{A.12}$$

This is a *second-order, central difference* approximation to d^2u/dx^2.

In the numerical solution of time-dependent partial differential equations involving one space dimension x and time t, $u = u(x,t)$ is a function of two independent variables. The Taylor series given by (A.6) becomes

$$u(x+h,t) = u(x,t) + \frac{h}{1!}\frac{\partial u(x,t)}{\partial x} + \frac{h^2}{2!}\frac{\partial^2 u(x,t)}{\partial x^2} + \dots \tag{A.13}$$

while (A.9) becomes

$$u(x-h,t) = u(x,t) - \frac{h}{1!}\frac{\partial u(x,t)}{\partial x} + \frac{h^2}{2!}\frac{\partial^2 u(x,t)}{\partial x^2} - \dots \tag{A.14}$$

In the same ways that (A.8), (A.10) and (A.11) were developed, it is easy to derive the following formulas with which to approximate $\partial u/\partial x$:

$$\frac{\partial u}{\partial x} = \frac{u(x+h,t) - u(x,t)}{h} + O(h) \ , \qquad \text{(first-order, forward difference)} \tag{A.15}$$

$$\frac{\partial u}{\partial x} = \frac{u(x,t) - u(x-h,t)}{h} + O(h) \ , \qquad \text{(first-order, backward difference)} \tag{A.16}$$

$$\frac{\partial u}{\partial x} = \frac{u(x+h,t) - u(x-h,t)}{2h} + O(h^2) \ , \quad \text{(second-order, central difference)} \tag{A.17}$$

Taking terms up to and including h^4 on the right-hand sides of (A.13) and (A.14) and adding leads to the second-order central difference approximant to $\partial^2u/\partial x^2$ given by

$$\frac{\partial^2 u}{\partial x^2} = \frac{u(x-h,t) - 2u(x,t) + u(x+h,t)}{h^2} + O(h^2) \ . \tag{A.18}$$

Formulas (A.15), (A.16), (A.17) and (A.18) were derived by allowing x to vary whilst t remained fixed. Keeping x constant and incrementing t by an amount $\pm l$ as appropriate, it is easy to derive the following formulas with which to approximate $\partial u/\partial t$ or $\partial^2u/\partial t^2$:

$$\frac{\partial u}{\partial t} = \frac{u(x,t+l) - u(x,t)}{l} + O(l) \ , \tag{A.19}$$

$$\frac{\partial u}{\partial t} = \frac{u(x,t) - u(x,t-l)}{l} + O(l) \ , \tag{A.20}$$

$$\frac{\partial u}{\partial t} = \frac{u(x,t+l) - u(x,t-l)}{2l} + O(l^2) \ , \tag{A.21}$$

$$\frac{\partial^2 u}{\partial t^2} = \frac{u(x,t+l) - 2u(x,t) + u(x,t-l)}{l^2} + O(l^2) \ . \tag{A.22}$$

Steady-state partial differential equations are discussed in Chapter 9 where $u = u(x,y)$ is a function of two space variables x and y. Formulas (A.19), (A.20), (A.21) and (A.22) can easily be adapted to give the derivatives of u with respect to y. Assuming that the increment in y is k, these derivatives are

$$\frac{\partial u}{\partial y} = \frac{u(x,y+k) - u(x,y)}{k} + O(k) \ , \tag{A.23}$$

$$\frac{\partial u}{\partial y} = \frac{u(x,y) - u(x,y-k)}{k} + O(k) \ , \tag{A.24}$$

$$\frac{\partial u}{\partial y} = \frac{u(x,y+k) - u(x,y-k)}{2k} + O(k^2) \ , \tag{A.25}$$

$$\frac{\partial^2 u}{\partial y^2} = \frac{u(x,y-k) - 2u(x,y) + u(x,y+k)}{k^2} + O(k^2) \ . \tag{A.26}$$

In section 9.4 of Chapter 9, the mixed second derivative $\partial^2 u/\partial x \partial y$ is required. In order to obtain this, Taylor's series in which both x and y receive increments is required. There are four such forms and they are given by

$$\begin{aligned} u(x+h,y+k) &= u(x,y) + \frac{1}{1!}\left(h\,\frac{\partial u(x,y)}{\partial x} + k\,\frac{\partial u(x,y)}{\partial y}\right) \\ &+ \frac{1}{2!}\left(h^2\,\frac{\partial^2 u(x,y)}{\partial x^2} + 2hk\,\frac{\partial^2 u(x,y)}{\partial x \partial y} + k^2\,\frac{\partial^2 u(x,y)}{\partial y^2}\right) + \dots \ , \end{aligned} \tag{A.27}$$

$$\begin{aligned} u(x+h,y-k) &= u(x,y) + \frac{1}{1!}\left(h\,\frac{\partial u(x,y)}{\partial x} - k\,\frac{\partial u(x,y)}{\partial y}\right) \\ &+ \frac{1}{2!}\left(h^2\,\frac{\partial^2 u(x,y)}{\partial x^2} - 2hk\,\frac{\partial^2 u(x,y)}{\partial x \partial y} + k^2\,\frac{\partial^2 u(x,y)}{\partial y^2}\right) + \dots \ , \end{aligned} \tag{A.28}$$

$$\begin{aligned} u(x-h,y+k) &= u(x,y) + \frac{1}{1!}\left(-h\,\frac{\partial u(x,y)}{\partial x} + k\,\frac{\partial u(x,y)}{\partial y}\right) \\ &+ \frac{1}{2!}\left(h^2\,\frac{\partial^2 u(x,y)}{\partial x^2} - 2hk\,\frac{\partial^2 u(x,y)}{\partial x \partial y} + k^2\,\frac{\partial^2 u(x,y)}{\partial y^2}\right) + \dots \ , \end{aligned} \tag{A.29}$$

$$u(x-h,y-k) = u(x,y) + \left(-h\,\frac{\partial u(x,y)}{\partial x} - k\,\frac{\partial u(x,y)}{\partial y}\right)$$

$$+\frac{1}{2!}\left(h^2\,\frac{\partial^2 u(x,y)}{\partial x^2} + 2hk\,\frac{\partial^2 u(x,y)}{\partial x\partial y} + k^2\,\frac{\partial^2 u(x,y)}{\partial y^2}\right) + \ldots . \quad \text{(A.30)}$$

It is easy to show that

$$u(x+h,y+k) - u(x+h,y-k) - u(x-h,y+k) + u(x-h,y-k)$$
$$= 4hk\,\frac{\partial^2 u}{\partial x\partial y} + O(h+k)^4 \ , \quad \text{(A.31)}$$

the terms involving first- and third-order derivatives and the derivatives $\partial^2 u/\partial x^2$ and $\partial^2 x/\partial y^2$ having vanished in this linear combination of the four functional values. It follows from (A.31) that

$$\frac{\partial u}{\partial x\partial y} = \frac{u(x+h,y+k) - u(x+h,y-k) - u(x-h,y+k) + u(x-h,y-k)}{h^2}$$
$$+ O\left(\frac{(h+k)^4}{hk}\right) . \quad \text{(A.32)}$$

In Chapter 9, the special case where $k = h$ is required so that (A.32) becomes

$$\frac{\partial^2 u}{\partial x\partial y} = \frac{u(x+h,y+h) - u(x+h,y-h) - u(x-h,y+h) + u(x-h,y-h)}{h^2}$$
$$+ O(h^2) . \quad \text{(A.33)}$$

The approximations in (A.32) and (A.33) are both second-order central difference replacements for $\partial^2 u/\partial x\partial y$.

A.3 CALCULUS OF VARIATIONS

The main problem in the calculus of variations is to find functions which maximize or minimize certain integrals, possibly under given extra conditions. In the context of second-order boundary value problems (section 6.3 of Chapter 6) these extra conditions are boundary conditions.

If a special function is introduced into the integral, a certain value will result and the independent variable can be thought of as a function while the dependent variable is a numerical value. Such a relationship is known as a *functional*.

In its simplest form the problem is to find a function $y = y(x)$ such that

$$I = \int_a^b F(x, y, y')\mathrm{d}x \tag{A.34}$$

becomes as small as possible. The classical approach in minimizing I is to introduce a *comparison function* $y + \varepsilon f(x)$ with $f(a) = f(b) = 0$. When $f(x)$ is a given function and y is the desired solution, I is a function of ε only, which has an extreme value for $\varepsilon = 0$. Thus

$$I = I(\varepsilon) = \int_a^b F(x, y + \varepsilon f(x), y' + \varepsilon f'(x))\mathrm{d}x$$

and so

$$\begin{aligned}\left[\frac{\mathrm{d}I}{\mathrm{d}\varepsilon}\right]_{\varepsilon=0} &= \int_a^b \left[\frac{\partial F}{\partial y} f(x) + \frac{\partial F}{\partial y'} f'(x)\right] \mathrm{d}x \\ &= \int_a^b \left[\frac{\partial F}{\partial y} - \frac{\mathrm{d}}{\mathrm{d}x}\left(\frac{\partial F}{\partial y'}\right)\right] f(x)\mathrm{d}x + \left[\frac{\partial F}{\partial y'} f(x)\right]_a^b \\ &= \int_a^b \left[\frac{\partial F}{\partial y} - \frac{\mathrm{d}}{\mathrm{d}x}\left(\frac{\partial F}{\partial y'}\right)\right] f(x)\mathrm{d}x \ ,\end{aligned} \tag{A.35}$$

since $f(a) = f(b) = 0$. At the extremum $\varepsilon = 0$, $\mathrm{d}I/\mathrm{d}\varepsilon$ vanishes; the right-hand side of (A.35) therefore vanishes provided

$$\frac{\partial F}{\partial y} - \frac{\mathrm{d}}{\mathrm{d}x}\left(\frac{\partial F}{\partial y'}\right) = 0 \ . \tag{A.36}$$

Equation (A.36) is known as the *Oiler* or *Euler-Lagrange equation*; its solutions are known as *extremals*. Euler's equation is a necessary condition for I to have an extreme value; it is not sufficient.

Example A.1 Let

$$I = \int_0^1 \{p(x)[y'(x)]^2 + q(x)[u(x)]^2 - 2g(x)y(x)\}\mathrm{d}x \tag{A.37}$$

and show that the Euler equation takes the form

$$-\frac{\mathrm{d}}{\mathrm{d}x}\left(p(x)\,\frac{\mathrm{d}y}{\mathrm{d}x}\right)+q(x)y(x) \;=\; g(x)\;. \tag{A.38}$$

In (A.37), $F(x,y,y')=p(x)[y'(x)]^2+q(x)[y(x)]^2-2g(x)y(x)$ so that

$$\frac{\partial F}{\partial y} \;=\; 2q(x)y(x) \quad \text{and} \quad \frac{\partial F}{\partial y'} \;=\; 2p(x)y'(x)$$

giving

$$\frac{\mathrm{d}}{\mathrm{d}x}\left(\frac{\partial F}{\partial y'}\right) \;=\; 2\frac{\mathrm{d}}{\mathrm{d}x}\left(p(x)\frac{\mathrm{d}y}{\mathrm{d}x}\right)\;.$$

It is then easy to show that (A.36) takes the form (A.38). The differential equation (A.38) arises in subsection 6.3.3 of Chapter 6.

Consider now a function $u=u(x,y)$ of the independent variables x,y, which takes on specified values on the boundary ∂R of some region R, and introduce the notation $u_x=\partial u/\partial x$ and $u_y=\partial u/\partial y$. The integral

$$I \;=\; \iint_R F(x,y,u,u_x,u_y)\mathrm{d}x\mathrm{d}y \tag{A.39}$$

is the natural extension of (A.34) and the Euler equation (A.36) becomes

$$\frac{\partial}{\partial x}\left(\frac{\partial F}{\partial u_x}\right)+\frac{\partial}{\partial y}\left(\frac{\partial F}{\partial u_y}\right)-\frac{\partial F}{\partial u} \;=\; 0\;; \tag{A.40}$$

equation (A.40) gives necessary conditions for I to have an extreme value.

Example A.2 Let

$$I=\iint_R\left[\left(\frac{\partial u}{\partial x}\right)^2+\left(\frac{\partial u}{\partial y}\right)^2+2uf(x,y)\right]\mathrm{d}x\mathrm{d}y \tag{A.41}$$

and show that the Euler equation takes the form

$$\frac{\partial^2 u}{\partial x^2}+\frac{\partial^2 u}{\partial y^2} \;=\; f(x,y)\;. \tag{A.42}$$

In (A.41), $F = u_x^2 + u_y^2 + 2uf(x,y)$ so that

$$\frac{\partial F}{\partial u_x} = 2u_x, \quad \frac{\partial f}{\partial u_y} = 2u_y \quad \text{and} \quad \frac{\partial F}{\partial u} = 2f(x,y)$$

giving

$$\frac{\partial}{\partial x}\left(\frac{\partial F}{\partial u_x}\right) = 2u_{xx} \quad \text{and} \quad \frac{\partial}{\partial y}\left(\frac{\partial F}{\partial u_y}\right) = 2u_{yy} \ .$$

Clearly (A.40) leads to (A.42) which is known as *Poisson's equation* or, if $f(x,y) \equiv 0$, *Laplace's equation* (see section 9.1 of Chapter 9).

The *variational formulations* just described arise in a more general context in Chapter 9 where they are discussed with respect to linear operators. The following definitions will facilitate understanding of section 9.6.2 of Chapter 9.

Definition A.8 The *inner product* of two functions f_1 and f_2 is defined by

$$(f_1,f_2) = \int_R f_1 f_2 \mathrm{d}R$$

If $\mathbf{f}_1$ and $\mathbf{f}_2$ are vector-value functions the inner product is defined by $\int_R \mathbf{f}_1^T \mathbf{f}_2 \mathrm{d}R$.

Definition A.9 The operator A is a *linear operator* if and only if

(1) A is a mapping of the Hilbert space H onto itself,
(2) $A(x+y) = A(x) + A(y)$ for all $x, y \in H$,
(3) $A(cx) = cA(x)$ where s is some scalar.

Definition A.10 The *norm* of a linear operator A is the smallest value of k for which A is bounded; the norm of A is denoted by $\|A\|$.

Definition A.11 A linear operator A is called a *bounded linear operator* if there exists a positive constant k such that

$$\|Ax\| \leqslant k\|x\| \text{ for all } x \in H \ .$$

Clearly, a bounded linear operator is continuous for, if the sequence $x_1, x_2, \ldots$ tends to the limit x, written $\{x_m\} \to x$, then $\{Ax_m\} \to Ax$.

Definition A.12 A linear operator A is said to be *positive definite* if and only if $(Ax,x) > 0$ for all $x \neq 0$ and to be *positive semi-definite* if $(Ax,x) \geqslant 0$ for all $x \neq 0$.

Definition A.13 The *adjoint* of linear operator A is the operator A^* for which $(Ax,y) = (x,A^*y)$ for all $x,y \in H$.

Definition A.14 The linear operator A is said to be *self-adjoint* if $(Tx,y) = (x,Ty)$ for all $x,y \in H$.

A.4 CLASSIFICATION OF SECOND-ORDER PARTIAL DIFFERENTIAL EQUATIONS

The general quasi-linear partial differential equation of the second order in two independent variables x and t has the form

$$a\frac{\partial^2 u}{\partial x^2} + b\frac{\partial^2 u}{\partial x \partial t} + c\frac{\partial^2 u}{\partial t^2} = e \tag{A.43}$$

where $u = u(x,t)$ and a,b,c,e are functions of $x,t,u,\partial u/\partial x,\partial u/\partial t$ but not of the second-order derivatives. In this section, the abbreviations p,q,r,s,w will be used to denote $\partial u/\partial x,\partial u/\partial t,\partial^2 u/\partial x^2,\partial^2 u/\partial x\partial t,\partial^2 u/\partial t^2$, respectively. Equation (A.43) can therefore be written in the alternative form

$$ar + bs + cw = e \ . \tag{A.44}$$

Suppose that the values of u,p,q are known at all points (x,t) on some smooth curve in the (x,t) plane. Then these values must satisfy the *total differential formula*

$$\mathrm{d}u = \frac{\partial u}{\partial x}\,\mathrm{d}x + \frac{\partial u}{\partial y}\,\mathrm{d}t = p\mathrm{d}x + q\mathrm{d}t \ . \tag{A.45}$$

Similarly p and q satisfy the formulas

$$\mathrm{d}p = \frac{\partial p}{\partial x}\,\mathrm{d}x + \frac{\partial p}{\partial t}\,\mathrm{d}t = r\mathrm{d}x + s\mathrm{d}t \ , \tag{A.46}$$

$$\mathrm{d}q = \frac{\partial q}{\partial x}\,\mathrm{d}x + \frac{\partial q}{\partial t}\,\mathrm{d}t = s\mathrm{d}x + w\mathrm{d}t \ . \tag{A.47}$$

Equations (A.44), (A.46) and (A.47) form a system of three equations in r,s,w which will not have a unique solution at any point at which the determinant of the coefficients of r,s,w vanish. That is, whenever

$$\begin{vmatrix} a & b & c \\ \mathrm{d}x & \mathrm{d}t & 0 \\ 0 & dx & \mathrm{d}t \end{vmatrix} = 0 \qquad (A.48)$$

holds. Expanding the determinant of (A.48) gives the equation

$$a\left(\frac{\mathrm{d}t}{\mathrm{d}x}\right)^2 - \mathrm{b}\left(\frac{\mathrm{d}t}{\mathrm{d}x}\right) + c = 0 \qquad (\mathrm{A}.49)$$

which has two real distinct, one real repeated, or conjugate complex roots in $\mathrm{d}t/\mathrm{d}x$ according as the discriminant $b^2 - 4ac$ is greater than, equal to, or less than zero. In a domain in the (x,t) plane in which u,p,q are defined, and throughout which equation (A.49) has two, one, or no real roots, the differential equation (A.43) is said to be *hyperbolic*, *parabolic* or *elliptic*, respectively.

In the case of elliptic equations which are steady-state, the time variable t is replaced by a second space derivative y, say, and (A.43) takes the form

$$a\,\frac{\partial^2 u}{\partial x^2} + b\,\frac{\partial^2 u}{\partial x \partial y} + c\,\frac{\partial^2 u}{\partial y^2} = e\ ; \qquad b^2 - 4ac < 0\ . \qquad (\mathrm{A}.50)$$

Example A.3 Classify the following equations as hyperbolic, parabolic, or elliptic, assuming that x,y,t,u are all real.

(1) $$\frac{\partial^2 u}{\partial x^2} + u\,\frac{\partial^2 u}{\partial x \partial y} + (1+u^2)\,\frac{\partial^2 u}{\partial y^2} = 0\ .$$

Here, $a = 1$, $b = u$, $c = 1 + u^2$, $e = 0$ so that

$$b^2 - 4ac = u^2 - 4(1+u^2) = -4 - 3u^2 = < 0\ .$$

Hence the equation is always elliptic.

(2) $$\frac{\partial^2 u}{\partial x^2} + 2\,\frac{\partial^2 u}{\partial x \partial t} + (1 - x^2 - t^2)\,\frac{\partial^2 u}{\partial t^2} = x^2 + t^2\ .$$

Here, $a = 1$, $b = 2$, $c = 1 - x^2 - t^2$, $e = x^2 + t^2$ so that

$$b^2 - 4ac = 4 - 4(1 - x^2 - t^2) = 4(x^2 + t^2)\ .$$

Hence the equation is hyperbolic except at the point (0,0) where it is parabolic.

Appendix B
NAG Fortran routines

The *Numerical Algorithms Group* (*NAG*) Fortran Library contains software directly related to many of the numerical methods outlined in this book. At the time of writing, the Library is at Mark 11 and full details of all available routines are contained in the NAG Fortran Library Manual. The three-character identification and title of the Library chapter, a description of the algorithm, and the six-character name of the routine are given below for a selection of programs which may be of use to the reader.

Chapter 1

F01 — Matrix Operations, Including Inversion

Matrix inversion:	
Real matrix, approximate inverse	F01AAF
Real symmetric positive definite matrix,	
approximate inverse	F01ADF
accurate inverse	F01ACF
accurate inverse (simplified parameter list)	F01ABF
Complex Hermitian positive-definite matrix, inverse	F01BPF
LU-factorization	
complex band matrix	F01NAF
real matrix	F01BTF
real band matrix	F01LBF
real tridiagonal matrix	F01LEF
real sparse matrix	F01BRF
real sparse matrix with known sparsity pattern	F01BSF
LL^T-factorization	
real symmetric positive definite matrix	F01BXF
real sparse symmetric positive definit matrix	F01MAF

F04 — Simultaneous linear equations

Solution of simultaneous linear equations (black box):
complex matrix,
approximate solution, multiple right-hand sides F04ADF
real matrix
accurate solution,
one right-hand side F04ATF
multiple right-hand sides F04AEF
approximate solution
one right-hand side F04ARF
multiple right-hand sides F04AAF
real tridiagonal matrix
approximate solution, one right-hand side F04EAF
real symmetric positive-definite matrix,
accurate solution
one right-hand side F04ASF
multiple right-hand side F04ABF
real symmetric positive definite band matrix,
approximate solution, multiple right-hand sides F04ACF
real symmetric positive-definite tridiagonal matrix
approximate solution, one right-hand side F04FAF
real sparse symmetric matrix F04MBF
Solution of simultaneous linear equations (after factorizing the matrix of coefficients):
complex matrix, approximate solution F04AKF
real matrix
accurate solution F04AHF
approximate solution F04AJF
real band matrix, approximate solution F04LDF
real tridiagonal matrix, approximate solution F04LEF
real sparse matrix, approximate solution F04AXF
real symmetric positive definite matrix,
accurate solution F04AFF
approximate solution (factorisation by F01BXF) F04AZF

F02 — Eigenvalues and eigenvectors
Matrix eigenvalue problems (black box):
Complex matrix:
all eigenvalues F02AJF
all eigenvalues and eigenvectors F02AKF
real matrix:
all eigenvalues F02AFF
all eigenvalues and eigenvectors F02AGF

Chapter 2

E01 — Interpolating functions
Interpolating functions:

polynomial interpolant (data may include derivative values)	E01AEF
cubic spline interpolant	E01BAF
rational interpolant	E01RAF
E02 — Curve and surface fitting	
Minimax curve fit by polynomials	E02ACF
Least squares curve fit:	
by polynomials, arbitrary datas points	E02ADF
L_1-approximation	
by general linear function	E02GAF
L_∞-approximation	
by general linear function	E02GCF
Padé approximants	E02RAF

Chapter 3

C02 — Zeros of Polynomials	
All zeros of polynomials	
complex coefficients	C02ADF
real coefficients	C02AEF
C05 — Roots of one or more transcendental equations	
Zero of continuous function of one variable	
in a given interval	C05ADF
from a given starting value	C05AGF
Solution of a system of nonlinear equations (easy-to-use)	
using function values only	C05NBE
Solution of a system of nonlinear equations (comprehensive)	
using function values only	C05NCF

Chapter 4

D04 — Numerical Differentiation	
Numerical differentiation of a function of one real variable, derivatives up to order 14	D04AAF
D01 — Quadrature	
One-dimensional quadrature	
adaptive integration of a function over a finite interval:	
badly-behaved integrands	D01AJF
well-behaved integrands	D01AHF
method suitable for oscillating functions	D01AKF
allowing for singularities at user-specified points	D01ALF
weight function $\cos(\omega x)$ or $\sin(\omega x)$	D01ANF
weight function with end-point singularities	
of algebraic–logarithmic type	D01APF
weight function $(x-c)^{-1}$ (Hilbert transform)	D01AQF

Gaussian quadrature rule-evaluation D01BAF
integrating of a function defined by data values only D01GAF
Weights and abscissae for Gaussian quadrature rules:
restricted choice of rule, using pre-computed weights and abscissae D01BBF
more general choice of rule, calculating the weights and abscissae D01BCF

Chapters 5 and 6

D02 — Ordinary Differential Equations

System of ordinary differential equations, initial value problems:
Simple driver routines:
Runge–Kutta–Merson method:
over a range D02BAF
Comprehensive routines:
Runge–Kutta–Merson method D02PAF
Interpolating solutions:
solution produced by D02PAF
all components D02XAF
one component D02XBF
Runge–Kutta–Merson method for integration over one step D02YAF
System of ordinary differential equations, boundary value problems:
finite difference technique with deferred correction:
simple nonlinear problem D02GAF
general linear problem D02GBF
general nonlinear problem, with continuation facility D02RAF
shooting and matching technique
boundary values to be determined D02HAF

Chapters 8 and 9

D03 — Partial Differential Equations

Partial differential equations, elliptic:
Laplace's equation in 2-D for an arbitrary domain D03EAF
Solution of finite difference equation by Strongly Implicit Procedure
for five-point 2-D molecule:
iterate to convergence D03EBF
one iteration D03UAF
Partial differential equations, parabolic, one space variable, method of lines:
single equation D03PAF
simple system D03PBF
general system D03PGF
Triangulation of a plane region D03MAF

References

Ackerman, E., Gatewood, L. C., Rosevear, J. W. and Molnar, G. D. (1965) 'Model studies of blood–glucose regulation', *Bull Math. Biophys.*, **27**, 21–38.

Bailey, P. B., Shampine, L. F. and Waltman, P. E. (1968) *Nonlinear Two-point Boundary Value Problems*, Academic Press, New York.

Barrodale, I. and Young, A. (1966) 'Algorithms for best L_1 and L_∞ linear approximations on a discrete set', *Numer. Math.*, **8**, 295–306.

Bartle, R. G. (1976) *The Elements of Real Analysis*, second edition, John Wiley, New York.

Besterman, J. M., Airhart, J. A., Woodworth, R. C. and Low, R. B. (1981) 'Exocytosis of pinocytosed fluid in culture cells: kinetic evidence for rapid turnover and compartmentation', *J. Cell Biol.*, **91**, 716–727.

Brown, M. S., Goldstein, J. L., Kreiger, M., Ho, Y. K. and Anderson, R. G. W. (1979) 'Reversible accumulation of cholesteryl esters in macrophages incubated with acetylated lipoproteins', *J. Cell Biol.*, **82**, 579–613.

Brusa, L. and Nigro, L. (1980) 'A one-step method for the direct integration of structural dynamic equations'. *Int. J. Num. Meth. Engng.*, **15**, 685–699.

Burden, R. L. and Faires, J. D. (1985) *Numerical Analysis*, third edition, Prindle, Weber and Schmidt, Boston, Mass.

Cheney, E. W. (1966) *Introduction to Approximation Theory*, McGraw-Hill, New York.

Cherruault, Y. (1986) *Mathematical Modelling in Biomedicine*, D. Reidel, Dordrecht, Holland.

Cullen, M. R. (1985) *Linear Models in Biology*, Ellis Horwood, Chichester.

Dahlquist, G. (1956) 'Convergence and stability in the numerical integration of ordinary differential equations', *Math. Scand.*, **4**, 33–53.

Davies, A. J. (1980) *The Finite Element Method (A First Approach)*, Oxford University Press.

Dean, R. T. and Jessup, W. (1982) 'Modulation of exocytosis of previously pinocytosed fluid by human fibroblasts', *Bioscience Reports*, **2**, 551–560 (The Biochemical Society).

Dinnar, U. (1981) *Cardiovascular Fluid Dynamics*, CRC Press, Boca Raton, FL.

FitzHugh, R. (1961) 'Impulses and physiological states in theoretical models of nerve membrane', *Biophys. J.*, **1**, 445–466.

Forsythe, G. E. and Moler, C. B. (1967) *Computer Solution of Linear Algebraic Systems*, Prentice-Hall, Englewood Cliffs, NJ.

Fröberg, C.-E. (1985) *Numerical Mathematics: Theory and Computer Applications*, Benjamin/Cummings, Menlo Park, CA.

George, A. and Liu J. W. H. (1981) *Computer Solution of Large Positive Definite Systems*, Prentice-Hall, Englewood Cliffs, NJ.

Gerald, C. F. and Wheatley, P. O. (1984), *Applied Numerical Analysis*, third edition, Addison-Wesley, Reading, MA.

Gourlay, A. R. and Morris, J. Ll. (1980) 'The extrapolation of first order methods for parabolic partial differential equations, II', *SINUM*, **17**, 641–655.

Greig, D. M. (1980) *Optimisation*, Longman, London.

Hastings, S. P. (1975) 'Some mathematical problems from neurobiology'. *Am. Math. Monthly*, **82**, 881–895.

Henrici, P. H. (1962) *Discrete Variable Methods in Ordinary Differential Equations*, John Wiley, New York.

Henrici, P. H. (1964) *Elements of Numerical Analysis*, John Wiley, New York.

Hodgkin, A. L. and Huxley, A. F. (1952) 'A quantitative description of membrane current and its application to conduction and excitation in nerve' *J. Physiol.* **117**, 500–544.

Hudetz, W. J. (1973) 'A computer simulation of the otolith membrane' *Comput. Biol. Med.*, **3**, 355–369.

Johnson, L. W. and Riess, R. D. (1982) *Numerical Analysis*, second edition, Addison-Wesley, Reading, MA.

Jones, D. S. and Sleeman, B. D. (1983) *Differential Equations and Mathematical Biology*, George Allen & Unwin, London.

Keller, H. B. (1968) *Numerical Methods for Two-point Boundary Value Problems*, Blaisdell, Waltham, MA.

Khaliq, A. Q. M. (1983) *Numerical Methods for Ordinary Differential Equations with Applications to Partial Differential Equation*, Ph.D. thesis, Brunel University.

Kubota, K. and Ishizaki, T. (1985) 'A theoretical consideration of percutaneous drug absorption', *J. Pharmacokin. Biopharm.*, **13**, 55–72.

Kubota, K. and Ishizaki, T. (1986a) 'A calculation of percutaneous drug absorption — I. Theoretical', *Comput. Biol. Med.*, **16**, 7–19.

Kubota, K. and Ishizaki, T. (1986b) 'A calculation of percutaneous drug absorption—II. Computation results', *Comput. Biol. Med.*, **16**, 21–37.

Kuznetz, L. H. (1979) 'A two dimensional transient mathematical model of human thermoregulation', *Am. J. Physiol.*, **237** (5), R266–R277.

Lambert, J. D. (1973) *Computational Methods in Ordinary Differential Equations*, John Wiley, Chichester.

Lambert, J. D. and Watson, I. A. (1976) 'Symmetric multistep methods for periodic initial value problems'. *J. Inst. Maths. Applics.*, **18,** 189–202.

Lawson, J. D. and Morris, J. Ll. (1977) 'A note on the efficient implementation of splitting methods in two space variables', *BIT,* **17,** 492–493.

Lawson, J. D. and Morris, J. Ll. 1978, 'The extrapolation of first order methods for parabolic partial differential equations. I', *SINUM*, **15**, 1212–1224.

Levenberg, K. (1944) 'A method for the solution of certain non-linear problems in least squares', *Q. Appl. Math.*, **2**, 164–168.

McDonald, D. A. (1974) *Blood Flow in Arteries*, second edition, Edward Arnold, London.

MacGregor, R. J. and Lewis, E. R. (1977) *Neural Modeling: Electrical Signal Processing in the Nervous System*, Plenum Press, New York.

Manoranjan, V. S., Mitchell, A. R., Sleeman, B. D. and Yu, K. P. (1984) 'Bifurcation studies in reaction-diffusion', *J. Comput. Appl. Math.*, **11,** 27–37.

Marquardt, D. W. (1963) 'An algorithm for least squares estimation of nonlinear parameters', *SIAM J.*, **11**(2), 431–441.

Mitchell, A. R. and Phillips, G. M. (1972) 'Construction of basis functions in the finite element method', *BIT*, **12**, 81–89.

Morris, J. Ll. (1983) *Computational Methods in Elementary Numerical Analysis*, John Wiley, Chichester.

Morton, K. W. (1980) 'Stability of finite difference approximations to a diffusion–convection equation', *Int. J. Num. Meth. Engng*, **15**, 677–683.

Nagumo, J., Arimoto, S. and Yoshizawa, S. (1962) 'An active pulse transmission line simulating nerve axon', *Proc IRE*, **50**, 2061–2071.

Ortega, J. M. (1972) *Numerical Analysis — a Second Course*, Academic Press, New York.

Peaceman, D. W. and Rachford, H. H. (1955) 'The numerical solution of parabolic and elliptic differential equations', *J. Soc. Indust. App. Math.*, **3**, 28–41.

Peskin, C. S. (1975) *Mathematical Aspects of Heart Physiology*, Courant Institute of Mathematical Sciences, New York.

Ralston, A. (1965) *A First Course in Numerical Analysis*, McGraw-Hill.

Rice, J. R. (1964) *The Approximation of Functions* (*Vol.* 1), Addison-Wesley, Reading, MA.

Rice, J. R. (1968) *The Approximation of Functions* (*Vol.* 2), Addison-Wesley, Reading, MA.

Sapirstein, L. A., Vidt, D. G., Mandel, M. J. and Hanusek, G. (1955) 'Volumes of distribution and clearances of intraveneously injected creatinine in the dog', *American J. Physiol.*, **181**, 330–336.

Schwan, J. P. (ed.) (1969) *Biological Engineering*, McGraw-Hill, New York.

Sharan, M. and Singh, M. P. (1984) 'Equivalence between one step kinetics and Hill's equation', *J. Biomed. Eng.*, **6**, 297–301.

Shrager, R. I. and Hill, E., Jr. (1980) 'Nonlinear curve fitting in the L_1 and L_∞ norms', *Math. Comp.*, **34**, 59–541.

Smirnova, O. A. and Stepanova, N. V. (1975) 'Mathematical model of autoimmunity', *Biofizika*, **20**, 1095–1098. [English translation: *Biophysics*, **20**(1975), 1117–1122.]

Smith G. D. (1985) *Numerical Solution of Partial Differential Equations: Finite Difference Methods*, third edition, Oxford University Press.

Smith, P. (1981) *Numerical Modelling of Human Thermoregulation*, Ph.D. Thesis, CNAA.

Smith, P. and Twizell, E. H. (1980) 'A finite element model of temperature distribution in the human torso', *Appl. Math. Mod.*, **4**, 146–154.

Smith, P. and Twizell, E. H. (1985) 'A transient two dimensional model of thermoregulation in a human subject', *IMA J. Maths Appl. Med. Biol.*, **2**, 161–181.

Spyropoulos, K., Kiountouzis, E. and Young, A. (1973) 'Discrete approximation in the L_1 norm', *Computer Journal*, **16**(2), 180–186.

Stear, E. B. (1975) 'Application of control theory to endocrine regulation and control'. *Ann. Biomed. Eng.*, **3**, 439–455.

Stolwijk, J. A. J. (1971) 'Mathematical model of thermoregulation'. In Hardy, J. D. *et al.* (eds), *Physiological and Behavioral Temperature Regulation*, Chapter 48, 703–721, C. C. Thomas, Springfield, IL.

Stolwijk, J. A. J. and Hardy, J. D. (1966) 'Temperature regulation in man: a theoretical study', *Pfluegers Arch.*, **291**, 129–162.

Strang, G. and Fix, G. J. (1973) *An Analysis of the Finite Element Method*, Prentice-Hall, Englewood Cliffs, NJ.

Swan, G. W. (1984) *Applications of Optimal Control Theory in Biomedicine*, Marcel Dekker, New York.

Swart, J. H. (1986) 'The stability of a nonlinear, age dependent population model, as applied to a biochemical reaction tank', *Math. Biosciences*, **80**, 47–56.

Talbot, S. A. and Gessner, U. (1973) *Systems Physiology*, John Wiley, New York.

Twizell, E. H. (1980) 'A variable gravity model of the otolith membrane', *Appl. Math. Modelling*, **4**(2), 82–86.

Twizell, E. H. (1984) *Computational Methods for Partial Differential Equations*, Ellis Horwood, Chichester.

Twizell, E. H. (1985) 'The extrapolation of implicit methods for the constant coefficient diffusion–convection equation', *Comm. Appl. Num. Math.*, **1**, 129–135.

Twizell, E. H. and Curran, D. A. S. (1977) 'A finite element model of the otolith membrane', *Comput. Biol. Med.*, **7**, 131–141.

Twizell, E. H. and Smith, P. (1982) 'Finite element solution of steady-state temperature distribution in the human torso', *Comput. Biol. Med.*, **12**(2), 119–132.

Ulanowicz, R. E. and Frazier, G. C., Jr. (1970) 'The transport of oxygen and carbon dioxide in haemoglobin systems', *Math. Biosciences*, **7**, 111.

Varga, R. (1962) *Matrix Interative Analysis*, Prentice-Hall, Englewood Cliffs, NJ.

Wachspress, E. L. (1975) *A Rational Finite Element Basis*, Academic Press, New York.

Wait, R. and Mitchell, A. R. (1986) *Finite Element Analysis and Applications*, John Wiley, Chichester.

Wilkinson, J. H. (1965) *The Algebraic Eigenvalue Problem*, Clarendon Press, Oxford.

Wissler, E. H. (1961) 'Steady state temperature distribution in man', *J. Appl. Physiol.*, **16**, 735–740.

Index

Mathematics and its Applications

Series Editor: G. M. BELL, Professor of Mathematics, King's College London (KQC), University of London

Faux, I.D. & Pratt, M.J.	**Computational Geometry for Design and Manufacture**
Firby, P.A. & Gardiner, C.F.	**Surface Topology**
Gardiner, C.F.	**Modern Algebra**
Gardiner, C.F.	**Algebraic Structures: with Applications**
Gasson, P.C.	**Geometry of Spatial Forms**
Goodbody, A.M.	**Cartesian Tensors**
Goult, R.J.	**Applied Linear Algebra**
Graham, A.	**Kronecker Products and Matrix Calculus: with Applications**
Graham, A.	**Matrix Theory and Applications for Engineers and Mathematicians**
Graham, A.	**Nonnegative Matrices and Applicable Topics in Linear Algebra**
Griffel, D.H.	**Applied Functional Analysis**
Griffel, D.H.	**Linear Algebra**
Hanyga, A.	**Mathematical Theory of Non-linear Elasticity**
Harris, D.J.	**Mathematics for Business, Management and Economics**
Hart, D. & Croft, T.	**Modelling with Projectiles**
Hoskins, R.F.	**Generalised Functions**
Hoskins, R.F.	**Standard and Non-standard Analysis**
Hunter, S.C.	**Mechanics of Continuous Media, 2nd (Revised) Edition**
Huntley, I. & Johnson, R.M.	**Linear and Nonlinear Differential Equations**
Jaswon, M.A. & Rose, M.A.	**Crystal Symmetry: The Theory of Colour Crystallography**
Johnson, R.M.	**Theory and Applications of Linear Differential and Difference Equations**
Johnson, R.M.	**Calculus: Theory and Applications in Technology and the Physical and Life Sciences**
Jones, R.H. & Steele, N.C.	**Mathematics of Communication**
Kelly, J.C.	**Abstract Algebra**
Kim, K.H. & Roush, F.W.	**Applied Abstract Algebra**
Kim, K.H. & Roush, F.W.	**Team Theory**
Kosinski, W.	**Field Singularities and Wave Analysis in Continuum Mechanics**
Krishnamurthy, V.	**Combinatorics: Theory and Applications**
Lindfield, G. & Penny, J.E.T.	**Microcomputers in Numerical Analysis**
Lord, E.A. & Wilson, C.B.	**The Mathematical Description of Shape and Form**
Malik, M., Riznichenko, G.Y. & Rubin, A.B.	**Biological Electron Transport Processes and their Computer Simulation**
Marichev, O.I.	**Integral Transforms of Higher Transcendental Functions**
Massey, B.S.	**Measures in Science and Engineering**
Meek, B.L. & Fairthorne, S.	**Using Computers**
Mikolas, M.	**Real Functions and Orthogonal Series**
Moore, R.	**Computational Functional Analysis**
Müller-Pfeiffer, E.	**Spectral Theory of Ordinary Differential Operators**
Murphy, J.A. & McShane, B.	**Computation in Numerical Analysis**
Nonweiler, T.R.F.	**Computational Mathematics: An Introduction to Numerical Approximation**
Ogden, R.W.	**Non-linear Elastic Deformations**
Oldknow, A.	**Microcomputers in Geometry**
Oldknow, A. & Smith, D.	**Learning Mathematics with Micros**
O'Neill, M.E. & Chorlton, F.	**Ideal and Incompressible Fluid Dynamics**
O'Neill, M.E. & Chorlton, F.	**Viscous and Compressible Fluid Dynamics**
Page, S. G.	**Mathematics: A Second Start**
Prior, D. & Moscardini, A.O.	**Model Formulation Analysis**
Rankin, R.A.	**Modular Forms**
Scorer, R.S.	**Environmental Aerodynamics**
Smith, D.K.	**Network Optimisation Practice: A Computational Guide**
Srivastava, H.M. & Karlsson, P.W.	**Multiple Gaussian Hypergeometric Series**
Srivastava, H.M. & Manocha, H.L.	**A Treatise on Generating Functions**
Shivamoggi, B.K.	**Stability of Parallel Gas Flows**
Stirling, D.S.G.	**Mathematical Analysis**
Sweet, M.V.	**Algebra, Geometry and Trigonometry in Science, Engineering and Mathematics**
Temperley, H.N.V. & Trevena, D.H.	**Liquids and Their Properties**
Temperley, H.N.V.	**Graph Theory and Applications**
Thom, R.	**Mathematical Models of Morphogenesis**
Toth, G.	**Harmonic and Minimal Maps**
Townend, M. S.	**Mathematics in Sport**
Towend, M.S. & Pountney, D.C.	**Computer-aided Engineering Mathematics**
Twizell, E.H.	**Computational Methods for Partial Differential Equations**
Twizell, E.H.	**Numerical Methods, with Applications in the Biomedical Sciences**
Vince, A. and Morris, C.	**Mathematics for Computer Studies**
Walton, K., Marshall, J., Gorecki, H. & Korytowski, A.	**Control Theory for Time Delay Systems**
Wheeler, R.F.	**Rethinking Mathematical Concepts**
Willmore, T.J.	**Total Curvature in Riemannian Geometry**
Willmore, T.J. & Hitchin, N.	**Global Riemannian Geometry**

Numerical Analysis, Statistics and Operational Research

Editor: B. W. CONOLLY, Professor of Operational Research, Queen Mary College, University of London

Beaumont, G.P.	**Introductory Applied Probability**
Beaumont, G.P.	**Probability and Random Variables**
Conolly, B.W.	**Techniques in Operational Research: Vol. 1, Queueing Systems**
Conolly, B.W.	**Techniques in Operational Research: Vol. 2, Models, Search, Randomization**
Conolly, B.W.	**Lecture Notes in Queueing Systems**
French, S.	**Sequencing and Scheduling: Mathematics of the Job Shop**
French, S.	**Decision Theory: An Introduction to the Mathematics of Rationality**
Griffiths, P. & Hill, I.D.	**Applied Statistics Algorithms**
Hartley, R.	**Linear and Non-linear Programming**
Jolliffe, F.R.	**Survey Design and Analysis**
Jones, A.J.	**Game Theory**
Kemp, K.W.	**Dice, Data and Decisions: Introductory Statistics**
Moscardini, A.O. & Robson, E.H.	**Mathematical Modelling for Information Technology**
Oliveira-Pinto, F.	**Simulation Concepts in Mathematical Modelling**
Oliveira-Pinto, F. & Conolly, B.W.	**Applicable Mathematics of Non-physical Phenomena**
Ratschek, J. & Rokne, J.	**Computer Methods for the Range of Functions**
Ratschek, J. & Rokne, J.	**New Computer Methods for Global Optimisation**
Schendel, U.	**Introduction to Numerical Methods for Parallel Computers**
Stoodley, K.D.C.	**Applied and Computational Statistics: A First Course**
Stoodley, K.D.C., Lewis, T. & Stainton, C.L.S.	**Applied Statistical Techniques**
Thomas, L.C.	**Games, Theory and Applications**
Whitehead, J.R.	**The Design and Analysis of Sequential Clinical Trials**

Ellis Horwood Series in
Mathematics and its Applications
Series Editor: G. M. BELL, Professor of Mathematics, King's College London (KQC), University of London
Statistics, Operational Research and Computational Mathematics Editor:
B. W. CONOLLY, Professor of Operational Research, Queen Mary College, University of London

NUMERICAL METHODS, WITH APPLICATIONS IN THE BIOMEDICAL SCIENCES

E. H. TWIZELL, Lecturer in Numerical Analysis, Brunel University

This timely book is the first (as far as the publisher is aware) to chart the rapidly-growing area of inter-disciplinary research in the mathematical and computational modelling of biomedical systems. Evidence of this is the wide sponsorship in Britain by the SERC, the MRC, the Wellcome Trust, and the pharmaceutical companies. Thus the book meets the needs of workers, both mathematical and biomedical, engaged in such research. The impact which high speed computers have made on scientific interpretation is recognised, and the book highlights the algorithmic approach in presenting the numerical methods. With its heavy emphasis on the detailed discussion of applications, it will interest the rising body of scientists researching or working in numerical modelling of natural phenomena, an exciting area of investigation which was stimulated by the evangelistic work of Lighthill in biomedical science, biomathematics and bioengineering.

This modelling/research invariably leads on to systems of algebraic, or ordinary or partial differential equations, linear or nonlinear, all of which require numerical methods for their solution.

An introductory chapter on background mathematics moves to matrix algebra, linear and nonlinear algebraic systems, interpolation and approximation, differentiation and integration, and ordinary and partial differential equations. It discloses the availability of a vast number of numerical methods for solving some of the equations arising from the mathematical modelling of biomedical systems. The text is liberally illustrated with case studies and worked examples from the biomedical sciences. These include mathematical models of predator/prey interactions, continuously-fed reaction tanks, therapy of auto-immune diseases, growth of oncological diseases, arterial blood pressure, nerve impulse transmissions, drug absorption, genetics, the otolith membrane, and heat flow in the human body. The numerical analysis is deliberately presented in a non-rigorous level yet sufficiently high to preserve its identification with the biomedical disciplines.

The computer programs can be written straight from the methods and algorithms as they are presented in the text. A first course of calculus and algebra and the ability to write programs in high level scientific languages are pre-requisites.

Readership: Senior undergraduate and postgraduate students of biomathematics, bioengineering, general biomedical sciences, mathematical modelling, general numerical analysis and mechanical engineering.